Advances in Industrial Control

Advances in Industrial Control is a series of monographs and contributed titles focusing on the applications of advanced and novel control methods within applied settings. This series has worldwide distribution to engineers, researchers and libraries.

The series promotes the exchange of information between academia and industry, to which end the books all demonstrate some theoretical aspect of an advanced or new control method and show how it can be applied either in a pilot plant or in some real industrial situation. The books are distinguished by the combination of the type of theory used and the type of application exemplified. Note that "industrial" here has a very broad interpretation; it applies not merely to the processes employed in industrial plants but to systems such as avionics and automotive brakes and drivetrain. This series complements the theoretical and more mathematical approach of Communications and Control Engineering.

Indexed by SCOPUS and Engineering Index.

Proposals for this series, composed of a proposal form downloaded from this page, a draft Contents, at least two sample chapters and an author cv (with a synopsis of the whole project, if possible) can be submitted to either of the:

Series Editors

Professor **Michael J. Grimble**
Department of Electronic and Electrical Engineering, Royal College Building, 204 George Street, Glasgow G1 1XW, United Kingdom
e-mail: m.j.grimble@strath.ac.uk

Professor **Antonella Ferrara**
Department of Electrical, Computer and Biomedical Engineering, University of Pavia, Via Ferrata 1, 27100 Pavia, Italy
e-mail: antonella.ferrara@unipv.it

or the

In-house Editor

Mr. **Oliver Jackson**
Springer London, 4 Crinan Street, London, N1 9XW, United Kingdom
e-mail: oliver.jackson@springer.com

Proposals are peer-reviewed.

Publishing Ethics

Researchers should conduct their research from research proposal to publication in line with best practices and codes of conduct of relevant professional bodies and/or national and international regulatory bodies. For more details on individual ethics matters please see: https://www.springer.com/gp/authors-editors/journal-author/journal-author-helpdesk/publishing-ethics/14214

More information about this series at http://www.springer.com/series/1412

Thivaharan Albin Rajasingham

Nonlinear Model Predictive Control of Combustion Engines

From Fundamentals to Applications

Springer

Thivaharan Albin Rajasingham
Institute for Dynamic Systems and Control
ETH Zurich
Zurich, Switzerland

ISSN 1430-9491 ISSN 2193-1577 (electronic)
Advances in Industrial Control
ISBN 978-3-030-68009-1 ISBN 978-3-030-68010-7 (eBook)
https://doi.org/10.1007/978-3-030-68010-7

MATLAB and Simulink are registered trademarks of The MathWorks, Inc. See https://www.mathworks.com/trademarks for a list of additional trademarks.

Mathematics Subject Classification: 49, 93, 80

This Springer imprint is published by the registered company Springer Nature Switzerland AG
The registered company address is: Gewerbestrasse 11, 6330 Cham, Switzerland

To Julin, Amma, Appa, and Elli

AIC Series Editors' Foreword

Control engineering is viewed rather differently by researchers and those that must implement and maintain control systems. Researchers develop general algorithms with a strong mathematical basis, while practitioners have more local concerns over the capabilities of equipment, quality of control, and plant downtime. The series *Advances in Industrial Control* attempts to bridge this divide and hopes to encourage the adoption of more advanced control techniques when they are likely to be beneficial.

The rapid development of new control theory and technology has an impact on all areas of control engineering and applications. This monograph series encourages the development of a more targeted control theory that is driven by the needs and challenges of applications. A focus on applications is essential if the different aspects of the control design problem are to be explored with the same dedication that control synthesis problems have received in the past. The series provides an opportunity for researchers to present an extended exposition of their new work on industrial control, raising awareness of the substantial benefits that can accrue, and exploring the challenges that can arise.

The author has made a number of contributions to the control of combustion engines using Model Predictive Control (MPC) methods and has also been involved in teaching the subject at various prestigious institutions. The text reflects this current research and the valuable practical experience gained. It begins by explaining the classical approaches to engine control and the major problem of engine calibration. When the number of calibration parameters was small, this was a practical process but for modern engines, the number of inputs and outputs and the number of calibration parameters have increased significantly and continue to do so. This and tighter emissions and fuel-consumption requirements led to substantial interest in model-based predictive control methods that are particularly suitable for complex nonlinear multivariable systems.

After the introductory chapter, the traditional decentralized PID control solution is discussed using look-up tables, and the problems of calibration are described. The problems encountered motivate the introduction of more systematic design procedures. These normally stem from optimal control or optimization-based methods covered in the following chapter. There is a very useful introduction to optimization techniques that are essential for predictive control where its constraint-handling capabilities are often the main feature.

The predictive control algorithms are based on optimization, which is a topic introduced from the first principles in Part I of the text. First, the linear MPC algorithms are introduced and then the use of hard constraints in a section on Constrained MPC. This is a relatively simple introduction, which is also valuable for the subsequent introduction to nonlinear MPC. This approach is a good candidate for engine control since engines are very nonlinear, and to improve performance some form of nonlinear control or compensation is essential.

Nonlinear optimization and numerical methods are introduced for the solution of the nonlinear predictive control problem that results. This is a topic that can often be hard to follow because of the mathematical tools required and the very large number of options available in modeling and optimization. This text provides a very straightforward approach of great practical use to the application. The feasibility of implementing the computational algorithm in real time is considered. This is a challenging and very practical problem in engine control, which seldom receives the attention it deserves in published work. It is also valuable that theoretical issues like stability questions are considered in this first part of the text.

The introduction to engine control problems including air path control, fuel and combustion control, and exhaust gas path control is followed by introductions to spark ignition and then compression-ignition engines in Part II. A good understanding of engines is needed before models can be produced and model-based techniques can be applied. The very clear introductions to the complicated control problems involved are therefore important. The final parts of the text are concerned with in-depth case studies that will probably be the most valuable chapters for engineers actually working in the automotive industry. These sections include useful practical details emphasizing the real engineering aspects of the problem not often available in research, rather than development-oriented texts.

This text concerns an area of advanced control where there is both a need and a desire for a practical solution. Senior management in the automotive industry has been very receptive to the use of models for the design of controllers that can provide higher performance and can cope with the complexity of the engine control problem. There is also a desire for systematic or formalized design procedures to be produced. This faith in model-based control has been justified recently in the first application of MPC to GM production engines. This text is therefore a very timely and a very welcome addition to the series on *Advances in Industrial Control*.

Glasgow, UK
Pavia, Italy
October 2020

Michael J. Grimble
Antonella Ferrara

Preface

Scope of the Book

The requirements on combustion engines used for transportation, i.e. automotive vehicles, ships, and airplanes, as well as for stationary energy supply, are constantly increasing. This concerns, for instance, the reduction of fuel consumption and pollutant emissions. An increasing number of sensors and actuators are implemented in combustion engines to improve their performance. To fully exploit the potentials of the system capabilities, complex control algorithms are required which allow for suitable process handling. The book describes the use of an advanced control algorithm, namely, the Model Predictive Control (MPC) method. The book aims to provide a comprehensive overview of the topics related to MPC for its application to engine systems.

Readers targeted are engineers and researchers in academia and industry working in the field of engine system control. The text is also suited for graduate students interested in this topic and who want to deepen their expertise. Readers should have a solid knowledge of control systems and the working principle of an engine system. Additional prerequisites are the fundamentals of mechanical engineering, such as mechanics and calculus.

Several good textbooks discuss the various topics that are needed for model predictive engine control. Textbooks exist that specifically detail the fundamentals of MPC, the fundamentals of numerical optimization, or the fundamentals of engine control and modeling. These textbooks treat the full scope of the specific topic in detail including their theoretical background. Rather than looking into any one topic in detail, the focus of this book is placed on the intersection of all the necessary topics. The algorithms and methods that are scattered over the various disciplines are brought together from a practical perspective. This allows the reader to become familiar with all the relevant aspects and to see how the various topics are interconnected. Instead of showing all available algorithms, particularly those are presented that have shown to work well in real vehicles and on real engine test benches. This concerns the characteristics of engine control, such as the necessity to consider nonlinearities and the small timescales of control that make the real-time feasibility a very demanding aspect. These algorithms are not only explained, but their application to engine control problems is detailed as well. After reading the book, readers have all the necessary fundamentals to implement their own MPC algorithms on real-world combustion engine control problems and advance the field of combustion engine control on their own.

Origin of the Book

The book originated from a graduate course that has been taught at various institutions. The courses are as follows:

- lecture series at ETH Zurich, Switzerland,
- lecture series at RWTH Aachen University, Germany,
- lecture at Peter the Great St. Petersburg Polytechnic University, Russia,
- summer school at Jilin University, Changchun, China and
- summer school at Sophia University, Tokyo, Japan.

Several Ph.D. students have been involved in the preparation of the lectures and the associated lab exercises, namely Martin Keller and Dennis Ritter at RWTH Aachen University as well as Richard Hutter, Severin Hänggi, Johannes Ritzmann, and Stijn van Dooren at ETH Zurich. I am thankful for all their support.

The content for the lectures and the book resulted from several research projects on the topic of MPC for engine system control. As a senior researcher and a group leader, I had the opportunity to work together with numerous Ph.D. students on this topic. I would like to especially mention the research unit FOR2401 as one project for which I had the honor to act as a spokesman. Numerous fruitful academic collaborations have resulted from the research unit, for instance, with Prof. Moritz Diehl on the topic of numerical optimization, with Prof. Heinz Pitsch on combustion modeling, with Prof. Katharina Kohse-Höinghaus on combustion chemistry, as well as with Prof. Stefan Pischinger and Prof. Jakob Andert on combustion engines. The book also benefited from publicly funded research projects with industrial collaborators such as General Electric and Ford AG.

The research work entered and shaped the book. Results from joint work with collaborators are properly cited in the various chapters. The focus is placed on application examples, where active research has been carried out—the application examples investigated all stem from research projects conducted. Within the research projects, the application examples described have been implemented by the author in real-world combustion engine test benches and in real-world vehicles which have been tested on public roads. Clearly, not all relevant and interesting application examples are treated. For instance, the energy management for hybrid drivetrains or the topic of exhaust gas aftertreatment is not treated within this book.

Acknowledgments

The financial support received from several funding agencies is acknowledged, which has made the research covered in this book possible. Part of the research was performed within the Research Unit (Forschungsgruppe) FOR2401 "Optimization based Multiscale Control for Low-Temperature Combustion Engines", which has been funded by the German Research Association (Deutsche Forschungsgemeinschaft, DFG).

Part of the research results presented was realized within the IGF research project 17733 N/1 "Prädiktive Steuerung für hochaufgeladene Ottomotoren" of the FVV. The IGF project 17733 N/1 of the research association Forschungskuratorium Maschinenbau e.V.—FKM, Lyoner Strasse 18, 60528 Frankfurt am Main, was financially supported by the AiF within the framework of the development program for Industrial Community Research (IGF) of the Federal Ministry of Economic Affairs and Energy based on a decision of the German Bundestag. The author thanks the AiF, the FVV, and the corresponding committee of the FVV for their financial support, expertise, and discussions during the working group meetings.

Part of the work presented was conducted within the research project "JB-X Clean—Model based Control of Dual-Fuel Combustion Engines". The project was funded by the Federal Ministry of Economic Affairs and Energy (BMWi) under the number BMBF 03SX375C. The author thanks the BMWi for its financial support and all project partners for the good cooperation.

In addition to the several funding agencies, several people have to be mentioned as well who have been essential in realizing the book. I would like to acknowledge Dennis Ritter with whom I worked together on many of the results presented. This concerns especially the topic of two-stage turbocharging and of combustion-rate shaping. Additionally, I would like to thank Severin Hänggi for his support on the EGR VTG case study and Jan Schilliger and Nils Keller who helped to streamline the book, for instance, with the implementations of several numerical examples. Brigitte Rohrbach helped to improve the quality of the book by very carefully proofreading the manuscript. I am very grateful to Prof. Chris Onder for hosting me for the last 2.5 years. He was very supportive in every regard and provided a research environment amenable to high-quality research. Finally, I would like to acknowledge Prof. Dirk Abel who has been a mentor for more than 10 years. It is a great honor to work with him and to learn from him.

Zurich, Switzerland
September 2020

Thivaharan Albin Rajasingham

Contents

Chapter 1
Introduction

Abstract There exists a high demand to improve combustion engines. The main focus is the reduction of carbon dioxide (CO_2) and pollutant emissions while sustaining the same performance, e.g. in terms of driveability, comfort, and durability. In this chapter, it is detailed that a lot of the improvements of the engine investigated come along with increased complexity of the process control. The state-of-the-art control algorithms are reaching the limits of their capabilities. To satisfy the rising requirements on the process control, a lot of calibration parameters are needed. The tuning of these parameters is very time consuming and cost-intensive, yet results in a suboptimal control behavior. Model predictive control (MPC) poses a very attractive alternative. It is suited to handle the complex system dynamic behavior of the engine in a systematic manner. This allows achieving better performance while reducing the time needed for the calibration of the controller.

1.1 Motivation to Improve Engine Performance

One of the central challenges in today's society is the ecological and economic energy supply for both mobile and stationary applications. In the upcoming decades, this topic will become even more important. Due to a significant increase in the population and prosperity worldwide, an over-proportional increase in primary energy consumption is predicted [4]. At present, a large portion of this energy is supplied by the combustion of hydrocarbon-based fuels [13]. Approximately 70% of the crude oil consumed, which amounts to 86 million barrels per day, is used in internal combustion engines (ICEs) [21]. They are used not only within a broad range of different applications such as passenger vehicles, heavy-duty trucks, and ships, but also for stationary energy supply, in order to drive electric generators.

The combustion of fossil fuels is associated with the emission of pollutants, e.g. particulate matter (PM), consisting mainly of soot, nitrogen oxides (NO_x), carbon

T. Albin Rajasingham, *Nonlinear Model Predictive Control of Combustion Engines*, Advances in Industrial Control,
https://doi.org/10.1007/978-3-030-68010-7_1

monoxide (CO), and unburned hydrocarbons (HC). These emissions contribute significantly to urban and rural air pollution. In addition, the greenhouse gas CO_2 is produced, which leads to a change in the global climate [12]. In the foreseeable future, ICEs will still continue to play a major role, although the relative contribution of ICEs to the global energy supply will decrease [22]. In the case of passenger cars, the partial or full electrification of the propulsion system is accelerated, which allows a significant reduction of CO_2 and pollutant emissions. The transformations will be an ongoing process over the next decades; see, for example, [4] where different outlooks for the share of electric vehicles by 2040 are highlighted. Besides their drawbacks, HC-based fuels still also have advantages compared to recently applied state-of-the-art batteries in electric vehicles. The fuels have a very high energy density, are easily storable, ubiquitously available, and are well-standardized. Especially for applications which require a high energy density such as ships, no technically reasonable alternatives are available at present. In the case of stationary applications, increasing shares of renewable energy will be deployed [4]. Still, also ICE-driven electric generators will be used in the future, as they allow for a load-flexible power supply which can stabilize the electrical grid by counter-balancing the varying levels of coverage given by renewable energies.

In the context of the depletion of resources and the increasing impact on the environment, the development of improved technologies for clean combustion in ICEs is extremely important. The requirements for more efficient and ecological combustion engines are enforced by both customer demands and political governments posing emission standards by law. In the future, emission regulations for various applications of ICEs will become even tighter. To illustrate exemplarily the tightening of requirements, the emission limits for automotive diesel engines and ship engines are shown in the following. For the automotive area, the development of emission regulations in the European Union (EU) is depicted in Fig. 1.1. It shows the constantly decreasing limits on PM, HC, and NO_x emissions for passenger cars with diesel engines [10]. For automotive engines, the so-called Real Driving Emissions (RDE) legislation is introduced, where a tougher determination of emissions is enforced. The vehicle's emissions are measured on the road using portable emission measuring systems (PEMS).

The same trends for emission legislation can be observed for maritime propulsion systems. The International Maritime Organization (IMO) sets emission standards for diesel and heavy-fuel engines depending on the maximum operating speed of the engine. Figure 1.2 depicts the NO_x limits subject to engine speed for engines with a displacement volume of 30l per cylinder or more [14]. Tier I and II limits are globally applied, while Tier III standards apply only in Emission Control Areas (ECA), including the North Sea, Baltic Sea, and most of the US and Canadian coasts.

All in all, ICEs will still play a central role in the energy supply for mobile and stationary applications in the upcoming decades. To reduce the negative impact of ICEs, their development is inevitable. The main drivers are the reduction of CO_2 and pollutant emissions. These reductions have to be achieved while customers simultaneously demand unchanged characteristics, e.g. concerning driveability, comfort, noise, vibration and harshness, and durability.

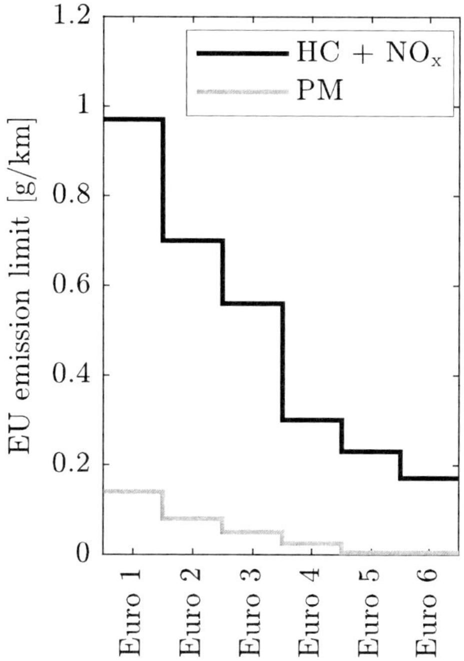

Fig. 1.1 EU emissions limits for HC+NO_x and PM in diesel vehicles

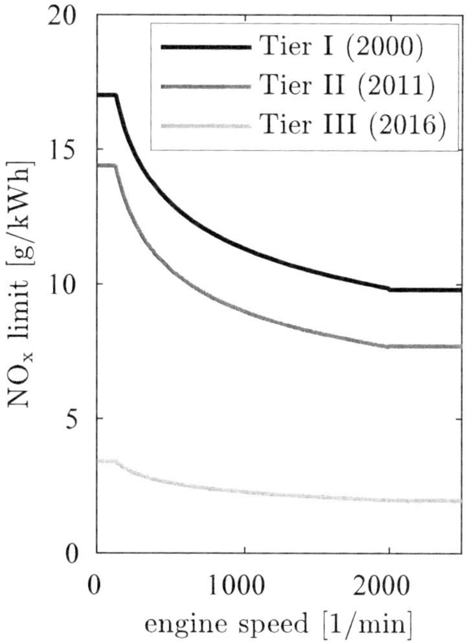

Fig. 1.2 IMO Tier emissions limits of NO_x for diesel and heavy-fuel maritime engines

1.2 Improving the Engine Performance by Advanced Control

In order to improve the performance of ICEs, e.g. in terms of fuel consumption or emission reduction, each component itself is revisited and improved. Examples include the combustion process, the air path, the exhaust gas aftertreatment, waste heat recovery, and auxiliary devices [18]. Many of the improvements investigated are accompanied by increased complexity of the process control which is detailed in the following.

A general development within the automotive area over the last decades has been the increase in electric/electronic parts, which often replace mechanical parts [5]. Within the engine area, a classic example is the throttle which used to be a purely mechanical element. With the mechanical throttle, no electronically implemented control algorithm is necessary. Today's standard is the use of electronic throttles which include a DC motor and a position sensor to actively control the throttle position. For the electronic throttle, one controller is needed to set the appropriate position of the throttle in dependence on parameters such as the required load and the ambient temperature. Additionally, at low level a controller is necessary to set the DC motor such that the desired throttle position results. The general trend of electrification of the components leads to increased degrees of freedom to control the entire process.

Additionally, more and more components are added to the engine system. One example is the introduction of turbocharging units for gasoline engines. The integration of the turbocharger unit allows for a downsizing of the ICE, thus increasing the efficiency in part-load operations. Often, turbochargers are equipped with a wastegate or a variable geometry turbine which allows for flexible use, but it adds a degree of freedom which has to be handled by the control unit.

The increased flexibility for operating the engine results in more possibilities to influence the processes involved. In order to fully exploit these capabilities, the number of sensors is increasing, which leads to better knowledge about the current system states. With these possibilities, the process can be adapted to handle varying conditions optimally, e.g. in terms of requested load, engine speed, and ambient conditions. Today's engines in automotive vehicles contain about 15 to 25 sensors and five to nine actuated variables that are considered in the process control [15]. In order to cope with these developments, improved electronic control units (ECU) are necessary. This concerns the hardware of the ECUs which are based on microcontrollers and also the software running on the ECUs. A crucial part of the software is the control algorithm which generates the appropriate actuator signals. Only if the control algorithm is able to fully exploit the given degrees of freedom, significant benefits in terms of fuel consumption and pollutant emissions can result.

For future engine systems, the requirements on the closed-loop system and the control algorithm itself will become increasingly complex. The engine controller has to handle the interaction between actuators and the engine process. For an appropriate control of the engine, several requirements have to be addressed. From a control

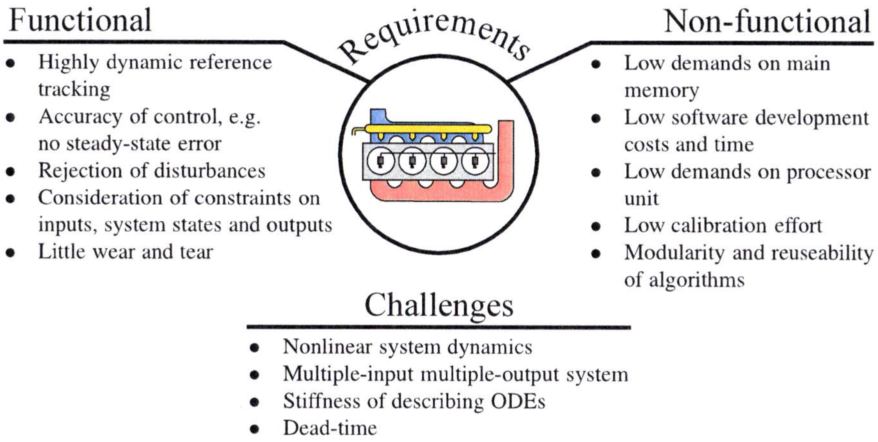

Fig. 1.3 Requirements and challenges on engine control algorithms

point of view, one can distinguish between the two categories, functional and non-functional requirements, as shown in Fig. 1.3. These requirements have to be fulfilled, while there are challenges arising from the open-loop dynamics of the system.

Functional Requirements

The control algorithm has to ensure that the engine provides the requested torque while satisfying numerous requirements such as emission legislation (NO_x, CO, HC, and PM), comfort demands, safety requirements, and reduction of CO_2 emissions. The control performance has a direct correlation to these engine characteristics. Relevant control specifications are the achieved closed-loop dynamics and the control accuracy, i.e. no steady-state error. The closed-loop dynamic of the intake manifold pressure in a gasoline engine, for instance, directly influences the performance for the acceleration of a vehicle. Additionally, the engine control should be able to reject disturbances. The engine is operated at various ambient conditions which concern the temperature, pressure, and humidity while not all of these influencing factors are measured or modeled. Disturbances to the nominal system also arise due to the aging of components over their lifetime. The engine control should be robust enough to cope with these (unmodeled) disturbances. Depending on the control task investigated, additional requirements occur, such as the prohibition of any considerable overshoot. On the other side, the control has to take into account the constraints of the system. They can be due to physical limitations, e.g. the throttle cannot be more than fully open. They can also occur due to safety requirements, i.e. the process control has to ensure that the engine (or its sub-components) is not operated in a safety-critical state. An example is the operation of components such as the exhaust gas aftertreatment at too high temperatures. Additionally, the control should reduce tear and wear. For example, high wear might appear when the actuator is used extensively.

Non-functional Requirements
The second category is given by the non-functional requirements. These arise especially in an industrial context, where the cost and development time are important factors. The control algorithm should have a low impact on the memory usage and a low computational burden to the control hardware. Concerning the software, the development costs and time are important. The development time is given by the time for coding and calibration of the software. To reduce the calibration effort, it is advantageous if the algorithm is as modular as possible. In such a case, a lot of software and calibration development can be reused for another control project, e.g. a slightly modified engine.

Due to the highly dynamic system, the sampling times of the control loops are chosen to be very short. Additionally, for realizing the requirements named, challenges arise from the (open-loop) system dynamics of the process to be controlled. Challenging system properties are present, such as

- nonlinearities,
- multiple-input multiple-output (MIMO) system dynamics,
- dead-time,
- stiffness of describing ordinary differential equations (ODE),
- highly dynamical systems, and
- noise on the measured signals.

1.3 Control Algorithms for Engine Control

In order to operate the engine, many diverse control tasks have to be tackled with various levels of complexity. Consequently, the most suitable control algorithm depends on the specific control task. In the following, the suitability of different control algorithms for use within engine control systems is examined. A very rough classification of control algorithms can be conducted as follows:

1. controller design and synthesis without any use of a process model,
2. controller design and synthesis (offline) by use of a process model, and
3. the process model is an integral part of the controller implemented.

For simple control tasks, the synthesis of the controller can often be conducted without the use of any process model, i.e. by heuristic tuning. In these cases, the use of a simple controller such as feedforward or a linear feedback controller is sufficient for the desired control quality. An application example for engines is the low-level control of actuators, e.g. the throttle position, which are often single-input single-output (SISO) control problems without strong impact of disturbance variables. In this simple setting, the heuristic tuning will even be quicker than a systematic approach. No in-depth system knowledge is necessary and also no process model has to be developed. Also, the computational demands for processing the algorithm are low.

In the case of more complex control tasks, heuristic tuning does not lead to satisfactory results, such that the use of a process model is necessary. This kind of control algorithm is referred to as model-based control. In general, there are two possibilities to use the process model for control synthesis. Either the process model is used to design a controller (offline) or the process model itself is an integral part of the controller implemented. The state of the art in terms of engine control is the use of a process model in order to calibrate the controller offline. The control algorithm typically consists of a combination of feedforward and PID-based feedback controllers. Look-up tables are applied to parametrize the various factors within the controller. Compared to heuristic tuning, a higher control quality can be achieved. Additionally, the control structure has low computational demands. For the systematic controller design, a model of the process is necessary, which can also be used for other purposes, such as state estimation. This approach comes along with drawbacks in terms of time effort for calibration, especially in the case of complex control tasks with high demands. The main reason lies in the inability to systematically consider various properties such as the handling of constraints. Due to this disadvantage, a lot of calibration parameters have to be included in order to still fulfill requirements such as constraint handling. As a result, the control algorithm is prone to the curse of dimensionality. An increase in degrees of freedom will lead to an exponential increase in the number of parameters which have to be calibrated, as discussed in Sect. 2.3 in more detail. In the context of engine control, these parameters are called calibration labels.

With the increase of actuators and sensors in modern engine systems, a controller based on look-up tables is reaching the limits of capability. Figure 1.4 shows the increase in the number of calibration labels over the number of degrees of freedom for different years [9].

In the case of complex control tasks with high demands, such as those given in modern engine systems, control algorithms are better suited where the controller itself contains a process model. Compared to the model-based control synthesis (offline), an even higher control performance can be achieved and more system properties can be handled in a systematic manner. Many different control algorithms exist where the model is contained in the controller itself, such as the Smith predictor [1], flatness-based control [16], and MPC. The Smith predictor is especially suitable for the dead-time compensation of SISO systems and of MIMO systems with dead-time directly associated with the outputs. The application of the Smith predictor for MIMO systems with arbitrary dead-times is far more complex [24]. Flatness-based control allows dealing with non-minimum phase systems. However, the flatness-based control approach can only be applied if the system fulfills the property of being flat. As MPC does not suffer from these limitations, it can be used for a wide range of system properties. On top of that, among all these control algorithms, MPC has an outstanding role as it is the only one that can inherently take into account constraints on actuated values, system states, and outputs. This is an important requirement in engine control, e.g. due to the given safety limitations. In the following, the fundamental concept of MPC will be explained along with advantages and disadvantages.

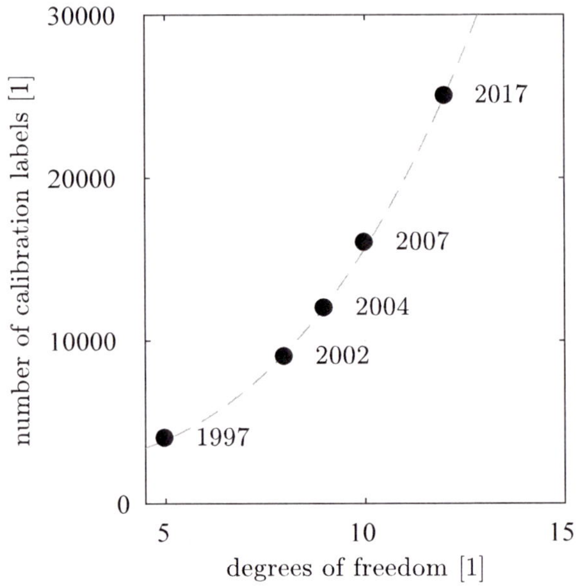

Fig. 1.4 Increase of engine management complexity based on data from [9]

1.4 Introduction to Model Predictive Control

MPC relates to a class of control strategies which uses optimization to calculate the actuated values. Figure 1.5 depicts the control structure using MPC. Compared to classical control structures, the feedforward and feedback action are combined in a single controller. Typically, MPC is used as a state-feedback controller, which means that the control action is based on the recent states of the system. These states are necessary, as they are used for the prediction of the output values in the process model. Either the system states are all measurable or, more commonly, a state observer has to be applied to estimate the system states.

Figure 1.6 depicts the working principle of MPC. The MPC algorithm solves an optimization problem in each sampling step, consisting of a cost function and constraints. In the optimization, a mathematical process model of the controlled application is explicitly utilized in order to predict the system behavior. Due to limitations on the maximum calculation time, the outputs of the model are predicted over a finite horizon. The result of the optimization algorithm is the optimal open-loop sequence of actuated values. The first values of the sequence are directly taken as actuated values that are applied to the system. In the next sampling step, the optimization problem is solved again and a new sequence of actuated values is calculated over a shifted horizon. This procedure is called the receding-horizon principle. Due to this principle, a feedback control algorithm is realized, as the most recent measurement is used as the basis for the new prediction.

The use of optimization along with a process model for predicting the system behavior has many benefits in terms of control performance and systematic design, which makes it attractive for engine control. A high control performance can be achieved, as the complex system dynamics are directly accounted for. Exemplary characteristics that can be handled in a systematic manner are nonlinearity, dead-time, non-minimum phase, and MIMO systems with strong couplings between the different inputs and outputs. The improvements in performance for engine control are reflected in better fuel economy or better vehicle acceleration performance, thus generating direct customer value. Its applicability to a wide range of system properties makes it possible to have a uniform approach for different control tasks instead of having individual solutions. This property is important for engine control. For instance, during the development process, the engine setup is changed from time to time. The change of the physical location of an actuator, for example, might result in a change of the system dynamics, going from no dead-time to a considerable dead-time. In the worst case, a control algorithm which was appropriate beforehand is no longer usable. For a systematic design, it is preferable if the control algorithm is employable as generically as possible – ideally independent of the system properties.

As the control behavior is directly determined by the optimization problem, further control requirements can be included. This concerns especially the inclusion of constraints which cannot be handled systematically by any other control algorithm. Additionally, the optimization allows considering preview information in the control concept. The preview can consider, for instance, the knowledge of future reference values or future disturbance variables. This becomes important for engine control as due to increased connectivity, e.g. by vehicle-to-vehicle information, preview information becomes more and more available [25]. The MPC algorithm can take the future information into account, in order to act in an anticipatory and thus acausal manner.

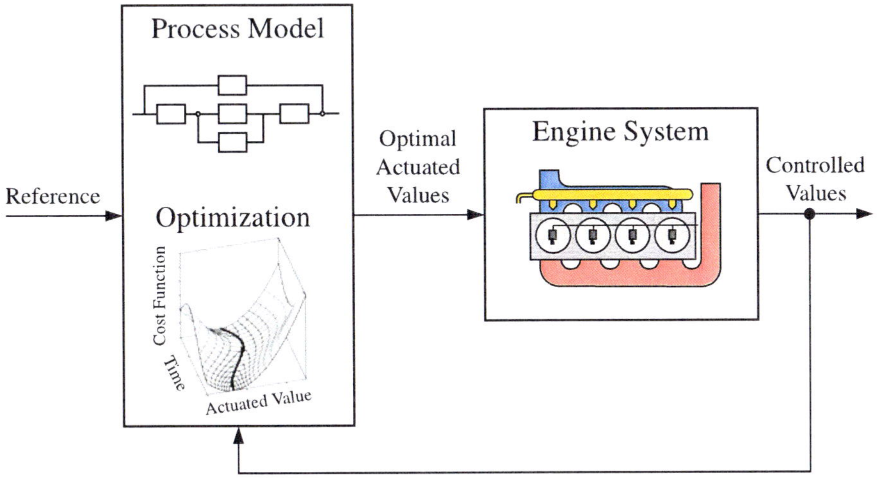

Fig. 1.5 Structure of a control loop with MPC

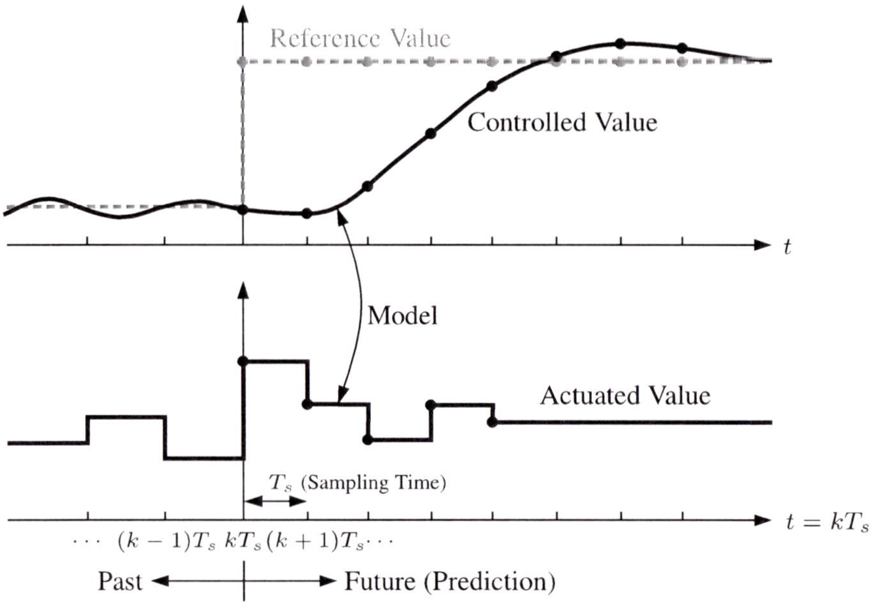

Fig. 1.6 Working principle of MPC

In addition to the advantage of better performance, the calibration time can be decreased. Some of the requirements, such as the consideration of constraints, are hard to tackle with conventional controllers that are based on look-up tables. A lot of heuristic tuning of calibration parameters has to be conducted and protection logics have to be included [8].

The MPC algorithm is able to consider the arising properties and requirements systematically. As a consequence, the process model has a crucial role. It is ideally built up in a very modular way, i.e. for each physical component of the system, there exists a separate submodel. For a change of a component, e.g. the injector, it is possible to only modify the submodel responsible for the injection system while all other submodels are reusable. If a new engine system is built, the controller can be designed via the use of submodels of each component, which might have already been developed in prior projects. This systematic design enables a fast adaptation to different engine hardware architectures – even for a mass production setting [3]. Consequently, the calibration procedure is shifted toward the use of measurement data to develop and validate the process model. As a result, the development and calibration time of the controller can be drastically reduced. A study conducted by a car manufacturer shows the reduction of calibration time from 4 months with an approach based on look-up tables to 15 d with an MPC-based approach for a turbocharging control problem [23]. Additionally, the mathematical models of the process are a good way to build up and save knowledge in a large organization. In the case of the calibration process based on look-up tables, the knowledge is directly linked to the experience of the people involved and thus is not easily accessible to all.

The main disadvantage of the MPC algorithm is the increased requirement for the hardware of the ECU. The memory storage and also the processor have to cope with the increased impact of the MPC algorithm. However, the hardware of ECUs is becoming more and more powerful, while the price for the hardware decreases [7]. Simultaneously, the algorithms and methods needed for MPC are improving such that the computational burden is reduced and its real-time feasibility is getting less critical. These improvements concern the numerical real-time optimization algorithms and also the reduced-order modeling of the components. Along with that, the MPC algorithm in general is much more complex than conventional controllers such as the PID-based controllers. Within MPC, a complex numerical algorithm is used, which usually relies on an iterative solution. The algorithm has to be designed very carefully, such that, for instance, a suitable solution is found always. In contrast, the PID-based control algorithm only calculates a few simple expressions. Besides these points, it has to be considered that the control engineers involved have to have a higher qualification level for the development of MPC-based algorithms. They need to be trained in the modeling of dynamic systems and in the topic of optimization-based control.

A successful implementation of MPC consists of three ingredients [6]. These are

1. a suitable formulation of the optimization problem,
2. a (reduced-order) process model, and
3. a real-time-feasible optimization algorithm.

In order for real-time-feasible control to be achieved and all the requirements on the closed-loop system to be fulfilled, the three parts have to match.

1.4.1 Formulation of the Optimization Problem

For a successful MPC-based control algorithm, a suitable optimization problem has to be formulated that allows fulfilling all requirements of the specific control task. The main advantage of MPC is that a large variety of control requirements can be covered by the possibilities given by the real-time optimization. In particular, the cost function can handle requirements on the time trajectory of the closed-loop system. One example is the performance requirement that the system states should follow the reference value as fast as possible, e.g. in order to allow a fast vehicle acceleration. However, the cost function can also consider other variables which have an impact on the process, e.g. the absolute value of the actuated values, the rate of change of actuated values, and economic considerations of the closed-loop system. The various elements can be included in the cost function and prioritized by weighting factors. The optimization is performed subject to constraints, which allows considering constraints on physics, performance, and safety. Physical constraints are given by the process setup, such as limits on the actuators, e.g. a valve has a minimum and maximum opening position. The performance constraints come into play if, for example, no overshoot is allowed, which can be formulated as a constraint on the output signal. Another important issue are the safety constraints, which have to be considered if system states or output values have to be kept in a certain operating region for safe operation.

Compared to classical controllers, the design is not conducted in the frequency domain but in the time domain which is more intuitive for the practitioner. However, a good physical understanding of the process and a systematic analysis of the process are still necessary for the formulation of a suitable optimization problem. Overall, the optimization problem allows sufficient freedom to account for all the requirements arising in engine control.

1.4.2 Control-Oriented Modeling

For MPC, control-oriented models have to be developed which capture the relevant system dynamics. The process model has a significant impact on the control performance, as the prediction capability of the model directly influences the quality of the control behavior. At the same time, the models have to be sufficiently simple such that they are suited for real-time optimization. Therefore, a trade-off has to be found between model accuracy and simplicity. This trade-off plays a fundamental role especially in engine control as on the one hand the sampling times are in the range of milliseconds and on the other hand, complex multi-physics processes have to be modeled.

For modeling the engine system, physics-based white-box models are available. These models allow for a detailed simulation of the processes involved, such as aerodynamics and chemical reaction kinetics. However, many of the physics-based models are far too complex to be used in a real-time optimization setting. An alternative approach uses data-based black-box models. The basis for these models consists of input/output data derived from measurements. With these experimental data, a candidate model is chosen that is usually given in a parametrized form. Through the use of the measured data set, the parameters are estimated. Black-box models allow reproducing the system dynamics with a model of low complexity. The drawback is that a large amount of measurement data is needed for the calibration of the black-box process model and that only a very limited extrapolation capability is given. In addition, no state estimation of physically meaningful quantities is possible. For this reason, the book will detail gray-box models [17]. In the approach taken, simplified, low-complexity physics-based models are used in addition to sub-components that are modeled using black-box models. One of the goals of model-based control is the reduction of measurements necessary or time at the test bench, respectively. Consequently, the advantages of model-based control are especially given if the portion of physics-based models is as high as possible in the gray-box approach.

1.4.3 Numerical Optimization

For systems with large sampling times, the real-time feasibility is not an issue, as the available time for solving the optimization problem is long enough. For systems with faster dynamics, the computational demand is the major bottleneck when applying MPC algorithms. This is the reason why the MPC algorithm was originally limited to relatively slowly varying applications as those present in the process and chemical industries [19]. The continuous improvement of methods and algorithms related to MPC allows the application of MPC to systems with ever-faster system dynamics [20].

Typical sampling times in combustion engines are in the order of milliseconds, which makes real-time feasibility a major challenge. As a consequence, the optimization methods applied play a major role. In general, several methods are available for solving the optimization problem. Depending on the class of system model, the type of cost function, and the constraints, different kinds of optimization problems result. In the case of a linear model, a quadratic cost function, and linear constraints, a Linear MPC (LMPC)[1] algorithm results. As the chapters below show, a quadratic program (QP) has to be solved in each sampling step. The algorithm is called Nonlinear MPC (NMPC), if, for example, the system model is nonlinear. In this case, a nonlinear program (NLP) results, which generally is harder to solve than the QP.

For linear and piecewise affine models, it is possible to calculate the control law resulting from the optimization-based controller already in advance offline. This approach is called explicit MPC [2]. In this case, the control laws are stored in maps and only the evaluation of the control law is conducted online. This approach is only feasible for systems with small dimensions. For this reason, it is usually not applicable to engine control tasks. Therefore, online methods are of interest for engine control, where the optimization problem is solved numerically during the runtime of the process. The numerical algorithm has to be fast in order to fulfill real-time feasibility and it has to be reliable such that a feasible solution is always available. The numerical methods applicable to these requirements are also called "embedded" optimization methods [11]. It is important to specifically design the optimization routine according to the given optimization problem. By proper choice of the optimization algorithm, a solution can be developed which can handle the given characteristics.

1.5 Aims and Outline of the Book

The present book provides a comprehensive overview of the application of MPC to complex control tasks arising in engine systems. The methods detailed in the book are presented from the algorithms to various in-depth application examples. The

[1] In the present book, the term MPC refers to the general case (Linear or Nonlinear MPC), whereas LMPC/NMPC refers specifically to the linear/nonlinear case.

application examples described have been implemented in real-world combustion engine test benches and in real-world vehicles which have been tested on public roads. Readers should have a solid knowledge of control systems and the working principle of an engine system. Additional prerequisites are the fundamentals of mechanical engineering, such as mechanics and calculus. Overall, the monograph consists of introductory chapters and four main parts.

Chapter 2 details the limitations of the current state-of-the-art control approach for ICEs. Current control approaches are often based on PID controllers which handle SISO control tasks in a decentralized manner. To take into account the nonlinearity of the plant, the parameters of the PID controller are adjusted via look-up tables. For complex control tasks, as present in engine systems, the control concept applied has some severe disadvantages. The chapter details for which circumstances the drawbacks arise. The last introductory chapter, Chap. 3, provides an overview of the fundamentals of optimization. The notations used in the book are introduced and classification of optimization problems is conducted. A special emphasis is placed on the optimality conditions for the typical optimization problems arising within engine MPC. These optimality conditions serve as fundamentals for numerical algorithms to solve the optimization problems.

Part I provides an overview of MPC methods that can be applied to achieve real-time-feasible engine control. First, the LMPC algorithm is introduced. It allows considering linear system models and linear constraints. The resulting optimization problem and its solution are presented. For the LMPC problem, a QP results which can be solved very reliably and very fast. Within engine control, the consideration of nonlinear system models and thus NMPC is often necessary. The fundamentals of NMPC are presented within the part. A focus is set on methods that allow for the real-time-feasible computation of the solution for the optimization problem formulated. This concerns discretization methods and also numerical solution algorithms. As one suitable solution algorithm, the so-called sequential quadratic programming (SQP) is introduced. Finally, suitable formulations of the optimization problem are discussed. This is a very crucial task as the optimization problem determines the control behavior. The formulation should allow to fully exploit the system capabilities and to fulfill the requirements on the engine controller. This concerns, for instance, the requirement to achieve offset-free control.

Part II provides an overview of the control tasks of an engine system. First, the general control structure is presented, along with an introduction of the control tasks within the fuel path, the air path, and the exhaust gas path. Based on this general overview, the specific requirements of a spark ignition (SI), a compression ignition (CI), and a low-temperature combustion (LTC) engines are explained. The control-relevant differences in the combustion concepts are detailed. For the SI and the CI engines, one typical setup is presented, along with its sensor and actuator architecture. For this setup, the various control loops are explained and typical sensitivities from actuated value to controlled value are demonstrated. For the LTC concept, three different possibilities for realizing LTC are revisited. Specifically the Gasoline Controlled Autoignition (GCAI), the diesel-based Premixed Charge Compression Ignition (PCCI), and the dual-fuel-based Reactivity Controlled Compression Ignition

(RCCI) are introduced. The control challenges that arise for the various realizations are described as well.

The purpose of Part III is to show in detail the application of MPC to air path control tasks, which is a very common and complex engine control problem. As a basis, the reduced-order modeling of the air path system is investigated. All common components of the air path, such as throttles, volumes, and turbochargers, thus are revisited to derive reduced-order models. A special focus is set on the suitability for the use within optimization-based control algorithms. The application of MPC is demonstrated by investigating a CI and an SI air path control problem. The CI engine air path consists of a combined exhaust gas recirculation and turbocharging with variable turbine geometry. The MPC approach is used to control the intake manifold pressure and the burnt gas ratio. For the SI engine, a two-stage turbocharging concept is investigated, where two wastegate-equipped turbochargers are placed in series to control the charging pressure. For both systems, the reduced-order modeling as well as a suitable MPC algorithm are detailed. The requirements and the setup of both systems are different. Nevertheless, a quite similar systematic procedure based on MPC can be used to develop appropriate control algorithms. This is possible due to the fact that MPC is inherently able to handle a broad range of requirements and system dynamic properties.

Part IV details the fundamentals of combustion control and modeling along with a case study. An overview of the in-cycle-resolved modeling of the CI engine process is presented first. The fundamental equations for the calculation of the cylinder pressure are detailed. Necessary submodels such as the ones for the determination of the cylinder volume and the heat transfer through cylinder walls are presented. The part details as well the state-of-the-art combustion control algorithms which rely on the use of cycle-integral parameters. Additionally, combustion rate shaping is introduced which is investigated in research for improved combustion control. The goal of combustion rate shaping is to shape the heat release highly resolved by the use of multiple-pulse fuel injections. Two aspects of the combustion rate shaping are investigated in the case study: the modeling of the combustion process and the so-called fuel injection rate digitalization. Algorithms are detailed that can be applied for both, conventional CI engines and PCCI engines. The combustion process is a complex multi-physics process, which is usually represented by very detailed models that are not suited for control purposes. Hence, data-based models of the process are investigated instead. Optimization-based methods can be used to determine the parameters of the model. As the second aspect, real-time optimization is used to realize the discretization of the continuous fuel injection rate. As a result, the discrete startpoints and durations of the various injection events can be determined that resemble the continuous reference fuel-rate flow as close as possible.

References

1. N. Abe and K. Yamanaka, Smith predictor control and internal model control – a tutorial, in *SICE Conference* (2003), pp. 1383–1387
2. A. Alessio, A. Bemporad, A survey on explicit model predictive control, in *Nonlinear Model Predictive Control* (Springer, 2009), pp. 345–369
3. A. Bemporad, D. Bernardini, R. Long, J. Verdejo, Model predictive control of turbocharged gasoline engines for mass production, in *SAE Technical Paper*, vol. 2018-01-0875 (2018)
4. BP p.l.c., Energy Outlook (2018). Accessed 2019-05-08. https://www.bp.com/content/dam/bp/en/corporate/pdf/energy-economics/energy-outlook/bp-energy-outlook-2018.pdf
5. M. Broy, I.H. Krüger, A. Pretschner, C. Salzmann, Engineering automotive software. Proc. IEEE **95**(2), 356–373 (2007)
6. L. Del Re, P. Ortner, D. Alberer, Chances and challenges in automotive predictive control, in *Automotive Model Predictive Control* (Springer, 2010), pp. 1–22
7. S. Di Cairano, An industry perspective on MPC in large volumes applications: potential benefits and open challenges. IFAC Proc. Vol. **45**(17), 52–59 (2012)
8. S. Di Cairano, I.V. Kolmanovsky, Automotive applications of model predictive control, in *Handbook of Model Predictive Control* (Springer, 2019), pp. 493–527
9. ETAS GmbH, ETAS calibration consulting flyer (2018). Accessed 2018-06-17. https://www.etas.com/download-center-files/company/Calibration_Consulting_Flyer_EN.pdf
10. European Union, Regulation (EC) No 715/2007 of the European Parliament and of the Council of 20 June 2007 on type approval of motor vehicles with respect to emissions from light passenger and commercial vehicles (Euro 5 and Euro 6) and on access to vehicle repair and maintenance information. Off. J. Eur. Union (2007)
11. H.J. Ferreau, S. Almér, R. Verschueren, M. Diehl, D. Frick, A. Domahidi, J.L. Jerez, G. Stathopoulos, C. Jones, Embedded optimization methods for industrial automatic control. IFAC-PapersOnLine **50**(1), 13 194–13 209 (2017)
12. J. Gregory, R.J. Stouffer, M. Molina et al., Climate change 2007: the physical science basis. *Intergovernmental Panel on Climate Change* (2007)
13. International Energy Agency, World Energy Outlook (2016). Accessed 2018-06-20. https://www.iea.org/publications/freepublications/
14. International Maritime Organization, *MARPOL Annex VI and NTC 2008: With Guidelines for Implementation* (IMO Publishing, 2013)
15. R. Isermann, *Engine Modeling and Control* (Springer, 2014)
16. J. Levine, *Analysis and Control of Nonlinear Systems: a Flatness-Based Approach* (Springer, 2009)
17. L. Ljung, Perspectives on system identification. Ann. Rev. Control **34**(1), 1–12 (2010)
18. F. Payri, J. Luján, C. Guardiola, B. Pla, A challenging future for the IC engine: new technologies and the control role. Oil Gas Sci. Technol. **70**(1), 15–30 (2014)
19. S.J. Qin, T.A. Badgwell, A survey of industrial model predictive control technology. Control Eng. Pract. **11**(7), 733–764 (2003)
20. S.V. Raković, W.S. Levine, *Handbook of Model Predictive Control* (Springer, 2018)
21. R.D. Reitz, Directions in internal combustion engine research. Combust. Flame **160**(1), 1–8 (2013)
22. U.S. Energy Information Administration, International Energy Outlook (2017). Accessed 2019-05-08. https://www.eia.gov/outlooks/ieo/pdf/0484(2017).pdf
23. D. von Wissel, A. Husson, V. Talon, L. Lansky, D. Pachner, M. Uchanski, Reducing engine calibration time and cost with model predictive control, in *IAV Automotive Powertrain Control Systems Conference* (2014)
24. Q.-G. Wang, *Decoupling ontrol* (Springer, 2002)
25. F. Willems, P. van Gompel, X. Seykens, S. Wilkins, Robust real-world emissions by integrated ADF and powertrain control development, in *Control Strategies for Advanced Driver Assistance Systems and Autonomous Driving Functions* (Springer), pp. 29–45 (2019)

Chapter 2
Model-Based Approach with PID Controllers

Abstract The state of the art in engine control is the use of decentralized PID controllers based on look-up tables. For complex engine tasks, the requirements on the controller can only be fulfilled by using a high number of calibration parameters. The tuning of the calibration parameters is very time-consuming and is associated with high costs as well as suboptimality. In this chapter, it will be shown why so many calibration parameters result. It will be exemplified by investigating two typical system dynamical properties which arise in combustion engines. In particular, multiple-input multiple-output (MIMO) system dynamics and actuator constraints are examined. The conventional control algorithms are not able to handle these system properties in a systematic manner. Instead, additional measures, such as decoupling terms or anti-windup mechanisms, have to be implemented to handle the system properties.

2.1 Multiple-Input Multiple-Output Systems

MIMO systems are characterized by several inputs and outputs, as opposed to single-input single-output (SISO). In the case of MIMO systems, the inputs and outputs may have cross-couplings, e.g. both inputs $\boldsymbol{u}(t)$ affect both outputs $\boldsymbol{y}(t)$ simultaneously; see Fig. 2.1. As the input and also the output are given as vectors rather than being scalar, additional properties are present, such as the concept of directionality. Thus, the gain of the system will not only depend on the frequency of the input but also on the direction of the input. The direction of the input can be expressed by singular value decomposition [11].

Within the present book, various examples will be investigated in detail where this MIMO characteristic is present. A common engine-related example is the air path control with exhaust gas recirculation (EGR) and turbocharging combined, as detailed in Chap. 10. The task is to control the intake manifold pressure and the burnt gas ratio with the actuated values being the guide vane position of the variable turbine geometry and the EGR valve position. However, changing the EGR valve

T. Albin Rajasingham, *Nonlinear Model Predictive Control of Combustion Engines*, Advances in Industrial Control,
https://doi.org/10.1007/978-3-030-68010-7_2

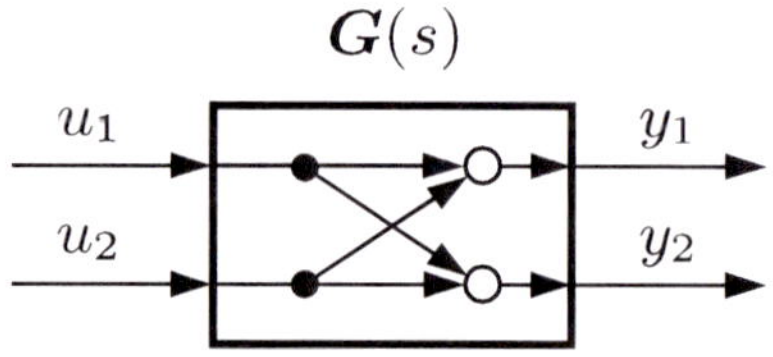

Fig. 2.1 Cross-couplings in a 2 × 2 MIMO system

position changes not only the burnt gas ratio but also the intake manifold pressure. As a consequence, the variable turbine geometry has to be adjusted simultaneously for appropriate closed-loop control.

In order to close-loop-control a MIMO system, two different approaches are possible: centralized and decentralized control; see Fig. 2.2. In the case of centralized control, one MIMO controller is used to control the entire system. Suitable control methods are H_∞, linear quadratic regulator (LQR), and model predictive control (MPC) [1, 11]. The main objective of decentralized control is to control the outputs of a MIMO system with distinct SISO controllers. In this case, standard SISO control concepts such as PID can be used. The applicability of the widespread PID controller makes the decentralized control the state of the art for engine control problems. In the case of weakly coupled systems, their treatment as several single-loop control problems might be sufficient. However, if the couplings are strong, the treatment of the MIMO system as several uncoupled SISO systems for control design can lead to unsatisfactory control performance. In the worst case, the negligence of the cross-couplings can lead to controllers which "work against each other". In this case, a suitable controller can be obtained only if the cross-couplings are considered. In general, the design of a decentralized controller is simpler. However, it is suboptimal as it does not take the cross-couplings into account. In the following, an overview will be given on decentralized control for MIMO systems. The fundamentals are outlined along with some of the challenges arising for the design of decentralized control.

2.1.1 Introduction to MIMO Systems

A common way to describe linear MIMO systems is the use of the transfer function matrix $\boldsymbol{G}(s)$. For a system with k inputs and l outputs, the transfer function matrix consists of the transfer functions G_{ij}, where $i = 1, \ldots, l$ and $j = 1, \ldots, k$. Each transfer function G_{ij} describes the relation between the input $U_j(s)$ and the output $Y_i(s)$. The inputs $U_j(s)$ and outputs $Y_i(s)$ define the Laplace transform of the corresponding scalar input $u_j(t)$ and scalar output $y_i(t)$ (for simplicity, vanishing initial conditions are assumed).

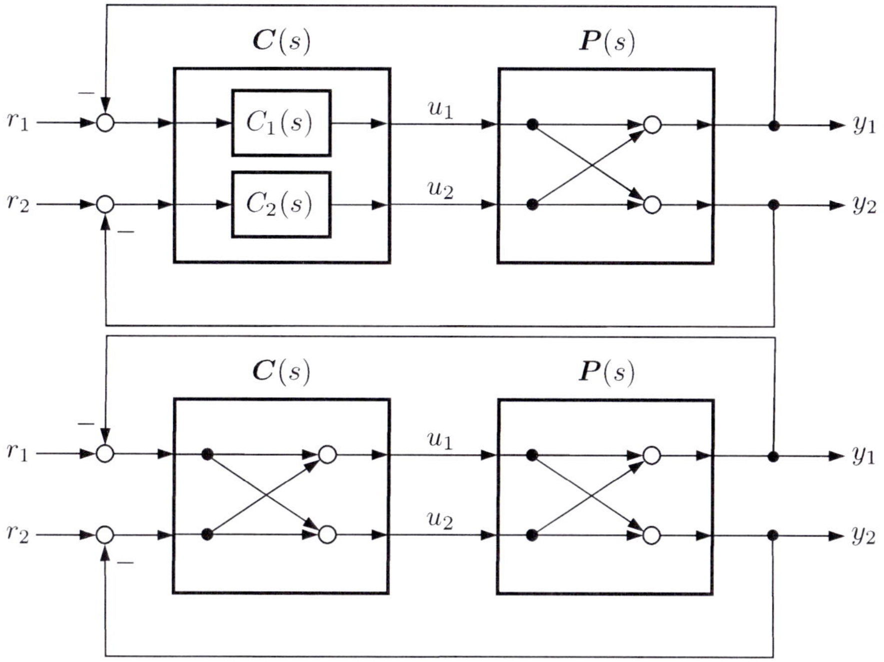

Fig. 2.2 Decentralized (upper plot) versus centralized (bottom plot) control for a 2×2 example

$$\begin{bmatrix} Y_1(s) \\ \vdots \\ Y_l(s) \end{bmatrix} = \begin{bmatrix} G_{11}(s) & \dots & G_{1k}(s) \\ \vdots & \ddots & \vdots \\ G_{l1}(s) & \dots & G_{lk}(s) \end{bmatrix} \begin{bmatrix} U_1(s) \\ \vdots \\ U_k(s) \end{bmatrix} \tag{2.1}$$

$$\boldsymbol{Y}(s) = \boldsymbol{G}(s)\boldsymbol{U}(s) \tag{2.2}$$

In the example of a linear system with two inputs and two outputs, four transfer functions result, as shown in Fig. 2.3.

$$\begin{bmatrix} Y_1(s) \\ Y_2(s) \end{bmatrix} = \begin{bmatrix} G_{11}(s) & G_{12}(s) \\ G_{21}(s) & G_{22}(s) \end{bmatrix} \begin{bmatrix} U_1(s) \\ U_2(s) \end{bmatrix} \tag{2.3}$$

The use of transfer function matrices allows to compute the input–output behavior of a complex system, for instance, when the system consists of several subsystems, each of which is described by a transfer function matrix. This is the case, for example, when the behavior of a MIMO closed-loop system is to be determined. For the calculation with transfer function matrices, the fact has to be considered that in general no commutativity is given. Thus, in contrast to SISO systems, the order of the transfer function matrices in the system does matter. Obviously, by changing the order of the transfer function matrices, the sizes of the matrices and thus their physical counterparts, i.e. the number of inputs, might no longer be applicable. In

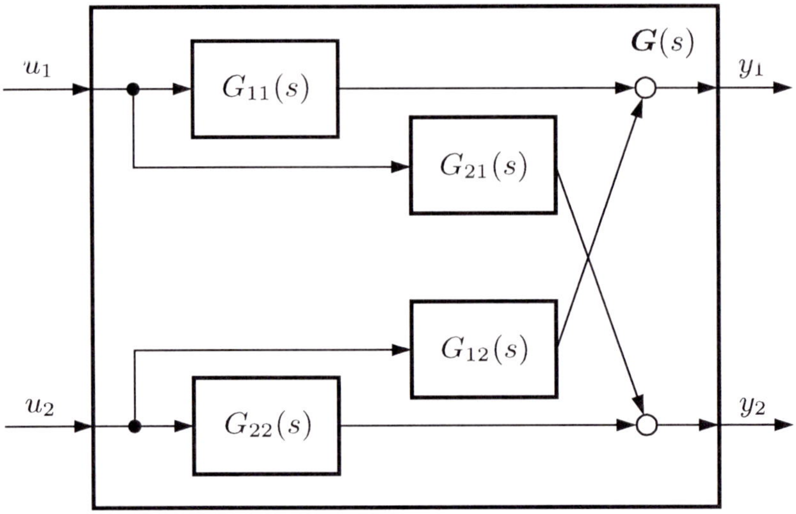

Fig. 2.3 Transfer functions from each input to each output for a linear 2×2 system

Fig. 2.4, the parallel, serial, and feedback connection of transfer function matrices is depicted. The general calculation rules for these fundamental operations are given in the following:

- Parallel connection
$$\boldsymbol{G} = \boldsymbol{G}_1 + \boldsymbol{G}_2 \tag{2.4}$$

- Series connection
$$\boldsymbol{G} = \boldsymbol{G}_2\boldsymbol{G}_1 \neq \boldsymbol{G}_1\boldsymbol{G}_2 \tag{2.5}$$

- Feedback
$$\boldsymbol{G} = \boldsymbol{G}_1(\boldsymbol{I} + \boldsymbol{G}_2\boldsymbol{G}_1)^{-1} = (\boldsymbol{I} + \boldsymbol{G}_1\boldsymbol{G}_2)^{-1}\boldsymbol{G}_1 \tag{2.6}$$

The transfer function matrix of the input–output behavior can be used to analyze the characteristics of the system investigated. The most important properties are the pole-zero locations, controllability, observability, and stability. Standard textbooks dealing with MIMO control, such as [10, 11], give an overview of these topics. They also describe in detail the differences between these concepts for the SISO and the MIMO case. Although the fundamental concepts are the same, differences result in calculation and interpretation, such as the directionality of zeros. Additionally, there are characteristics which are specific to MIMO systems and are directly related to the concept of directionality. Examples are the singular value decomposition and the relative gain array (RGA) matrix [1]. As an example, the RGA methodology is detailed in the following.

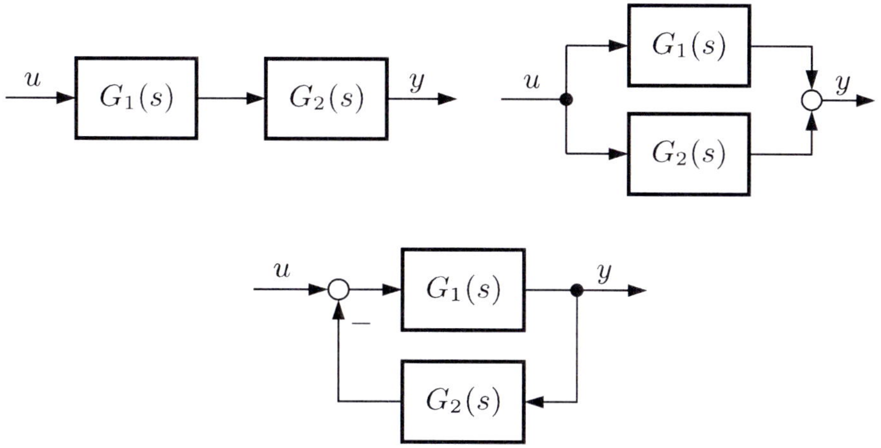

Fig. 2.4 Connections of subsystems: serial (upper left), parallel (upper right), and feedback (below)

2.1.2 *Relative Gain Array*

An example of a closed-loop system with a decentralized controller and a 2×2 plant is depicted in Fig. 2.5. In this case, the decentralized controller consists of the transfer functions $C_1(s)$ and $C_2(s)$. The first step to design decentralized controllers is to determine which combination of input and output is the best choice for the feedback loops. This question is sometimes called the *pairing problem*. For this purpose, the RGA can be used [1]. It allows quantifying the effectiveness of a certain input to influence a certain output in a closed-loop control setting. This influence consists of two portions: One is the direct influence occurring when the entire system is operated in an open-loop fashion ($u_1(t) \rightarrow y_1(t)$ via P_{11} in Fig. 2.5), and the other one is given by the influence through the cross-couplings when the process is operated in a closed-loop system ($u_1(t) \rightarrow y_1(t)$ via P_{21} and subsequent subsystems in Fig. 2.5).

First, with all other inputs being constant, the open-loop gain from $U_j(s)$ to $Y_i(s)$ can be determined.

$$\left.\frac{\partial Y_i(s)}{\partial U_j(s)}\right|_{U_{k\neq j}=\text{const.}} = G_{ij}(s) \tag{2.7}$$

Next, the gain is quantified from U_j to Y_i when all other loops are closed. The ideal actuated values for perfect closed-loop tracking are given by

$$\boldsymbol{U}(s) = \boldsymbol{G}^{-1}(s)\boldsymbol{Y}(s) \tag{2.8}$$

When all other outputs $Y_{k\neq i}$ are perfectly closed-loop tracked, the sensitivities can be calculated.

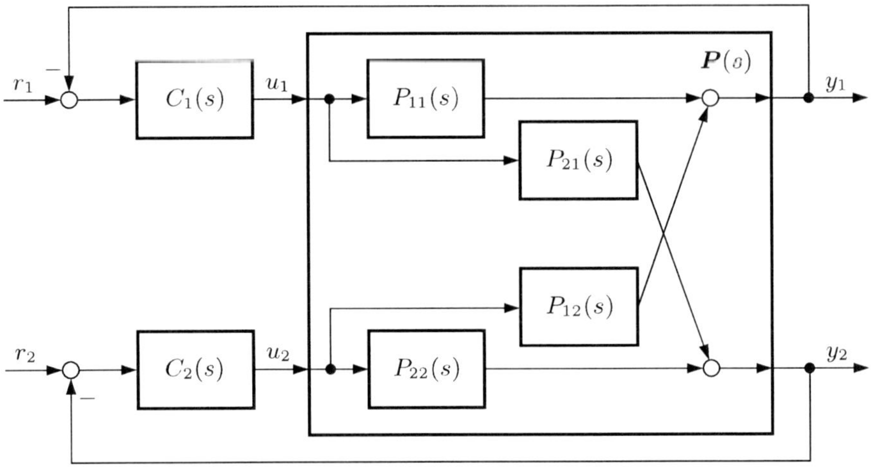

Fig. 2.5 Decentralized control for a 2×2 system

$$\left.\frac{\partial U_j(s)}{\partial Y_i(s)}\right|_{Y_{k\neq i}=\text{const.}} = [\boldsymbol{G}^{-1}(s)]_{ji} \quad \Rightarrow \quad \left.\frac{\partial Y_i(s)}{\partial U_j(s)}\right|_{Y_{k\neq i}=\text{const.}} = \frac{1}{[\boldsymbol{G}^{-1}(s)]_{ji}} \tag{2.9}$$

The RGA element $\Lambda_{ij}(s)$ is given by the ratio of these two gains:

$$\Lambda_{ij}(s) = \frac{\left.\frac{\partial Y_i}{\partial U_j}\right|_{U_{k\neq j}=const.}}{\left.\frac{\partial Y_i}{\partial U_j}\right|_{Y_{k\neq i}=const.}} = G_{ij}(s)[\boldsymbol{G}^{-1}(s)]_{ji} \tag{2.10}$$

Alternatively, a matrix representation can be used for $\boldsymbol{\Lambda}(s)$.

$$\boldsymbol{\Lambda}(s) = \boldsymbol{G}(s) \times (\boldsymbol{G}^{-1}(s))^{\mathsf{T}} \tag{2.11}$$

Here, the $\times$ operator represents an element-wise multiplication, also called a Schur product. The matrix $\boldsymbol{\Lambda}(s)$ is a transfer function matrix with some specific properties: It is symmetric, and all rows and columns add up to one. As ratios are calculated, the RGA matrix $\boldsymbol{\Lambda}(s)$ is independent of scaling and thus also independent of the physical units used.

Example 2.1 (*Calculation of the RGA matrix*) The RGA matrix $\boldsymbol{\Lambda}(s)$ shall be calculated for a system with two inputs and two outputs (the dependence on s is omitted for readability).

$$\boldsymbol{G}(s) = \begin{bmatrix} G_{11} & G_{12} \\ G_{21} & G_{22} \end{bmatrix}, \quad \boldsymbol{\Lambda}(s) = \frac{1}{G_{11}G_{22} - G_{12}G_{21}} \begin{bmatrix} G_{11}G_{22} & -G_{12}G_{21} \\ -G_{12}G_{21} & G_{11}G_{22} \end{bmatrix} \tag{2.12}$$

The RGA matrix $\mathbf{\Lambda}(s)$ can be evaluated to determine suitable parings of inputs and outputs. As $\mathbf{\Lambda}(s)$ is a transfer function matrix, the evaluation has to be conducted at a certain frequency. Typically, the steady-state properties $s = 0$ and the crossover frequency region $s \in [j\omega_{c,low}, j\omega_{c,up}]$ are of interest. In the case of $\Lambda_{ij}(s) = 1$, the input–output combination is only affected by the open-loop gain and is not affected by any cross-couplings. Consequently, this would be a perfect choice for pairing. In general, it is advisable to pair inputs and outputs where the RGA elements are close to one. The more closely the RGA matrix is to the identity matrix $\boldsymbol{I}$, the better the system is suited for decentralized control. For the quantification of "closeness", any distance norm can be used. Pairings with very high or very low values in the RGA matrix should be avoided, as this indicates a very strong interaction with other variables than the paired variables. For $\Lambda_{ij}(s) = 0$, there is no open-loop effect of the input on the output, thus it is a bad choice for pairing. Pairings with $\Lambda_{ij}(s) < 0$ should also be avoided, as these can lead to problems with closed-loop stability [11].

Example 2.2 (*Design of a decentralized controller*) In the following, an example for decentralized control design is investigated. The system to be controlled is given by the transfer function matrix $\boldsymbol{P}(s)$:

$$\boldsymbol{Y}(s) = \boldsymbol{P}(s)\boldsymbol{U}(s) \tag{2.13}$$

$$\begin{bmatrix} Y_1 \\ Y_2 \end{bmatrix} = \begin{bmatrix} \frac{1}{s+1} & \frac{4}{s+8} \\ \frac{0.5}{s+1} & \frac{1}{s+1} \end{bmatrix} \begin{bmatrix} U_1 \\ U_2 \end{bmatrix} \tag{2.14}$$

The considered system has similarities with the Rosenbrock system which is classically used as an example for a system that looks simple to control, but has some fundamental limitations [3]. In order to choose the pairing of inputs and outputs, the RGA matrix is calculated. With the equations given in (2.12), the RGA matrix can be calculated exemplarily for $s = 0$.

$$\mathbf{\Lambda}(0) = \begin{bmatrix} 1.33 & -0.33 \\ -0.33 & 1.33 \end{bmatrix} \tag{2.15}$$

The resulting $\mathbf{\Lambda}$ matrix suggests to pair input y_1 with u_1 and, for the second control loop, to pair y_2 with u_2. For an initial control synthesis, the cross-couplings are ignored, i.e. $P_{12} = P_{21} = 0$. The effects of cross-couplings are considered as external disturbances d_1 and d_2, which can ideally be rejected by the closed-loop controller. As a consequence, two SISO control loops remain, which can easily be designed; see Fig. 2.6. Ignoring the cross-couplings, results in the system $\tilde{\boldsymbol{P}}(s)$ with two first-order elements on the diagonal.

$$\boldsymbol{Y}(s) = \tilde{\boldsymbol{P}}(s)\boldsymbol{U}(s) \tag{2.16}$$

$$\begin{bmatrix} Y_1 \\ Y_2 \end{bmatrix} = \begin{bmatrix} \frac{1}{s+1} & 0 \\ 0 & \frac{1}{s+1} \end{bmatrix} \begin{bmatrix} U_1 \\ U_2 \end{bmatrix} \tag{2.17}$$

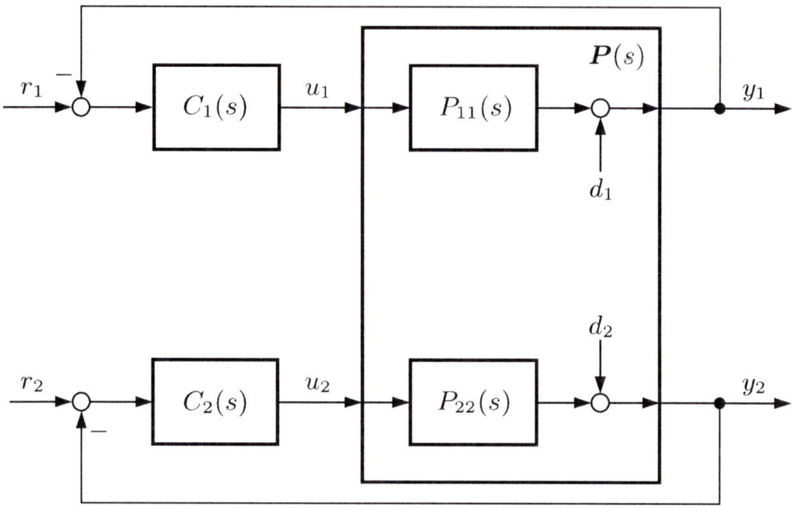

Fig. 2.6 The decentralized controller for the system investigated

Simple proportional controllers with $C_1 = 18,\ C_2 = 20$ result in suitable closed-loop control for the two SISO loops.

$$\begin{bmatrix} U_1 \\ U_2 \end{bmatrix} = \begin{bmatrix} 18 & 0 \\ 0 & 20 \end{bmatrix} \begin{bmatrix} R_1 - Y_1 \\ R_2 - Y_2 \end{bmatrix} \tag{2.18}$$

To determine the closed-loop dynamics of the entire system, the complementary sensitivity functions can be calculated.

$$\begin{aligned} T_1(s) &= \frac{Y_1(s)}{R_1(s)} = \frac{C_1(s)P_{11}(s)}{1 + C_1(s)P_{11}(s)} = \frac{18}{s+19} \\ T_2(s) &= \frac{Y_2(s)}{R_2(s)} = \frac{C_2(s)P_{22}(s)}{1 + C_2(s)P_{22}(s)} = \frac{20}{s+21} \end{aligned} \tag{2.19}$$

The left-hand side plot in Fig. 2.7 shows the step response of the two single closed loops without any further cross-couplings considered, i.e. $\tilde{\boldsymbol{P}}(s)$ is used as a plant. A suitable closed-loop control performance is achieved. However, the picture changes as the full system, i.e. $\boldsymbol{P}(s)$ as a plant, is investigated. The step response of the closed-loop system is shown in the right-hand plot of Fig. 2.7. Including the cross-couplings, the system turns out to be unstable. Figure 2.8 shows the pole-zero map of the open-loop system $\boldsymbol{P}(s)$. The difficulty in controlling the system arises due to the existence of a right half-plane transmission zero. Consequently, cross-coupling can lead to instability, even if all SISO closed loops are stable. In order to achieve suitable control performance, the closed-loop couplings have to be considered in the design. One possibility to still rely on a SISO control synthesis while considering the

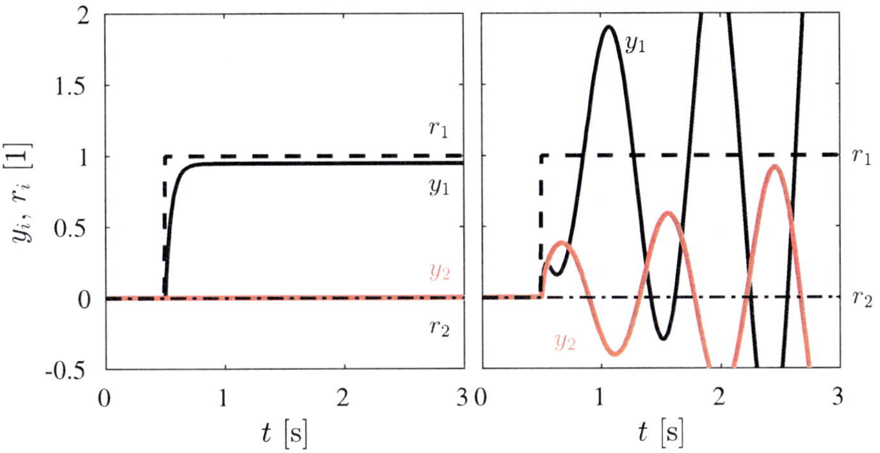

Fig. 2.7 Step response of the closed-loop system with $\tilde{\boldsymbol{P}}(s)$ as a plant (left) versus $\boldsymbol{P}(s)$ as a plant (right)

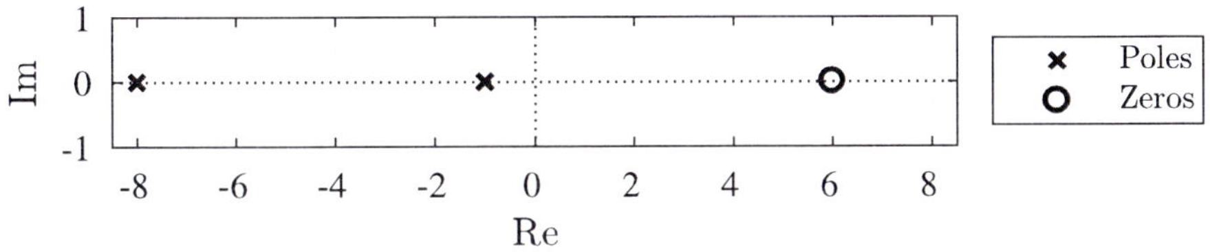

Fig. 2.8 Pole-zero map of the open-loop system $\boldsymbol{P}(s)$

couplings in the design process can be achieved by using system decoupling, which is explained in the following.

2.1.3 *System Decoupling*

If the controller design with separate SISO controllers for the decentralized system is insufficient, the cross-couplings have to be considered in the design. One approach for dealing with the cross-couplings while still using SISO control synthesis is the application of decoupling in addition to the decentralized controller. The idea is to introduce a decoupling matrix $\boldsymbol{D}(s)$ to achieve a sufficient decoupling of the system to allow for decentralized control synthesis. Ideally, after the introduction of $\boldsymbol{D}(s)$, no cross-coupling effects are present anymore.

The diagonal controller matrix $\boldsymbol{C}(s)$ is designed in a separate step once the system is decoupled. The structure of the closed-loop system including the decoupling matrix is given in Fig. 2.9. The dynamics of the closed-loop system can be calculated by the

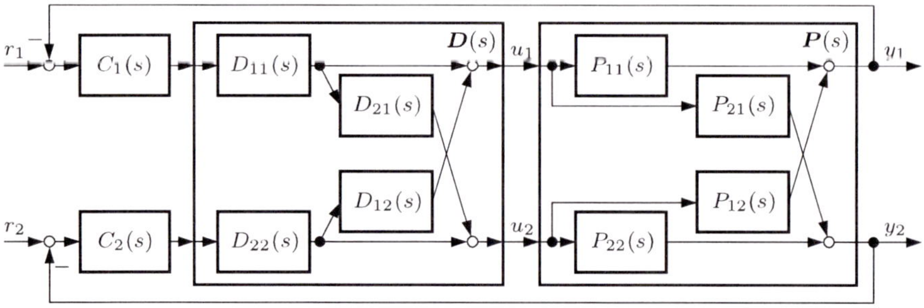

Fig. 2.9 Principle of decoupling for a 2×2 plant

transfer function matrix $\boldsymbol{T}(s)$. For readability, the dependency on s is omitted when appropriate.

$$\boldsymbol{T} = \boldsymbol{PDC}\Big(\boldsymbol{I} + \boldsymbol{PDC}\Big)^{-1} \tag{2.20}$$

The closed-loop system is perfectly decoupled if $\boldsymbol{T}(s)$ is diagonal and nonsingular [12]. The matrix $\boldsymbol{C}(s)$ is diagonal, as a decentralized controller is investigated. If $\boldsymbol{T}(s)$ is diagonal, the inverse $\boldsymbol{T}^{-1}(s)$ has to be diagonal as well. The inverse is given by

$$\begin{aligned} \boldsymbol{T}^{-1} &= (\boldsymbol{I} + \boldsymbol{PDC})\Big(\boldsymbol{PDC}\Big)^{-1} \\ &= \Big(\boldsymbol{PDC}\Big)^{-1} + \boldsymbol{I} \end{aligned} \tag{2.21}$$

As a consequence, the closed-loop system $\boldsymbol{T}(s)$ is diagonal (and thus decoupled), if and only if the open-loop system $\boldsymbol{P}(s)\boldsymbol{D}(s)$ is decoupled. The matrix $\boldsymbol{D}(s)$ has to be designed such that $\boldsymbol{P}(s)\boldsymbol{D}(s)$ is diagonal. In this case, the following open-loop behavior is achieved.

$$\boldsymbol{P}(s)\boldsymbol{D}(s) = \text{diag}\big\{Q_{11}(s), Q_{22}(s), \dots\big\} \tag{2.22}$$

In a perfectly decoupled system, all off-diagonal terms are zero. The diagonal entries of the new decoupled system matrix are represented by $Q_{11}(s)$, $Q_{22}(s)$, and so on. For a 2×2 plant, the following equations result with the structure corresponding to Fig. 2.9.

$$\boldsymbol{D}(s) = \begin{bmatrix} D_{11} & D_{22}D_{12} \\ D_{11}D_{21} & D_{22} \end{bmatrix} \tag{2.23}$$

Using $\det(\boldsymbol{P}(s))$ for the determinant of $\boldsymbol{P}(s)$, the elements of the decoupling matrix $\boldsymbol{D}(s)$ are given by

$$D_{11} = \frac{P_{22}}{\det(\boldsymbol{P})} Q_{11} = \frac{P_{22}}{P_{11}P_{22} - P_{12}P_{21}} Q_{11}, \quad D_{12} = -\frac{P_{12}}{P_{11}},$$
$$D_{22} = \frac{P_{11}}{\det(\boldsymbol{P})} Q_{22} = \frac{P_{11}}{P_{11}P_{22} - P_{12}P_{21}} Q_{22}, \quad D_{21} = -\frac{P_{21}}{P_{22}} \tag{2.24}$$

The control engineer designs the transfer functions $Q_{ii}(s)$ on the diagonal of the resulting decoupled MIMO system. The appropriate choice of the transfer functions for $Q_{ii}(s)$ depends strongly on the original system. One possibility is to use the diagonal elements of the original system, i.e. $Q_{ii}(s) = P_{ii}(s)$.

This general decoupling approach relies on inverting the transfer functions of the original system. The inversion-based design has limitations for many cases, such as the presence of model uncertainty, dead-times, right half-plane zeroes, and poles. These properties can have a major (negative) impact on the performance of the decoupled system. Furthermore, an exact decoupling may be rendered impossible by the fact that all terms in the decoupling matrix have to be proper to be applicable in an online control system. For more information on how the decoupling matrix is calculated, as well as which constraint exists on the performance achievable, the reader is referred to [12].

For certain systems, a transfer function matrix $\boldsymbol{D}(s)$ exists for perfect decoupling. However, this approach often fails. In general, perfect decoupling is difficult or impossible to achieve for a given system. In practice, perfect decoupling is often not even necessary, because an approximate decoupling is sufficient. Depending on the system characteristics present, various approaches for approximate decoupling exist [12]. For instance, the issue of improper terms in $\boldsymbol{D}(s)$ can be solved by introducing additional poles until the terms are proper. Doing so, the original decoupling terms are recovered for low frequencies. Another alternative for approximate decoupling is to use decoupling at one specific frequency, i.e. at $s = j\omega_{des}$. If the determinant $\det(\boldsymbol{D}(j\omega_{des})) \neq 0$, then $\boldsymbol{D}(j\omega_{des})$ can be chosen as a constant, i.e. a frequency-independent decoupling matrix. A common choice is $\omega_{des} = 0\frac{\text{rad}}{\text{s}}$ such that the system is statically decoupled, i.e. at steady state. Another alternative is a decoupling at the desired bandwidth. The use of static decoupling is often better than no decoupling. If approximate decoupling is applied, it is important to check if the closed-loop system fulfills the desired properties, of which at least stability is critical. There are no theoretical guarantees on performance. However, quite good results are usually observed in practice.

Example 2.3 (*Closed-loop control with decoupling*) The closed-loop control system from Example 2.2 is to be extended by a decoupling term. Figure 2.8 showed that the system has a non-minimum phase zero at a frequency of $6\frac{\text{rad}}{\text{s}}$, which limits the feasible choices for the diagonal terms of the decoupled system. The zero has to occur in the chosen transfer functions Q_{ii}. Additional poles are introduced for Q_{ii}, such that the resulting decoupling terms are strictly proper. One feasible choice is given by

$$Q_{11} = Q_{22} = -\frac{0.1(s-6)}{s^2 + 9s + 8} \tag{2.25}$$

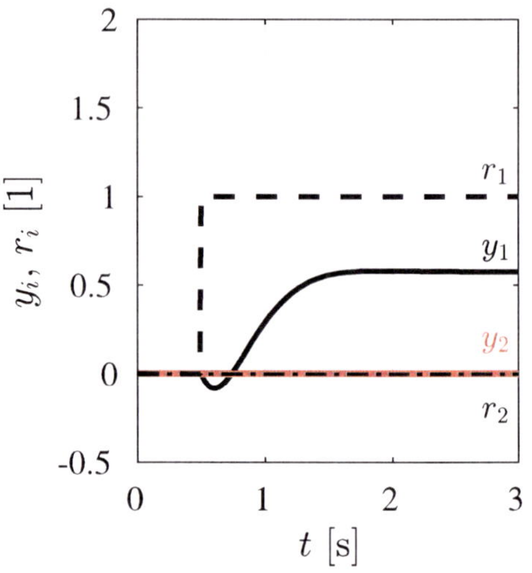

Fig. 2.10 Closed-loop control including decoupling matrix with simple feedback controller

With (2.24), the decoupling matrix can be calculated as

$$
\begin{aligned}
D_{11} &= D_{22} = 0.1 \\
D_{12} &= -\frac{4(s+1)}{s+8}, \qquad D_{21} = -\frac{1}{2}
\end{aligned}
\tag{2.26}
$$

Example 2.2 showed that the controllers $C_1 = 18$ and $C_2 = 20$ lead to unstable closed-loop system behavior when no decoupling is used. When the decoupling term is added, the closed-loop system becomes stable. Figure 2.10 shows the step response for the resulting system with the calculated decoupling matrix and the same controllers $C_1 = 18$ and $C_2 = 20$. The closed-loop system is stable, and even perfect decoupling can be achieved. For a step response on y_1, there is no impact on y_2. For the given system, the MIMO non-minimum phase transmission zero cannot be canceled: It is still present in the decoupled system. By using this decoupling approach, any SISO control design can be applied to shape the closed-loop control response. With the simple proportional controller, for instance, a significant steady-state error results. This can be eliminated by more advanced controllers.

2.2 Actuator Constraints

There is extensive theory available about linear controllers. The closed-loop control performance remains good as long as the plant dynamics remain close to the linear system dynamics that was used for synthesizing the controller. In reality, in all plants nonlinear effects are present. Constraints are one example of a nonlinear effect that

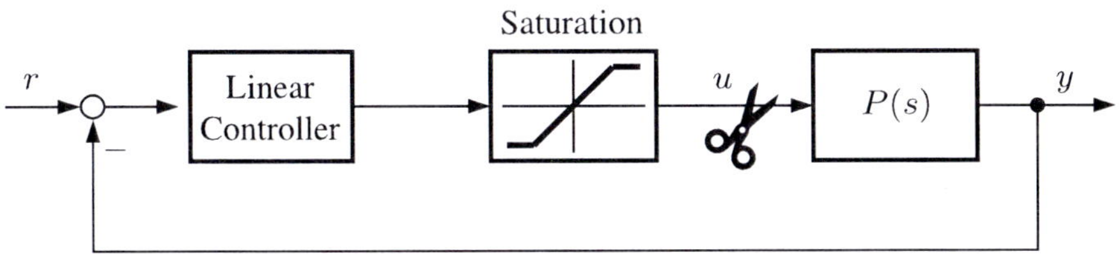

Fig. 2.11 System becomes open-loop if saturation is active

is always present. The constraints concern, for instance, the actuator: Due to limited energy, the actuators do have an upper and a lower limit on the absolute value of the actuated value or on the rate of change. Often, these constraints limit the closed-loop control performance. Usually, the actuator cannot be changed for economic, size, or weight reasons.

In Fig. 2.11, the closed-loop system is shown including actuator constraints. The presence of these actuator saturations can give rise to undesired behavior in the closed-loop system. The saturation can lead to closed-loop instability or a sluggish response with temporary large control errors. As soon as the saturation is active, the closed-loop system is not present anymore. Instead, an open-loop behavior is present. For an unstable plant, this inevitably leads to an unstable closed-loop system. Additionally, problems result with integrator windup and MIMO systems, as detailed in the following.

2.2.1 Integrator Windup

If an integrator is present within the controller or the plant, the problem of integrator windup can occur. A typical example is the combination of a controller with integral action and an actuator that has saturations. When the actuated value is saturated, the integral part will continue to integrate the error, as the error is typically not zero. Due to the constrained actuated value, the integration is decoupled from the actual system behavior, i.e. the increasing integral part does not have an influence on the actuated value. The calculated actuated value of the controller can become very large due to the term of the integrator. Even when the error changes, e.g. due to a change of the reference value, it takes a long time until the non-saturated region is reached again. Consequently, a sluggish behavior results in transients.

One possibility to counteract this undesired behavior is the use of an anti-windup mechanism. A well-known example is the so-called back-calculation scheme [4]. Figure 2.12 shows the functional diagram of the back-calculation scheme. The anti-windup scheme has no effect when the actuator is not saturating, i.e. when $e_t(t) = 0$. The time constant T_t determines how quickly the integrator of the PID controller is reset. Thus, T_t serves as a tuning parameter. For calculating the difference $e_t(t) = v(t) - u(t)$, the actual (saturated) output $u(t)$ of the actuator has to be

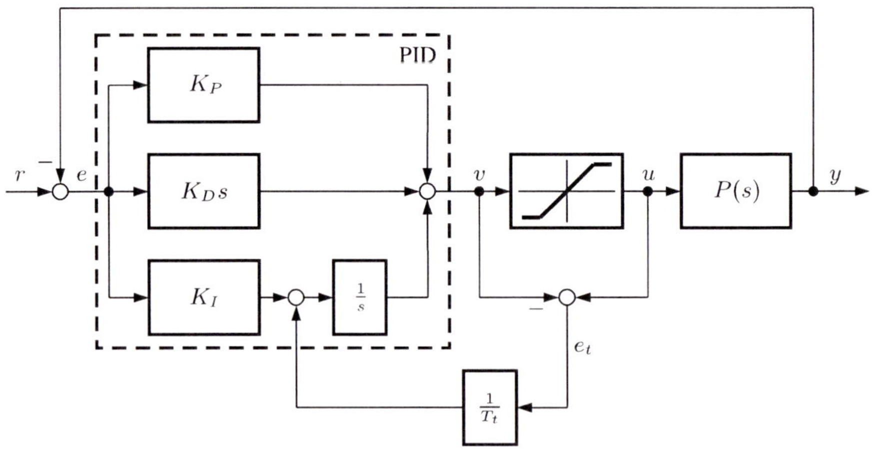

Fig. 2.12 Anti-windup with back calculation

either measured or determined by a mathematical model. A typical example for such a mathematical model is given in (2.27).

$$u(t) = \text{sat}(v) = \begin{cases} u_{max}, & \text{for } v > u_{max} \\ u, & \text{for } u_{min} \leq v \leq u_{max} \\ u_{min}, & \text{for } v < u_{min} \end{cases} \tag{2.27}$$

2.2.2 *Directionality in Constrained MIMO Systems*

Not only the magnitude and frequency of the input are important in MIMO systems, but also their direction due to cross-couplings. Active input constraints may change the direction of the input and a nonlinear system behavior of the plant results. In a closed-loop system, the active input constraints can lead to an undesirable system response. This problem is independent of the integrator windup discussed previously. The problem with directionality can also occur when no integrators are involved, e.g. when a simple proportional feedback controller is used.

To further illustrate this problem, a closed-loop system with a decentralized controller is to be treated. In Sect. 2.1.3, the usefulness of decoupling was introduced. A system is considered that is in nominal operation, i.e. without any active saturations, perfectly decoupled. As soon as input saturations are active, the directionality of the inputs is changed, such that for saturated inputs the decoupling is lost. Example 2.3 showed that the decoupling has a significant effect on the closed-loop stability. As a result, the closed-loop system can become unstable as soon as the input constraints are active, and thus decoupling is no longer given. Additional measures have to be introduced to counteract this behavior. One example is the introduction of a reference shaping filter [8].

Example 2.4 (*Actuator constraints for a MIMO system*) Suppose a 2×2 MIMO system is given. The transfer function matrix of the system is given by $\boldsymbol{P}(s)$.

$$\boldsymbol{P}(s) = \frac{\boldsymbol{Y}(s)}{\boldsymbol{V}(s)} = \begin{bmatrix} \frac{2}{s+4} & \frac{4}{s^2+3s+4} \\ 0 & \frac{1}{s+2} \end{bmatrix} \tag{2.28}$$

The input to the system is denoted by $\boldsymbol{v}$. It is saturated at $v_{1,max} = 0.8$ and $v_{2,max} = 1$. In order to decouple the system, the following decoupling matrix $\boldsymbol{D}(s)$ is introduced.

$$\boldsymbol{D}(s) = \frac{\boldsymbol{V}(s)}{\boldsymbol{U}(s)} = \begin{bmatrix} -1 & \frac{2(s+4)}{s^2+3s+4} \\ 0 & 1 \end{bmatrix} \tag{2.29}$$

The open-loop transfer function without considering the input saturations is given by $\boldsymbol{P}(s)\boldsymbol{D}(s)$.

$$\boldsymbol{P}(s)\boldsymbol{D}(s) = \begin{bmatrix} \frac{2}{s+4} & 0 \\ 0 & \frac{1}{s+2} \end{bmatrix} \tag{2.30}$$

The left-hand side of Fig. 2.13 shows the response to a step in u_2 for the system $\boldsymbol{P}(s)\boldsymbol{D}(s)$ without considering the input constraints. The system is perfectly decoupled, i.e. the input u_2 only affects y_2 but not y_1. However, if the step response is simulated for the system including the input constraints, the decoupling is not effective any more; see the right-hand side plot of Fig. 2.13. Due to saturation, the

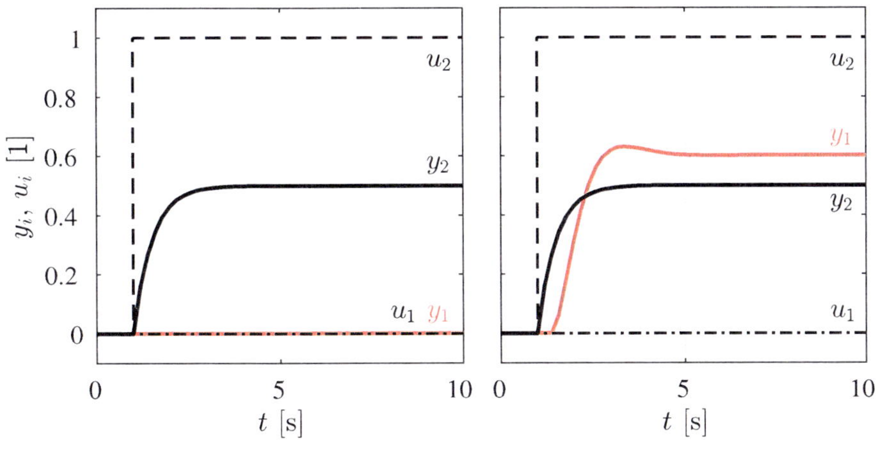

Fig. 2.13 Step response of the MIMO system without input saturations (left) and with input saturations (right)

directionality changes and the nonlinear system behavior no longer lead to a decoupling. In the case of a closed-loop control system, this can affect the stability of the system.

2.3 Control Approach Based on Look-Up Tables

Since the engine is characterized by nonlinear behavior, a linear controller is often not sufficient for appropriate control performance. In practice, the nonlinearity is tackled by the use of look-up tables. In a control approach based on look-up tables, depending on the control task, either feedforward control, feedback control, or a combination of both is applied. Figure 2.14 shows the resulting closed-loop control structure when feedforward and feedback control are used simultaneously. The feedforward control action is directly calculated with the use of a look-up table. For this case, at least one signal is present which serves as an input to the look-up table. This signal might either be the reference value or a disturbance variable which is measured or estimated. Based on this input signal, the look-up table calculates the output signal. For this reason, the look-up table is segmented into different regions, depending on the input value(s). For each cell, a feedforward control value is assigned. In between these values, a linear interpolation is conducted [6]. In the case of multiple input variables, a bilinear area interpolation is performed [9].

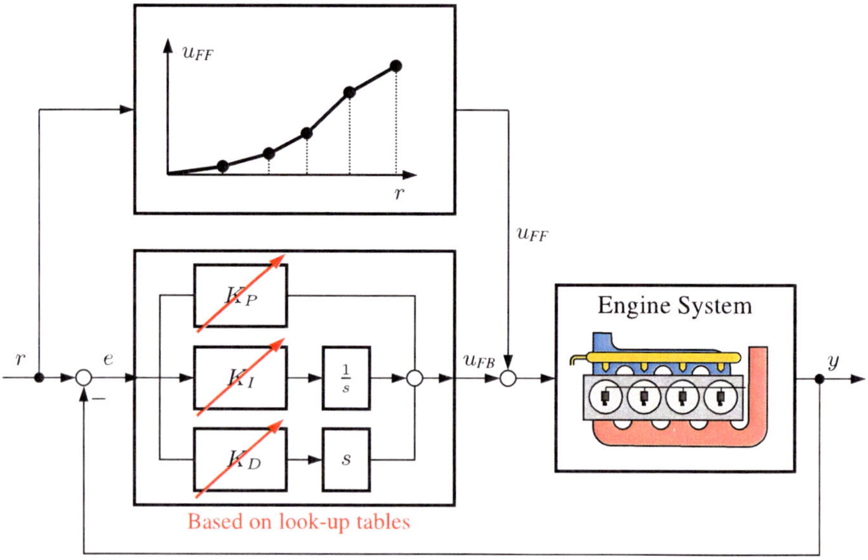

Fig. 2.14 Controller based on look-up tables

As a feedback control algorithm, most commonly SISO control loops with PID controllers are used, resulting in decentralized controllers. A standard PID controller is designed with three parameters, K_P, K_I, and K_D [2]. However, if a nonlinear process is controlled, where the system behavior depends strongly on the operating point, a standard PID controller is insufficient to achieve a highly dynamic reference tracking. Consequently, the parameters K_P, K_I, and K_D are parametrized depending on the operating point based on look-up tables. A possible approach is to use linearization of the nonlinear model to design the values K_P, K_I, and K_D for the different operating points. The look-up tables work equivalently to the description given in the case of the feedforward part. The main advantage of the approach based on look-up tables is the fast processing of the algorithm due to its simplicity. Thus, only low requirements for the control hardware in terms of memory storage and processor result. The simple structure of the controller also allows for easy recalibration. If the control behavior is to be adapted only at a certain operating point, this can be done in a straightforward manner.

Drawbacks arise especially in the case of a high number of inputs and outputs as a high amount of calibration work results. Figure 2.15 shows an example of a feedforward control calibration based on look-up tables. In this example, the fuel mass injected per cylinder is plotted on the left in terms of engine speed and torque. On the right-hand side, in addition to the engine speed and torque, the ambient temperature is used as an input for the look-up table. Changing from two inputs to three inputs, each input considered with four segments, leads to an increase in number of calibration parameter from $4^2 = 16$ to $4^3 = 64$ values. Clearly, the curse of dimensionality is coming into full play. For a small number of inputs, a relatively small number of calibration parameters is needed. However, already for a system that

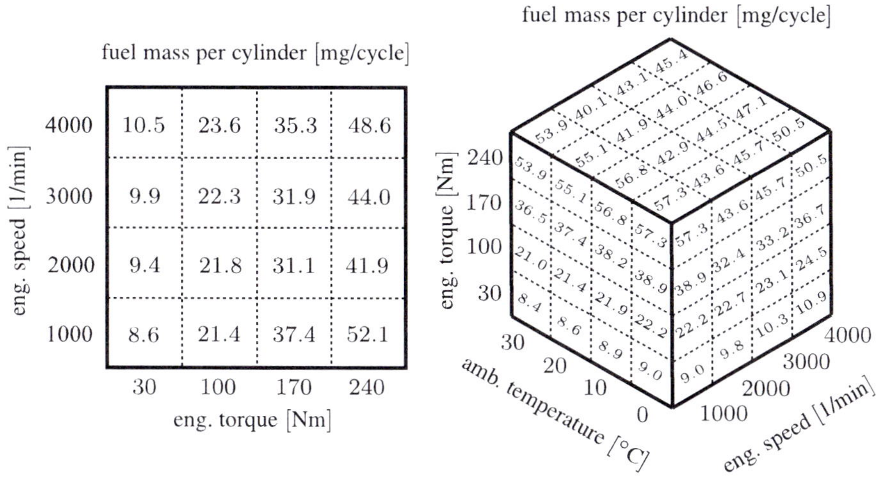

Fig. 2.15 Structure of 2D and 3D look-up tables

is just slightly more complex, large numbers of calibration parameters are required. At the same time, the decentralized PID controllers are unable to account for the MIMO nonlinear dynamics in a systematic way. As shown in the previous sections, additional measures are needed. Constraints on the actuators pose problems as well, e.g. additional anti-windup schemes have to be implemented. The constraints can also exist on the system states. This might be due to safety limitations, e.g. because the temperature has to stay within a certain region for a safe operation. The PID-based controllers handle constraints on system states by choosing setpoints sufficiently far from constraints and by adding protection logics [5]. Therefore, suboptimal plant operation results.

As several system dynamic characteristics cannot be handled in a systematic way, the calibration process can be very time-consuming. For proof of validity, a lot of tests, e.g. at the test bench or in the vehicle have to be conducted. This results in high costs for the control development. Further restrictions of the approach based on look-up tables arise due to its structure, which does not allow high reusability of parts of the controller. Consequently, a new setup, e.g. due to a change of individual components, often leads to a new determination of the calibration labels.

In conclusion, the control approach based on look-up tables is well suited for control tasks that are characterized by a low number of inputs and outputs, no strong nonlinearity, and no strong interconnections with other components such that they can be treated as SISO systems. However, in the case of modern engines, the number of inputs and outputs is increasing, causing the number of calibration parameters to be drastically increasing as well. This makes the calibration of controllers with look-up tables very tedious. In Fig. 1.4, the development of the number of calibration labels from 1997 to 2017 was already shown [7]. The calibration labels increase by a factor of more than six for an increase in the number of degrees of freedom by approximately 2.5. Future engines will likely have even a higher number of degrees of freedom. Consequently, increasingly more calibration labels have to be determined by heavy use of measurements at a test bench or in the vehicle. In [7], the work share during the calibration process of current engine controllers is detailed as follows:

1. 5% calibration time in a virtual environment,
2. 25% calibration time at the test bench, and
3. 70% calibration time in the vehicle.

Due to the strong use of late development stages (such as vehicle testings) in the calibration procedure, the calibration procedure is not only very time-consuming but also very expensive. A possible alternative to overcome these drawbacks is given by a switch to new advanced control algorithms, such as MPC.

References

1. P. Albertos, S. Antonio, *Multivariable Control Systems: An Engineering Approach* (Springer, 2006)
2. K.J. Aström, T. Hägglund, Advanced PID control, in *ISA – The Instrumentation, Systems, and Automation Society*, vol. 461 (2006)

3. K.J. Aström, K.H. Johansson, Q.-G. Wang, Design of decoupled PID controllers for MIMO systems, in *American Control Conference* (2001), pp. 2015–2020
4. K.J. Aström, R.M. Murray, *Feedback Systems: An Introduction for Scientists and Engineers* (Princeton University Press, 2010)
5. S. Di Cairano, I.V. Kolmanovsky, Automotive applications of model predictive control, in *Handbook of Model Predictive Control* (Springer, 2019), pp. 493–527
6. L. Eriksson, L. Nielsen, *Modeling and Control of Engines and Drivelines* (Wiley, 2014)
7. ETAS GmbH, ETAS calibration consulting flyer (2018). Accessed 2018-06-17. https://www.etas.com/download-center-files/company/Calibration_Consulting_Flyer_EN.pdf
8. P. Hippe, *Windup in Control: Its Effects and Their Prevention* (Springer, 2006)
9. R. Isermann, *Engine Modeling and Control* (Springer, 2014)
10. J.M. Maciejowski, *Multivariable Feedback Design* (Addison-Wesley, 1989)
11. S. Skogestad, I. Postlethwaite, *Multivariable Feedback Control: Analysis and Design* (Wiley, 2007)
12. Q.-G. Wang, *Decoupling Control* (Springer, 2002)

Chapter 3
Mathematical Fundamentals of Optimization

Abstract The fundamental principle of model predictive control (MPC) is the solution of an optimization problem in real time. The optimization problem is designed in such a way that it reflects the goals of the control algorithm. For the development of MPC controllers for complex engine tasks, a solid knowledge of optimization is required. This chapter provides a brief overview of the fundamentals of optimization. Rather than an overview of the entire field, the optimization fundamentals are described that are particularly needed for the application of MPC in the field of combustion engines. The reader also is familiarized with the notation used. The chapter is to discuss various classifications of optimization problems. Furthermore, the concept of convexity is introduced. Additionally, the optimality conditions for nonlinear programs (NLP) are examined.

3.1 Introduction to Optimization Problems

The optimization tasks that are to be treated within this book can be seen as the mathematical search for the best solution. Within MPC, optimization is used to find the most suitable actuated values to control the process. For this purpose, various criteria are taken into account, such as deviations from the predicted output to the reference value. Optimization tasks obviously are not only found in the context of closed-loop control but also in many other applications. Examples are portfolio management, where the investment in different assets is optimized [4], routing for autonomous-mobility-on-demand, where the traffic throughput through a city is investigated [9], or system identification, where the model parameters are explored for the best fit to measurement data [7]. This chapter provides a brief overview of the fundamentals of optimization that are necessary for the MPC of combustion engines and introduces the notation used. Readers who are interested in further details, such as mathematical

T. Albin Rajasingham, *Nonlinear Model Predictive Control of Combustion Engines*, Advances in Industrial Control,
https://doi.org/10.1007/978-3-030-68010-7_3

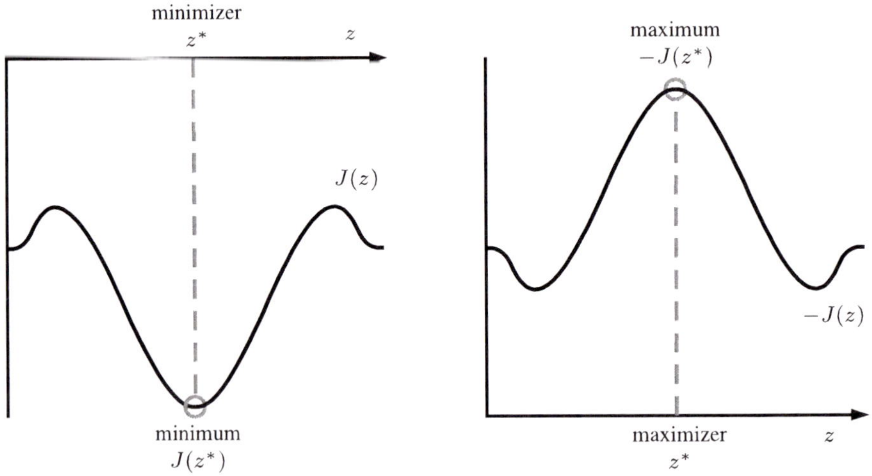

Fig. 3.1 Equivalence of maximization and minimization problems

proofs of the presented statements, are referred to the many good textbooks which suit this purpose [2, 3, 8].

To find the best solution of an optimization problem, there has to be a quantification of the objective. This is done by the so-called cost function $J(z)$ (sometimes also called objective function) where z represents the optimization variables. The output of the cost function is a scalar value which enables a comparison of different costs. Depending on the application, the goal is to minimize or maximize this cost function under certain constraints. A minimization is desired to reduce the material usage for the production of a certain product, for instance, whereas the maximization is requested for the throughput of a product in a plant. However, they are equivalent: the minimization of $J(z)$ corresponds to the maximization of $-J(z)$, as illustrated in Fig. 3.1. From now on, the focus will be on the minimization of the cost function.

One specific class of optimization problems is investigated first in order to familiarize the reader with the notation used. It is the so-called NLP, which is an important class of optimization problems for engine MPC. For an NLP, the cost function is defined by $J(z) : \mathbb{R}^n \rightarrow \mathbb{R}$, where $J(z)$ is a general nonlinear function. The cost function is optimized for the optimization variables $z \in \mathbb{R}^n$, which are given as column vector $z = [z_1, \ldots, z_n]^\mathsf{T}$. The minimal cost is called optimum or alternatively minimum and is denoted as $J^* = J(z^*)$. To denote the optimizer, also called minimizer, of the optimization problem, z^* is used. In the case of an unconstrained optimization problem, the cost function is minimized without taking into account any additional constraints. The unconstrained optimization problem for an NLP reads as follows:

$$\min_{z \in \mathbb{R}^n} \quad J(z) \tag{3.1}$$

Besides the cost function, additional constraints can be added, such that a constrained optimization problem results. The constraints can be divided into equality constraints, which have to be fulfilled exactly, and inequality constraints, where the inequality has to be fulfilled. The general constrained NLP is denoted by the following equations:

$$\min_{z \in \mathbb{R}^n} \quad J(z) \tag{3.2a}$$

$$\text{s.t.} \quad \boldsymbol{h}(z) = \boldsymbol{0}, \quad \boldsymbol{h} : \mathbb{R}^n \to \mathbb{R}^p, \tag{3.2b}$$

$$g(z) \leq \boldsymbol{0}, \quad g : \mathbb{R}^n \to \mathbb{R}^q \tag{3.2c}$$

In (3.2b), the equality constraints are given, whereas (3.2c) shows the inequality constraints. Both, the equality and the inequality constraints have to be fulfilled, which is why they are also called hard constraints. The constraints of the optimization problem define the feasible set Ω, i.e. the set of all points which satisfy the constraints.

$$\Omega = \{z \in \mathbb{R}^n | g(z) \leq \boldsymbol{0}, \boldsymbol{h}(z) = \boldsymbol{0}\} \tag{3.3}$$

With Ω, the NLP can be rewritten in an equivalent form.

$$\min_{z \in \Omega} \quad J(z) \tag{3.4}$$

For the solution of the optimization problem, various types of minimizers z^* can be distinguished. Examples of the various types of solutions are depicted in Fig. 3.2.

- The minimizer z^* is a *local* minimizer if $z^* \in \Omega$ and if there is a neighborhood $\mathcal{N}(z^*)$ of z^*, such that

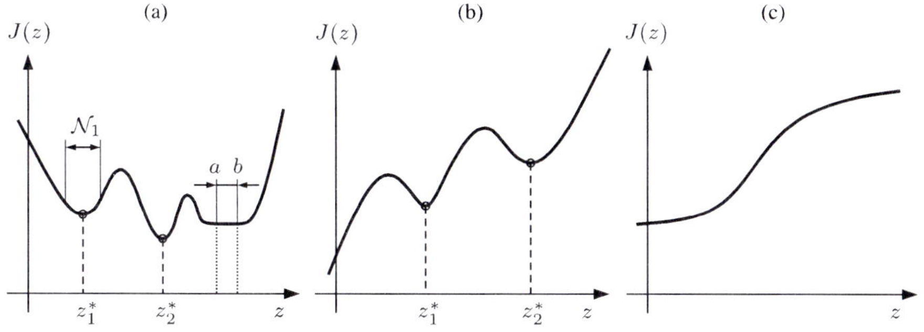

Fig. 3.2 Illustrations of the various types of solutions: **a** z_1^* is a strict local solution, z_2^* is a unique global solution, and each $z^* \in [a, b]$ is a non-strict local solution; **b** z_1^* and z_2^* are strict local solutions, there exists no global solution; **c** there exists no solution

$$J(z^*) \leq J(z)\ \forall z \in \mathcal{N}(z^*) \subset \Omega \tag{3.5}$$

- The minimizer z^* is a *strict local* minimizer if $z^* \in \Omega$ and if there is a neighborhood $\mathcal{N}(z^*)$ of z^*, such that

$$J(z^*) < J(z)\ \forall z \in \mathcal{N}(z^*) \subset \Omega, z \neq z^* \tag{3.6}$$

- The minimizer z^* is a *global* minimizer if $z^* \in \Omega$ and

$$J(z^*) \leq J(z)\ \forall z \in \Omega \tag{3.7}$$

- The minimizer z^* is an *unique global* minimizer if $z^* \in \Omega$ and

$$J(z^*) < J(z)\ \forall z \in \Omega \tag{3.8}$$

The influence of the constraints on the location of the minimizer is visualized in Fig. 3.3, which shows one cost function with iso-contour lines for the same cost and where the arrow indicates an increasing cost. If the unconstrained minimizer is within the feasible set of the constrained optimization problem, the minimizer does not change. An inequality constraint i is called active if $g_i(z^*) = 0$ and inactive in the case of $g_i(z^*) < 0$. Disregarding the inactive inequality constraints in the optimization problem does not change the minimizer. However, usually it is not known in advance as to which constraints are inactive. In contrast to the inequality constraints, all equality constraints are active, as they always have to be fulfilled.

Within the numerical optimization algorithms, derivatives of functions play an important role. These are especially the gradient, Jacobian, and the Hessian of a function, which are introduced in the following.

Gradient

For a continuously differentiable function $f(z) : \mathbb{R}^n \rightarrow \mathbb{R}$, the *gradient* is given by the column vector ∇f.

$$\nabla f = \begin{bmatrix} \frac{\partial f}{\partial z_1} \\ \vdots \\ \frac{\partial f}{\partial z_n} \end{bmatrix} \tag{3.9}$$

Jacobian

The *Jacobian* matrix of a (vector-valued) continuously differentiable function $f(z) : \mathbb{R}^n \rightarrow \mathbb{R}^m$ is given by $\frac{\partial f}{\partial z}$.

$$\frac{\partial f}{\partial z} = \begin{bmatrix} \frac{\partial f_1}{\partial z_1} & \cdots & \frac{\partial f_1}{\partial z_n} \\ \vdots & \ddots & \vdots \\ \frac{\partial f_m}{\partial z_1} & \cdots & \frac{\partial f_m}{\partial z_n} \end{bmatrix} \tag{3.10}$$

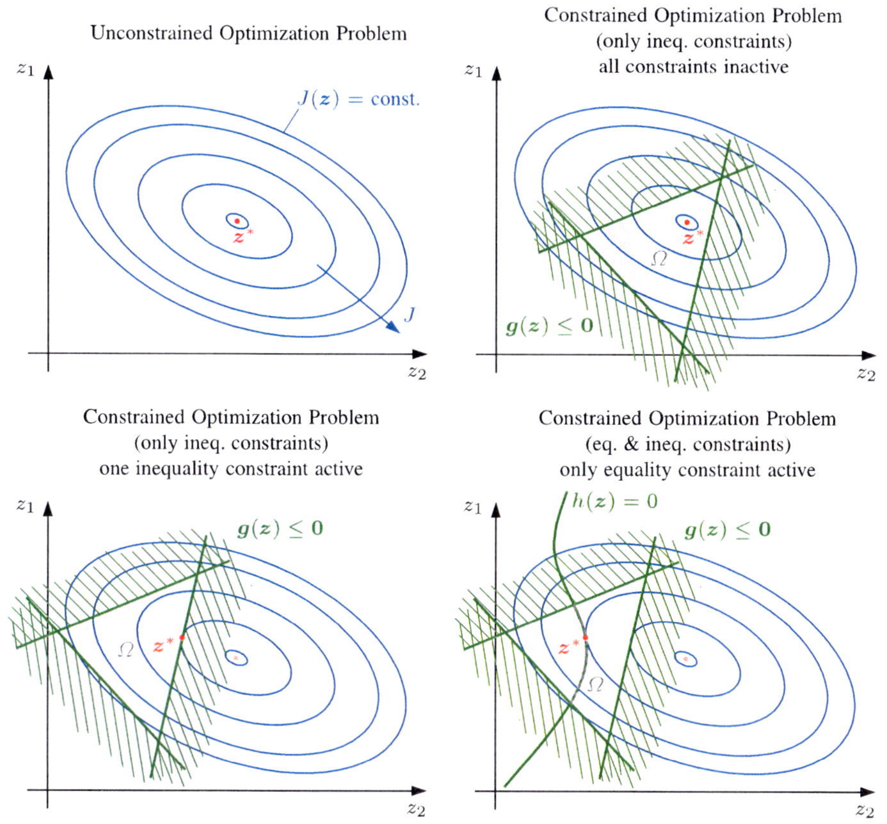

Fig. 3.3 Visualizations of the influence of constraints on the location of the minimizer

As commonly done in optimization literature, the nabla operator will also be used for vector-valued functions. The matrix $\nabla \boldsymbol{f}(z)$ defines the transpose of the Jacobian matrix

$$\nabla \boldsymbol{f} = \begin{bmatrix} \frac{\partial f_1}{\partial z_1} & \cdots & \frac{\partial f_m}{\partial z_1} \\ \vdots & \ddots & \vdots \\ \frac{\partial f_1}{\partial z_n} & \cdots & \frac{\partial f_m}{\partial z_n} \end{bmatrix} \tag{3.11}$$

which can also be denoted by

$$\nabla \boldsymbol{f}(z) = [\nabla f_1(z), \nabla f_2(z), \ldots, \nabla f_m(z)] \tag{3.12}$$

Hessian

For a twice continuously differentiable function $f(z) : \mathbb{R}^n \to \mathbb{R}$, the *Hessian* defines a matrix-valued function. The Hessian will be denoted by $\nabla^2 f$. Due to the commutative property of the differentiation, the Hessian matrix is symmetric.

$$\nabla^2 f = \begin{bmatrix} \frac{\partial^2 f}{\partial z_1^2} & \cdots & \frac{\partial^2 f}{\partial z_1 z_n} \\ \vdots & \ddots & \vdots \\ \frac{\partial^2 f}{\partial z_n z_1} & \cdots & \frac{\partial^2 f}{\partial z_n^2} \end{bmatrix} \tag{3.13}$$

Within optimization algorithms, the definiteness of the Hessian is an important characteristic. A symmetric n×n matrix $\boldsymbol{A}$ is called *positive (semi)definite* if the following condition holds:

$$\boldsymbol{s}^\mathsf{T} \boldsymbol{A} \boldsymbol{s} > (\geq)\, 0,\ \forall \boldsymbol{s} \in \mathbb{R}^n, \boldsymbol{s} \neq \boldsymbol{0} \tag{3.14}$$

A practical way to determine whether a matrix is positive (semi)definite is the evaluation of the eigenvalues. The eigenvalues $\sigma_i, i = 1, ..., n$ can be calculated by the solution of

$$\det(\boldsymbol{A} - \sigma \boldsymbol{I}) = 0 \tag{3.15}$$

Here, $\boldsymbol{I}$ is the n×n identity matrix. The matrix $\boldsymbol{A}$ is positive (semi)definite, if it holds

$$\sigma_i > (\geq)\, 0,\ \forall i \in \{1, \ldots, n\} \tag{3.16}$$

If $(-\boldsymbol{A})$ is positive (semi)definite, then $\boldsymbol{A}$ is called negative (semi)definite. The notation $\boldsymbol{A} \succ \boldsymbol{0}$, $\boldsymbol{A} \succeq \boldsymbol{0}$, $\boldsymbol{A} \prec \boldsymbol{0}$, $\boldsymbol{A} \preceq \boldsymbol{0}$ denotes a positive-definite, positive-semidefinite, negative-definite, and a negative-semidefinite matrix $\boldsymbol{A}$.

3.2 Convex Optimization Problems

Convexity is an important characteristic for the distinction of optimization problems. When the optimization problem is convex, a lot of beneficial properties are present, such that it becomes easier to find the solution numerically. In order to classify optimization problems as convex, the notions of a convex set and a convex function have to be known.

Convex Sets

A set $\mathcal{Z} \subseteq \mathbb{R}^n$ is convex if the following condition is fulfilled:

$$z = k\boldsymbol{x} + (1 - k)\boldsymbol{y} \in \mathcal{Z},\ \forall \boldsymbol{x}, \boldsymbol{y} \in \mathcal{Z}, k \in [0, 1] \tag{3.17}$$

The geometric interpretation is that the set $\mathcal{Z}$ is convex if the straight line connecting any pair of points from within the set lies completely inside the set, as shown in Fig. 3.4. One property that can be utilized for MPC is the property that the intersection set of convex sets is a convex set again. This property holds true for an infinite number of sets. Scaling does not influence the convexity properties, i.e. $\alpha\mathcal{Z}$, where $\alpha \in \mathbb{R}$ is

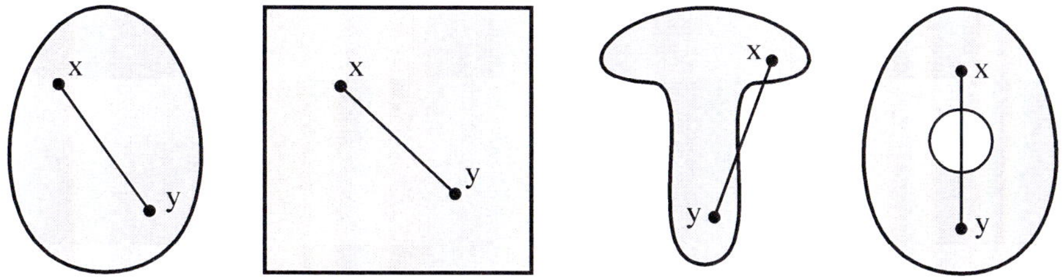

(a) convex (b) convex (c) non-convex (d) non-convex

Fig. 3.4 Examples for convex and non-convex sets

a convex set if $\mathcal{Z}$ is convex. There are also many other properties that are associated with convex sets and operations that preserve convexity [3].

Convex Function

A function $f(z) : \mathcal{Z} \rightarrow \mathbb{R}$ is convex if $\mathcal{Z} \subseteq \mathbb{R}^n$ is a convex set and the function satisfies the following condition for all points $\boldsymbol{x}, \boldsymbol{y} \in \mathcal{Z}, \boldsymbol{x} \neq \boldsymbol{y}$:

$$f(z) \leq kf(\boldsymbol{x}) + (1-k)f(\boldsymbol{y}) \quad \forall z = k\boldsymbol{x} + (1-k)\boldsymbol{y},\ k \in [0, 1] \tag{3.18}$$

This condition can be interpreted in a geometrical sense that the line segment connecting any two points on the graph of the function lies above (or on) the graph. Furthermore, the function f is called strictly convex if the following inequality holds:

$$f(z) < kf(\boldsymbol{x}) + (1-k)f(\boldsymbol{y}) \quad \forall z = k\boldsymbol{x} + (1-k)\boldsymbol{y},\ k \in (0, 1) \tag{3.19}$$

The function f is called (strictly) concave if $-f$ is (strictly) convex.

If the function $f(z) : \mathcal{Z} \rightarrow \mathbb{R}$ is twice continuously differentiable, the convexity can be examined in the following practical way:

- $f(z)$ is convex if the Hessian $\nabla^2 f(z)$ is positive semidefinite $\forall z \in \mathcal{Z}$.
- $f(z)$ is strictly convex if the Hessian $\nabla^2 f(z)$ is positive definite $\forall z \in \mathcal{Z}$.
- $f(z)$ is concave if the Hessian $\nabla^2 f(z)$ is negative semidefinite $\forall z \in \mathcal{Z}$.
- $f(z)$ is strictly concave if the Hessian $\nabla^2 f(z)$ is negative definite $\forall z \in \mathcal{Z}$.

Exemplary convex and non-convex functions are demonstrated in Fig. 3.5. For convex functions, there are also operations that preserve convexity [3]. For instance, the sum of two convex functions is a convex function again $f_3(z) = f_1(z) + f_2(z)$.

Convex Optimization Problem

Assume that the following NLP is given:

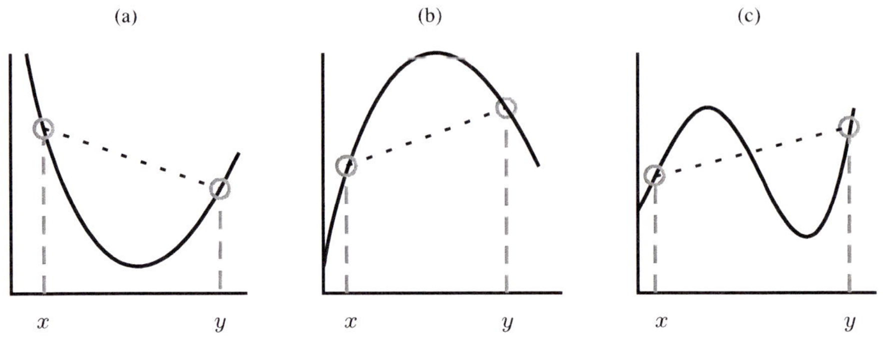

Fig. 3.5 **a** Convex function, **b** concave function, and **c** function is neither convex nor concave

$$\begin{aligned} \min_{z \in \mathbb{R}^n} \quad & J(z) \\ \text{s.t.} \quad & h_i(z) = \mathbf{0}, \quad i = 1, ..., p, \\ & g_i(z) \leq \mathbf{0}, \quad i = 1, ..., q \end{aligned} \tag{3.20}$$

The individual constraint functions are given by $h_i(z) : \mathbb{R}^n \to \mathbb{R}$ and $g_i(z) : \mathbb{R}^n \to \mathbb{R}$ and define the feasible set $z \in \Omega$. An optimization problem is (strictly) convex if the cost function J is a (strictly) convex function and the feasible set Ω is a convex set.

Convexity of the feasible set is achieved, for instance, when each single constraint defines a convex set, as the intersection of convex sets becomes a convex set again. In this case, each equality constraint function $h_i(z)$, $i = 1, \ldots, p$ has to be a linear or an affine function. For a nonlinear function, the equality constraint directly translates to a non-convex set. Additionally, for each inequality constraint $g_i(z) \leq 0$, $i = 1, \ldots, q$, the functions g_i have to be convex functions, as these define convex sets [3].

Extensive theory exists on the subject of convex optimization [3]. Convex optimization problems have properties that make them attractive for numerical solutions. In general, they can be solved more easily and faster than non-convex optimization problems. Additionally, conclusions can be drawn about global optimality which is generally not possible for non-convex optimization problems. For convex optimization problems, every local minimum is also a global minimum. As a consequence, if there exists an isolated local minimum, it is also the unique global minimum. Optimization problems with a strictly convex cost function and a convex feasible set have either one or no minimum. If there exists a minimum, it is directly the unique global minimum.

Example 3.1 (*Convex functions*) In the following, the connection between the definiteness of the Hessian and the convexity of the function is demonstrated. Figure 3.6 shows plots of functions with $f(z) : \mathbb{R}^2 \to \mathbb{R}$ as well as their projection of the surface onto the plane $z_1 z_2$ ($\mathbb{R}^2$). The left plot shows a strictly convex function. The positive definiteness of $\nabla^2 f(z)\ \forall z \in \mathcal{Z}$ renders it strictly convex. If this function

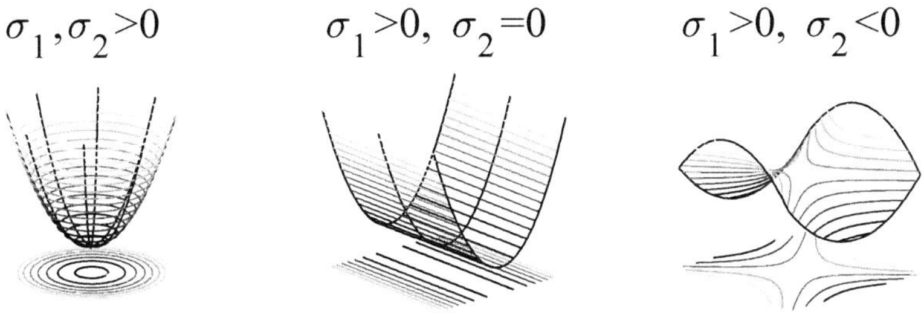

Fig. 3.6 Function: **a** strictly convex, **b** convex, and **c** neither convex nor concave

was a cost function to be minimized, one unique global minimum of the function would result. In the case of positive semidefinite Hessian matrices $\nabla^2 f(z)\ \forall z \in \mathcal{Z}$ with one eigenvalue equal to zero, a convex but not strictly convex function would result. The corresponding function is shown in the middle. The function is constant in the direction of the eigenvector associated with $\sigma_2 = 0$. If this function was a cost function to be minimized, no unique minimum would be present anymore. In the case where $\nabla^2 f(z)\ \forall z \in \mathcal{Z}$ has one positive and one negative eigenvalue, the function is neither convex nor concave. The function is shown on the right plot. It is characterized by a saddle point.

3.3 Classes of Optimization Problems

There are many possibilities to classify optimization problems. Depending on the class of optimization problem, the representable challenges vary, the characteristics change, and, accordingly, different suitable solution methods are available. In the following, some important optimization classes for engine MPC are introduced. First, the difference between static and dynamic optimization problems will be outlined. Based on that outline, various examples of static optimization problems, specifically linear, quadratic, nonlinear, integer, and mixed-integer programs are described.

3.3.1 Dynamic Optimization

The goal of dynamic optimization is to determine the function of an independent variable (often time t) that solves the optimization problem given. For that reason, the optimization is conducted over a function space and can thus be interpreted as an optimization problem with *infinite* number of optimization variables. One specific class within dynamic optimization is the class of optimal control problems (OCP) which will be relevant for the engine MPC. For an OCP problem, a possibly nonlinear dynamic system is given with an input $\boldsymbol{u} \in \mathbb{R}^m$ and the system states $\boldsymbol{x} \in \mathbb{R}^n$. The dynamic system is described by an ordinary differential equation (ODE) and its initial value.

$$\begin{aligned} \dot{\boldsymbol{x}}(t) &= \boldsymbol{f}(\boldsymbol{x}(t), \boldsymbol{u}(t), t) \\ \boldsymbol{x}(t_0) &= \boldsymbol{x}_0 \end{aligned} \tag{3.21}$$

The task within the OCP is to find a trajectory $\boldsymbol{u}(t) \in \mathbb{R}^m, t \in [t_0, t_f]$, such that the resulting state trajectory $\boldsymbol{x}(t) \in \mathbb{R}^n, t \in [t_0, t_f]$ minimizes the cost function $J(\boldsymbol{x}(t), \boldsymbol{u}(t))$ while satisfying the constraints. The cost function is defined as J_{ocp}.

$$J_{ocp}(\boldsymbol{x}(t), \boldsymbol{u}(t)) = l_f(\boldsymbol{x}(t_f)) + \int_{t_0}^{t_f} l_s(\boldsymbol{x}(t), \boldsymbol{u}(t))\mathrm{d}t \tag{3.22}$$

The cost function J_{ocp} is often divided in two parts, the stage cost resulting from the integration of l_s called Lagrange term and the final cost l_f called Mayer term. A typical example of an OCP then reads as follows:

$$\min_{\boldsymbol{x}(t), \boldsymbol{u}(t)} \quad J_{ocp}(\boldsymbol{x}(t), \boldsymbol{u}(t)) \tag{3.23a}$$

$$\text{s.t.} \quad \dot{\boldsymbol{x}}(t) = \boldsymbol{f}(\boldsymbol{x}(t), \boldsymbol{u}(t), t), \forall t \in [t_0, t_f], \tag{3.23b}$$

$$\boldsymbol{x}(t_0) = \boldsymbol{x}_0, \tag{3.23c}$$

$$\boldsymbol{h}(\boldsymbol{x}(t_f)) = \boldsymbol{0}, \tag{3.23d}$$

$$g(\boldsymbol{x}(t), \boldsymbol{u}(t)) \leq \boldsymbol{0}, \forall t \in [t_0, t_f] \tag{3.23e}$$

In this optimization problem, the system dynamics are considered in (3.23b), while (3.23c) considers the initial state of the system at $t = t_0$. The function $\boldsymbol{h}(\boldsymbol{x}(t_f)) \in \mathbb{R}^n$ represents the equality constraints defining the final state, i.e. the states where the system is steered to. The path constraints are defined by the inequality constraints with $g(\boldsymbol{x}(t), \boldsymbol{u}(t)) \in \mathbb{R}^q$.

3.3.2 *Static Optimization*

In static optimization, the optimization variables are characterized by a *finite* number of optimization variables. Within static optimization, numerous classifications exist. They depend, for example, on the type of cost function and constraints (e.g. convex vs. non-convex) and the domain of the optimization variables (e.g. discrete vs. continuous variables). In the following, an overview is presented on NLPs, Quadratic Programs (QP), Linear Programs (LP), Integer Programs (IP), and Mixed-Integer Programs (MIP).

Nonlinear Program (NLP)

For the NLP, the cost function, or at least one constraint, is nonlinear. In general, it can be a non-convex optimization problem. NLPs occur in many practical applications, for example, in the production planning [11]. An NLP is defined by the following optimization problem:

$$\begin{aligned} \min_{z \in \mathbb{R}^n} \quad & J(z) \\ \text{s.t.} \quad & \boldsymbol{h}(z) = \mathbf{0}, \\ & g(z) \leq \mathbf{0} \end{aligned} \tag{3.24}$$

Quadratic Program (QP)

For a QP, the cost function is quadratic and the constraint functions are affine. The optimization problem is (strictly) convex if the Hessian $\boldsymbol{H}$ is positive semidefinite (positive definite). A typical example for QPs is the parameter fitting via a least-squares approach [5]. The QP is defined by

$$\begin{aligned} \min_{z \in \mathbb{R}^n} \quad & \frac{1}{2} z^\mathsf{T} \boldsymbol{H} z + \boldsymbol{x}^\mathsf{T} \boldsymbol{q} \\ \text{s.t.} \quad & \boldsymbol{A} z - \boldsymbol{b} = \mathbf{0}, \\ & \boldsymbol{C} z - \boldsymbol{d} \leq \mathbf{0} \end{aligned} \tag{3.25}$$

Linear Program (LP)

For a LP, the cost function and the constraint functions are both affine. The optimization problem is always convex; as a consequence, every local solution is a global solution. An example where LPs are used is the balancing of supply and demand points, for instance, in transportation problems [10]. The LP is given in standard form as

$$\begin{aligned} \min_{z \in \mathbb{R}^n} \quad & \boldsymbol{c}^\mathsf{T} z \\ \text{s.t.} \quad & \boldsymbol{A} z - \boldsymbol{b} = \mathbf{0}, \\ & z \geq \mathbf{0} \end{aligned} \tag{3.26}$$

Integer Program (IP)

Within an IP, the optimization variables are integer variables. The traveling salesman problem, for instance, can be formulated as an IP [6]. An example for this class of optimization problems is the Integer Nonlinear Program which can be formulated as

$$\begin{aligned} \min_{z_a} \quad & J(z_a) \\ \text{s.t.} \quad & \boldsymbol{h}(z_a) = \mathbf{0}, \\ & g(z_a) \leq \mathbf{0}, \\ & z_a \in \mathbb{Z}^n \end{aligned} \tag{3.27}$$

Mixed-Integer Program (MIP)

Within an MIP, some optimization variables are integer variables and others are continuous. For instance, the energy management of hybrid vehicles can be posed as a MIP, as it involves the discrete decision to turn on or off the combustion engine [1]. One example for this class of optimization problems is the Mixed-Integer Nonlinear Program which is given as

$$\begin{aligned} \min_{z_a, z_b} \quad & J(z_a, z_b) \\ \text{s.t.} \quad & h(z_a, z_b) = \mathbf{0}, \\ & g(z_a, z_b) \le \mathbf{0}, \\ & z_a \in \mathbb{Z}^{n_a}, \\ & z_b \in \mathbb{R}^{n_b} \end{aligned} \tag{3.28}$$

3.4 Optimality Conditions for NLPs

In the following, an overview of optimality conditions for NLPs is provided. As QPs and LPs can be seen as a subset of NLPs, the optimality conditions also hold for these classes. These conditions offer the possibility to find the minimizers z^* of an optimization problem, and they allow to check if a certain candidate is indeed a minimizer. In general, the conditions provide information about *local* optimality. Conclusions about global optimality can only be drawn for special cases, such as convex optimization problems. The functions involved, i.e. the cost function, the equality constraints, and inequality constraints, are assumed to be twice continuously differentiable functions. First, the unconstrained case is considered. It is then extended by equality and inequality constraints.

3.4.1 Unconstrained Optimization Problems

For the unconstrained case, optimization problems are considered that can be put in the following form:

$$\min_{z \in \mathbb{R}^n} \quad J(z) \tag{3.29}$$

The cost function J can either be a linear, a quadratic, or a general nonlinear function. There exists a necessary condition for optimality which is also called the *first-order necessary condition* (FONC). The FONC is given by the following equation:

$$\nabla J(z^*) = \mathbf{0} \tag{3.30}$$

All points that satisfy this equation are called stationary points. Every local minimizer fulfills the FONC. However, the reverse is not true: Not every stationary point is a local minimizer. A stationary point can also be a maximum or a saddle point. Thus, the

FONC can be used to determine all stationary points which are possible candidates for minimizers.

For determining whether a candidate is indeed a local minimizer, the sufficient condition can be checked. This is the so-called *second-order sufficient condition* (SOSC) for optimality. The criteria are based on evaluation of the positive definiteness of the Hessian of the cost function at the candidate point. The SOSC is fulfilled when the Hessian matrix of the cost function is positive definite:

$$\nabla^2 J(z^*) \succ \mathbf{0} \tag{3.31}$$

If the candidate is a stationary point and the SOSC holds, the point z^* is a strict local minimizer. If the SOSC is not fulfilled, the stationary point could be a saddle point, a maximum, or still a minimum, e.g. $J(z) = z^4, z \in \mathbb{R}$.

For a convex cost function J, the SOSC does not need to be checked: All stationary points are local minima. Additionally, every local minimizer is directly also a global minimizer. For strictly convex optimization problems, there exists either no stationary point or exactly one, which is simultaneously the unique global minimizer.

Example 3.2 (*Unconstrained NLP*) A simple example for minimizing an unconstrained NLP by application of the optimality conditions is demonstrated. The cost function J_1 investigated is the well-known Rosenbrock function [8]. Figure 3.7 shows a plot of the function J_1.

$$J_1(z_1, z_2) = (1 - z_1)^2 + 100(z_2 - z_1^2)^2 \tag{3.32}$$

The following unconstrained NLP defines the optimization task:

$$\min_{z \in \mathbb{R}^2} \quad J_1(z) \tag{3.33}$$

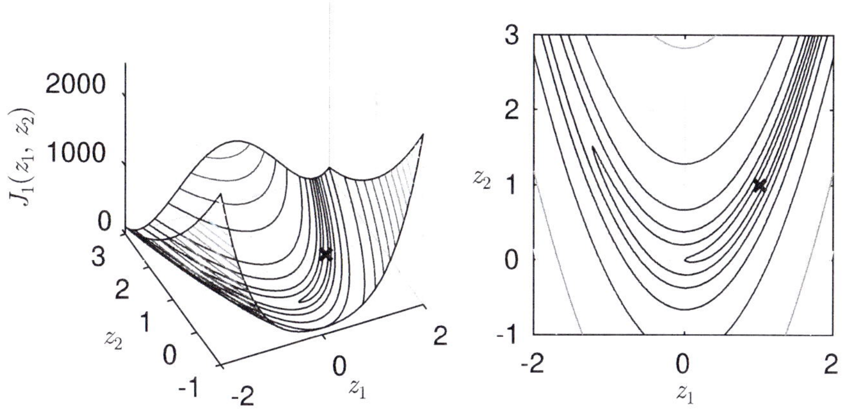

Fig. 3.7 Visualizing the Rosenbrock function from Example 3.2

The FONC can be set up for this optimization problem.

$$\nabla J_1(z_1, z_2) = \begin{bmatrix} 2z_1 - 400z_1(-z_1^2 + z_2) - 2 \\ -200z_1^2 + 200z_2 \end{bmatrix} = \mathbf{0} \tag{3.34}$$

The stationary point is given by $z_1^* = 1,\ z_2^* = 1$. The SOSC is evaluated to determine if the point is indeed a local minimizer:

$$\begin{aligned} \nabla^2 J_1(z_1, z_2) &= \begin{bmatrix} 1200z_1^2 - 400z_2 + 2 & -400z_1 \\ -400z_1 & 200 \end{bmatrix} \\ \nabla^2 J_1(z_1^*, z_2^*) &= \begin{bmatrix} 802 & -400 \\ -400 & 200 \end{bmatrix} \end{aligned} \tag{3.35}$$

The matrix $\nabla^2 J_1(z_1^*, z_2^*)$ has eigenvalues $\sigma_1 = 1002$ and $\sigma_2 = 0.4$, which shows that it is positive definite and that therefore SOSC is fulfilled. Consequently, the stationary point $z^* = [1, 1]^\mathsf{T}$ calculated by the FONC is indeed a strict local minimizer.

3.4.2 Constrained Case: Equality Constraints

The optimality conditions for an equality-constrained problem is to be summarized next. The optimization problem investigated is defined by the following equations:

$$\begin{aligned} \min_{z \in \mathbb{R}^n} \quad & J(z) \\ \text{s.t.} \quad & h_i(z) = 0, \quad i = 1, \ldots, p \end{aligned} \tag{3.36}$$

The cost function is defined by $J(z) : \mathbb{R}^n \rightarrow \mathbb{R}$ and the p equality constraints are given by $h_i(z) : \mathbb{R}^n \rightarrow \mathbb{R}$. For setting up the optimality conditions, the so-called *Lagrange* function (also known as *Lagrangian*) is introduced. The Lagrange function for an equality-constrained problem is defined by $L(z, \boldsymbol{\lambda}) : \mathbb{R}^n \times \mathbb{R}^p \rightarrow \mathbb{R}$.

$$L(z, \boldsymbol{\lambda}) = J(z) + \sum_{i=1}^{p} \lambda_i h_i(z) \tag{3.37}$$

$$= J(z) + \boldsymbol{\lambda}^\mathsf{T} \boldsymbol{h}(z) \tag{3.38}$$

The variables $\boldsymbol{\lambda} = [\lambda_1, .., \lambda_p]^\mathsf{T}$ are called Lagrange multipliers or dual variables. To obtain a more compact form, the equality constraints are summarized to $\boldsymbol{h}(z) = [h_1(z), .., h_p(z)]^\mathsf{T}$, with $\boldsymbol{h}(z) \in \mathbb{R}^p$. To each constraint $h_i(z)$, there exists an associated scalar value λ_i.

First-Order Necessary Conditions
The FONC can be defined using the Lagrange function. For an equality-constrained problem, the FONC are defined by the following equations:

$$\begin{aligned}\nabla L(z^*, \boldsymbol{\lambda}^*) = \begin{bmatrix}\nabla_z L(z^*, \boldsymbol{\lambda}^*)\\ \nabla_{\boldsymbol{\lambda}} L(z^*, \boldsymbol{\lambda}^*)\end{bmatrix} &= \begin{bmatrix}\nabla J(z^*) + \sum_{i=1}^{p} \lambda_i^* \nabla h_i(z^*)\\ \boldsymbol{h}(z^*)\end{bmatrix} \\ &= \begin{bmatrix}\nabla J(z^*) + \nabla \boldsymbol{h}(z^*)\boldsymbol{\lambda}^*\\ \boldsymbol{h}(z^*)\end{bmatrix} = \mathbf{0}\end{aligned} \tag{3.39}$$

In these equations, $\nabla \boldsymbol{h}(z)$ is defined to be $\nabla \boldsymbol{h}(z) = \left[\nabla h_1(z), \nabla h_2(z), \dots, \nabla h_p(z)\right]$, the transpose of the Jacobian matrix of the equality-constrained function $\boldsymbol{h}(z)$. If the point $(z^*, \boldsymbol{\lambda}^*)$ fulfills $\nabla_{\boldsymbol{\lambda}} L(z^*, \boldsymbol{\lambda}^*) = \mathbf{0}$, it is said to be primal feasible. The FONC for equality-constrained problems are defined by $n + p$ equations for $n + p$ unknown variables.

Constraint Qualification
For constrained optimization problems, there might exist a minimum even though the FONC conditions are not fulfilled. To ensure that the FONC conditions hold at a local minimizer z^*, the problem has to fulfill some regularity conditions, the so-called constraint qualifications. There exist several constraint qualifications, which differ in their ease of application and also in how restrictive they are [8]. A commonly used constraint qualification is the *linear independence constraint qualification* (LICQ). This constraint qualification holds if $\nabla \boldsymbol{h}(z^*)$ has full column rank, i.e. rank p. If LICQ is fulfilled, each local minimizer $(z^*, \boldsymbol{\lambda}^*)$ fulfills the FONC, i.e. it solves $\nabla L(z, \boldsymbol{\lambda}) = \mathbf{0}$. Additionally, there exist a unique set of Lagrange multipliers to fulfill $\nabla L(z, \boldsymbol{\lambda}) = \mathbf{0}$. If the constraints are not qualified at z^*, then $\nabla L(z^*, \boldsymbol{\lambda}^*) = \mathbf{0}$ is not a necessary condition for local optimality, which means that local minima might exist which do not fulfill the FONC. An example is shown in Sect. 3.4.4 below.

Second-Order Sufficient Optimality Conditions
Let $z^* \in \mathbb{R}^n$ be a point at which LICQ holds, together with the Lagrange multipliers $\boldsymbol{\lambda}^*$, such that the FONC conditions are satisfied. In this case, $z^* \in \mathbb{R}^n$ is a candidate for a local minimizer. However, it can still be a point which is no local minimizer. The SOSC can be applied to check if it is indeed a minimizer. For equality-constrained problems, the SOSC are given by the following equations:

$$\boldsymbol{s}^\mathsf{T} \nabla_{zz}^2 L(z^*, \boldsymbol{\lambda}^*) \boldsymbol{s} > 0, \qquad \forall \boldsymbol{s} \neq \mathbf{0} \tag{3.40}$$

$$\nabla \boldsymbol{h}^\mathsf{T}(z^*) \cdot \boldsymbol{s} = 0 \tag{3.41}$$

If the LICQ is fulfilled along with the FONC and the SOSC, the candidate z^* is a strict local minimizer. The variables $\boldsymbol{s}$ represent the constraint tangent space and are defined by the dot product in (3.41) [8]. The constraint tangent space yields the feasible directions of the optimization problem at the point $(z^*, \boldsymbol{\lambda}^*)$. Thus, the SOSC checks whether in all the feasible directions, a positive curvature is present. For the unconstrained case, a positive curvature is demanded for all directions. If the SOSC is not fulfilled, no conclusion about optimality can be drawn.

Note that for the constrained optimization problem, the Hessian matrix with respect to z of the Lagrange function – and not of the cost function – is checked for the SOSC. If the Hessian of the Lagrange function with respect to z at the candidate point is positive definite, i.e. $\nabla^2_{zz} L(z^*, \boldsymbol{\lambda}^*) \succ \mathbf{0}$, the SOSC is automatically fulfilled. For a convex optimization problem, the Hessian matrix $\nabla^2_{zz} L(z^*, \boldsymbol{\lambda}^*) \succ \mathbf{0}$ is inherently positive definite. Consequently, for a convex optimization problem, SOSC does not need to be checked as the FONC is a necessary and sufficient condition.

Example 3.3 (*Equality-constrained QP*) The optimality conditions shall be applied for an equality-constrained QP. The QP considers n optimization variables $z \in \mathbb{R}^n$ and p equality constraints defined by the matrices $\boldsymbol{A} \in \mathbb{R}^{p \times n}$, $\boldsymbol{b} \in \mathbb{R}^p$. The optimization problem is defined by

$$\begin{aligned} \min_{z \in \mathbb{R}^n} \quad & \frac{1}{2} z^{\mathsf{T}} \boldsymbol{H} z + z^{\mathsf{T}} \boldsymbol{q} \\ \text{s.t.} \quad & \boldsymbol{A} z - \boldsymbol{b} = \mathbf{0} \end{aligned} \tag{3.42}$$

First, the Lagrange function of the optimization problem is formulated as

$$L(z, \boldsymbol{\lambda}) = \frac{1}{2} z^{\mathsf{T}} \boldsymbol{H} z + z^{\mathsf{T}} \boldsymbol{q} + \boldsymbol{\lambda}^{\mathsf{T}} (\boldsymbol{A} z - \boldsymbol{b}) \tag{3.43}$$

The constraints have to fulfill the LICQ, in order for the FONC to be a necessary condition for the local minimizers. For the given optimization problem, the matrix $\boldsymbol{A}$ has to have full row rank, i.e. rank p, to fulfill the LICQ. The FONC are given by the following equation:

$$\nabla L(z^*, \boldsymbol{\lambda}^*) = \mathbf{0} \tag{3.44}$$

The application of this condition to a QP results in a square system of linear equations. This system is called the Karush–Kuhn–Tucker (KKT) system, with the KKT matrix on the left-hand side. It consists of $n + p$ linear equations with $n + p$ unknown variables, which even allows for an analytic solution.

$$\begin{bmatrix} \boldsymbol{H} & \boldsymbol{A}^{\mathsf{T}} \\ \boldsymbol{A} & \mathbf{0} \end{bmatrix} \begin{bmatrix} z^* \\ \boldsymbol{\lambda}^* \end{bmatrix} = \begin{bmatrix} -\boldsymbol{q} \\ \boldsymbol{b} \end{bmatrix} \tag{3.45}$$

For finding the possible candidates, this system of linear equations has to be solved. In order for a unique solution to exist, the KKT matrix has to be nonsingular. One prerequisite for the existence of a unique solution is that LICQ is fulfilled. If LICQ does not hold (because $\boldsymbol{A}$ does not have full row rank), the KKT matrix becomes singular. If the Hessian matrix is positive definite, i.e. $\boldsymbol{H} \succ \mathbf{0}$, and LICQ holds, the optimization problem is strictly convex, the KKT matrix becomes nonsingular, and a unique solution exists [8]. Actually, even in the case that a positive curvature exists in the constraint tangent space and LICQ is fulfilled, a unique solution exists.

Example 3.4 (*Constrained optimization of a QP by substitution*) A numerical example for the application of optimality conditions on an optimization problem including equality constraints is investigated. The cost function is given by J_2.

$$J_2(z) = z_1^2 + z_2^2 + z_3^2 + z_1 z_3 - 4z_2 \tag{3.46}$$

For the optimization task, one equality constraint is considered.

$$\begin{aligned} \min_{z \in \mathbb{R}^3} \quad & J_2(z) \\ \text{s.t.} \quad & z_3 - 2 = 0 \end{aligned} \tag{3.47}$$

Obviously, this implies that $z_3^* = 2$. This can be substituted into the cost function J_2 which yields another unconstrained minimization problem named $\tilde{J}_2$, with $\tilde{z} = [z_1, z_2]^\mathsf{T}$.

$$\tilde{J}_2(\tilde{z}) = z_1^2 + z_2^2 + 2z_1 - 4z_2 + 4 \tag{3.48}$$

The FONC yields the candidate $\tilde{z}^*$ for the cost function $\tilde{J}_2$.

$$\tilde{z}^* = \begin{bmatrix} -1 & 2 \end{bmatrix}^\mathsf{T} \tag{3.49}$$

For checking the SOSC, the Hessian of $\tilde{J}_2(\tilde{z})$ can be calculated as

$$\nabla^2 \tilde{J}_2(\tilde{z}) = \begin{bmatrix} 2 & 0 \\ 0 & 2 \end{bmatrix} \tag{3.50}$$

The Hessian of $\tilde{J}_2(\tilde{z})$ is a positive definite and constant matrix. Therefore, the optimization problem is strictly convex and the candidate is a unique global minimizer. Together with $z_3 = 2$, the result is $z^* = \begin{bmatrix} -1 & 2 & 2 \end{bmatrix}^\mathsf{T}$.

Example 3.5 (*Constrained optimization of a QP using the Lagrange function*) The problem of minimizing J_2 as shown in Example 3.4 can also be solved by means of the Lagrange function. The Lagrange function for this problem is defined by $L(z, \lambda)$:

$$L(z, \lambda) = J_2(z) + \lambda(z_3 - 2) \tag{3.51}$$

For the simple equality constraint, the LICQ is fulfilled. To obtain possible candidates, the FONC is evaluated.

$$\nabla L(z, \lambda) = \begin{bmatrix} \nabla_z L(z, \lambda) \\ \nabla_\lambda L(z, \lambda) \end{bmatrix} = \begin{bmatrix} 2z_1 + z_3 \\ 2z_2 - 4 \\ 2z_3 + z_1 + \lambda \\ z_3 - 2 \end{bmatrix} = \mathbf{0} \tag{3.52}$$

$$\Rightarrow z^* = \begin{bmatrix} -1 & 2 & 2 \end{bmatrix}^\mathsf{T}, \quad \lambda^* = -3 \tag{3.53}$$

By evaluation of the FONC, the candidate $z^* = [-1 \ 2 \ 2]^\mathsf{T}$ can be found. For checking if the candidate is indeed a local minimizer, the Hessian matrix of the cost function is evaluated.

$$\nabla^2 J_2(z) = \begin{bmatrix} 2 & 0 & 1 \\ 0 & 2 & 0 \\ 1 & 0 & 2 \end{bmatrix} \succ \mathbf{0} \tag{3.54}$$

The cost function J_2 is a strictly convex function and the equality constraint defines a convex feasible set, such that the optimization problem is actually a convex optimization problem. As a result, the FONC is also a sufficient condition. Consequently, $z^* = [-1 \ 2 \ 2]^\mathsf{T}$ is the unique global minimizer.

3.4.3 Constrained Case: Inequality and Equality Constraints

In the following, optimization problems with equality and inequality constraints are treated. The optimization problem is defined by the cost function $J(z) : \mathbb{R}^n \to \mathbb{R}$, the p equality constraints are given by $h_i(z) : \mathbb{R}^n \to \mathbb{R}$, and the q inequality constraints by $g_i(z) : \mathbb{R}^n \to \mathbb{R}$.

$$\begin{aligned} \min_{z \in \mathbb{R}^n} \quad & J(z) \\ \text{s.t.} \quad & h_i(z) = \mathbf{0}, \quad i = 1, \ldots, p, \\ & g_i(z) \leq \mathbf{0}, \quad i = 1, \ldots, q \end{aligned} \tag{3.55}$$

First-Order Necessary Condition

The FONC can be set up using the Lagrange function. When equality and inequality constraints are present, it is given by $L(z, \boldsymbol{\lambda}, \boldsymbol{\mu}) : \mathbb{R}^n \times \mathbb{R}^p \times \mathbb{R}^q \to \mathbb{R}$.

$$L(z, \boldsymbol{\lambda}, \boldsymbol{\mu}) = J(z) + \sum_{i=1}^{p} \lambda_i h_i(z) + \sum_{i=1}^{q} \mu_i g_i(z) \tag{3.56}$$

$$= J(z) + \boldsymbol{\lambda}^\mathsf{T} \boldsymbol{h}(z) + \boldsymbol{\mu}^\mathsf{T} g(z) \tag{3.57}$$

The Lagrange multipliers are given by the vectors $[\lambda_1, \ldots, \lambda_p]^\mathsf{T} = \boldsymbol{\lambda} \in \mathbb{R}^p$ and by $[\mu_1, \ldots, \mu_q]^\mathsf{T} = \boldsymbol{\mu} \in \mathbb{R}^q$. The equality constraints are summarized to be $\boldsymbol{h}(z) = \begin{bmatrix} h_1(z), h_2(z), \ldots, h_p(z) \end{bmatrix}^\mathsf{T}$ and the inequality constraints to be $g(z) =$

$\left[g_1(z), g_2(z), \ldots, g_q(z)\right]^\mathsf{T}$. Each scalar value λ_i is associated with the equality constraint $h_i(z)$ and each scalar value μ_i is associated with the inequality constraint $g_i(z)$.

For constrained optimization problems, the FONC are named after their inventors, KKT conditions. A point $(z^*, \boldsymbol{\lambda}^*, \boldsymbol{\mu}^*)$ that satisfies the KKT conditions and LICQ is also called a KKT point. The KKT conditions for a constrained problem are given by the following equations:

$$\nabla_z L \quad (z^*, \boldsymbol{\lambda}^*, \boldsymbol{\mu}^*) = \mathbf{0} \qquad \text{Stationarity} \tag{3.58}$$

$$\begin{cases} h_i(z^*) = 0, \ \forall i = 1, \ldots, p \\ g_i(z^*) \leq 0, \ \forall i = 1, \ldots, q \end{cases} \qquad \text{Primal Feasibility} \tag{3.59}$$

$$\mu_i^* \quad \geq 0, \ \forall i = 1, \ldots, q \qquad \text{Dual Feasibility} \tag{3.60}$$

$$\mu_i^* g_i(z^*) \quad = 0, \ \forall i = 1, \ldots, q \qquad \text{Complementary Slackness} \tag{3.61}$$

The expression $\nabla_z L(z^*, \boldsymbol{\lambda}^*, \boldsymbol{\mu}^*)$ can be calculated by

$$\nabla_z L(z^*, \boldsymbol{\lambda}^*, \boldsymbol{\mu}^*) = \nabla J(z^*) + \sum_{i=1}^{p} \lambda_i^* \nabla h_i(z^*) + \sum_{i=1}^{q} \mu_i^* \nabla g_i(z^*) \tag{3.62}$$

$$= \nabla J(\boldsymbol{z}^*) + \nabla \boldsymbol{h}(\boldsymbol{z}^*)\boldsymbol{\lambda}^* + \nabla \boldsymbol{g}(\boldsymbol{z}^*)\boldsymbol{\mu}^* \tag{3.63}$$

By inclusion of the inequality constraints in the optimization problem, additional conditions have to be fulfilled. The condition $\boldsymbol{\mu}^* \geq \mathbf{0}$ prohibits a reduction of the cost function toward the infeasible directions of the inequality constraints. A graphical example will be shown in Sect. 3.4.4 below. The gradients of the equality constraints $\nabla \boldsymbol{h}_i(z^*)$ can point in any direction, which is why $\boldsymbol{\lambda}^*$ can be negative as well.

The condition (3.61) is called the *complementary slackness* condition. It is an elegant way to formulate that one of three scenarios is present:

1. The inequality constraint i is strictly active with $g_i(z^*) = 0$ and $\mu_i^* > 0$.
2. The inequality constraint i is inactive with $g_i(z^*) < 0$ and the associated Lagrange multiplier has to be $\mu_i^* = 0$.
3. The inequality constraint i is weakly active with $g_i(z^*) = 0$ and $\mu_i^* = 0$.

In the case of an optimization problem with solely equality constraints, the FONC leads to a system of $n + p$ equations with $n + p$ unknowns. Usually, this system of equations can be solved very effectively. Sometimes, it is even possible to find an analytic solution for this system of equations, as shown in Example 3.3 for the equality-constrained QP. The KKT conditions for inequality-constrained optimization problems generate a system of equations and inequalities. As a consequence, an analytic solution is even for simple systems not possible. Instead, numerical solution techniques are necessary. The solution manifold resulting from the KKT conditions is non-smooth which renders the problem challenging even for numerical solution techniques. Consequently, in general, optimization problems with inequality constraints are harder to solve than purely equality-constrained optimization problems.

Constraint Qualification
Analogous to the equality constrained case, a regularity condition has to be fulfilled to ensure that a local minimizer z^* fulfills the KKT conditions. The LICQ for an optimization problem with equality and inequality constraints requires that the gradients of both the strictly and weakly active inequality constraints and of all equality constraints are linearly independent. This is done by stacking all equality and active inequality constraints together. For this purpose, the active set $\mathcal{A}(z)$ of the optimization problem is defined as the sum of the indices of the inequality constraints i for which $g_i(z) = 0$ holds.

$$\mathcal{A}(z) = \{i \in \{1, \ldots, q\} \mid g_i(z) = 0\} \tag{3.64}$$

The stacked active inequality constraints are denoted by $g_{\mathcal{A}}(z)$. Now, the gradient of all active constraints can be computed. To fulfill LICQ, this matrix has to have full column rank.

$$\operatorname{rank}\left[\nabla \boldsymbol{h}(z), \nabla g_{\mathcal{A}}(z))\right] = p + |\mathcal{A}| \tag{3.65}$$

Suppose that z^* is a local minimizer of the constrained optimization problem and LICQ is satisfied. Then there is a unique Lagrange multiplier vector $\left[\boldsymbol{\lambda}^*, \boldsymbol{\mu}^*\right]^\mathsf{T}$ such that the KKT conditions hold at $(z^*, \boldsymbol{\lambda}^*, \boldsymbol{\mu}^*)$. A graphical interpretation of LICQ will be shown in Sect. 3.4.4 below.

Second-Order Sufficient Optimality Conditions
If LICQ holds and the KKT conditions are fulfilled at a point z^*, the point z^* can still be a point which is not a local minimizer. For checking if the candidate z^* is indeed a local minimizer, the SOSC can be examined which, together with LICQ and the KKT conditions, is a sufficient condition for optimality. Let $z^* \in \mathbb{R}^n$ be a point at which LICQ holds, together with the Lagrange multipliers $\boldsymbol{\lambda}^*$, $\boldsymbol{\mu}^*$ such that the KKT conditions are satisfied. The additional fulfillment of SOSC ensures that z^* is a local minimizer. The SOSC checks if the curvature is positive in all feasible directions:

$$\boldsymbol{s}^\mathsf{T} \nabla_{zz}^2 L(z^*, \boldsymbol{\lambda}^*, \boldsymbol{\mu}^*) \boldsymbol{s} > 0, \qquad \forall \boldsymbol{s} \neq \boldsymbol{0} \tag{3.66}$$

$$\nabla h_i^\mathsf{T}(z^*) \cdot \boldsymbol{s} = 0, \qquad \forall i = 1, \ldots, p \tag{3.67}$$

$$\nabla g_i^\mathsf{T}(z^*) \cdot \boldsymbol{s} = 0, \qquad \forall i \in \mathcal{A}(z^*), \text{ for } \mu_i^* > 0 \tag{3.68}$$

$$\nabla g_i^\mathsf{T}(z^*) \cdot \boldsymbol{s} \leq 0, \qquad \forall i \in \mathcal{A}(z^*), \text{ for } \mu_i^* = 0 \tag{3.69}$$

The feasible directions $\boldsymbol{s}$ are given by the dot products in (3.67)–(3.69). For the inequality constraints, a distinction of cases has to be made. If the inequality constraint is strictly active, i.e. $\mu_i^* > 0$, the feasible direction is given by the constraint tangent space as in the case of equality constraints. However, if the inequality constraint is weakly active $\mu_i^* = 0$, the entire half-space, defined by (3.69), is a feasible direction.

If the Hessian of the Lagrange function with respect to zz is positive definite at the point z^*, $\boldsymbol{\lambda}^*$, $\boldsymbol{\mu}^*$, the SOSC is automatically fulfilled. If a convex optimization problem is given, the FONC and the fulfillment of LICQ are necessary and sufficient conditions, such that SOSC does not need to be checked.

3.4.4 Graphical Interpretation

In the following, the KKT conditions and the LICQ condition are interpreted in a graphical way.

Graphical Interpretation of KKT Conditions for Equality Constraints
Both plots in Fig. 3.8 show the iso-contour lines of a cost function with two optimization variables z_1, z_2, along with an affine equality constraint of the form $h(z) = \boldsymbol{A}z - b = 0$. According to the FONC, ∇J and ∇h are parallel at the minimizer z^*. If these are parallel, there is no feasible direction in which a decrease in the cost function is possible while satisfying the constraints. The left-hand plot of Fig. 3.8 shows additionally a point $\boldsymbol{z}_\alpha$ as a counterexample where the FONC is not fulfilled. At the point $\boldsymbol{z}_\alpha$, the two gradients ∇J and ∇h are not parallel. Consequently, there is a direction $\boldsymbol{p}_\alpha$ in which it is possible to decrease the cost function while the constraint is still fulfilled. For an equality constraint, the Lagrange multiplier λ^* can have a positive or a negative sign. The sign is always such that the weighted sum of the gradients of the cost function and of the constraints equal to $\mathbf{0}$; see (3.39). In the left-hand plot, the Lagrange multiplier λ^* has a positive sign. As an alternative example, the right-hand plot shows a case where the Lagrange multiplier λ^* has a negative sign.

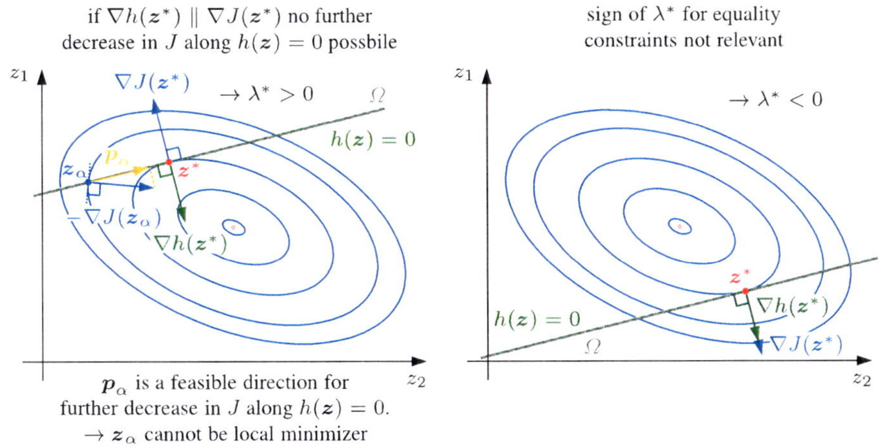

Fig. 3.8 Graphical interpretation of KKT conditions for equality constraints

In general, the single Lagrange multipliers λ_i^* can be interpreted as a sensitivity. High absolute values of λ_i^* mean that if the constraint is relaxed by a small amount, the cost can be decreased by a high value [3].

Graphical Interpretation of KKT Conditions for Inequality Constraints

Figure 3.9 shows the iso-contour lines of a cost function with the two optimization variables z_1, z_2. One affine inequality constraint of the form $g(z) = \boldsymbol{A}z - b \leq 0$ is considered. In the left graph, the unconstrained minimizer lies within the feasible set, such that it is also the constrained minimizer. In this case, the inequality constraint is inactive. At the point $\boldsymbol{z}_\alpha$, the gradients of the cost function $\nabla J(z^*)$ and of the constraint $\nabla g(z^*)$ point in the same direction. According to (3.58), the sum of the two gradients would result in 0 for $\mu_\alpha \leq 0$. For inequality constraints, the feasible region is restricted to only one direction. For negative values of μ, there exists a direction where the inequality constraints are fulfilled and the cost can be decreased. This is the reason why $\boldsymbol{z}_\alpha$ is no local minimizer. For an actual minimizer, the inequality $\mu^* \geq 0$ has to hold. The right-hand graph of Fig. 3.9 shows an example of an active inequality constraint. In this case, the inequality $\mu^* > 0$ holds, such that the KKT conditions are fulfilled.

Linear Independence Constraint Qualification

The LICQ poses a regularity condition for the optimization problem. For LICQ to hold, all active constraints need to be linearly independent. In this case, the KKT conditions as given in (3.58)–(3.61) are necessary conditions for the minimizer. If they are not fulfilled, there might exist a minimizer where the FONC is not satisfied. Figure 3.10 shows an example of the LICQ not being fulfilled. A cost function with the two optimization variables z_1, z_2 is present, along with one inequality constraint $g(z) \leq 0$ and one equality constraint $h(z) = 0$. There exists a minimum of this optimization problem at z^*. Although it is a minimizer, the KKT conditions are not

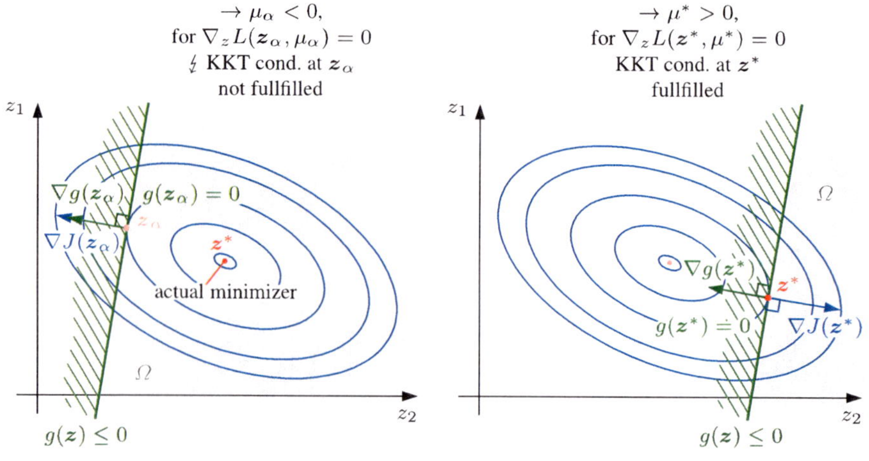

Fig. 3.9 Graphical interpretation of KKT conditions for inequality constraints

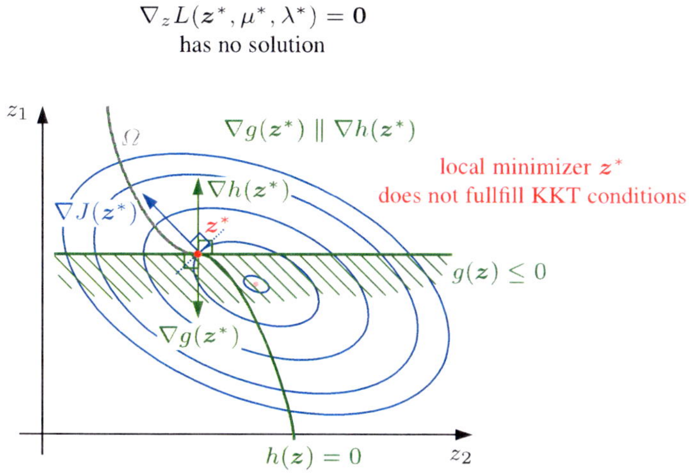

Fig. 3.10 Graphical interpretation of the LICQ

fulfilled at this point z^*. The LICQ is not fulfilled as the gradients of $h(z)$ and $g(z)$ are parallel. As a consequence, they do not span a basis to represent the gradient $\nabla J(z^*)$. Consequently, the stationarity condition $-\nabla J(z^*) = \lambda^* \nabla h(z^*) + \mu^* \nabla g(z^*)$ cannot be fulfilled for any choice of λ^*, μ^*. As the LICQ is not fulfilled, the KKT conditions do not pose a necessary condition for the minimizer.

References

1. R. Beck, A. Bollig, D. Abel, Comparison of two real-time predictive strategies for the optimal energy management of a hybrid electric vehicle. Oil & Gas Science and Technology-Revue de l'IFP **62**(4), 635–643 (2007)
2. J.T. Betts, *Practical methods for optimal control and estimation using nonlinear programming* (SIAM, 2010)
3. S. Boyd, L. Vandenberghe, *Convex optimization* (Cambridge University Press, 2004)
4. P.N. Kolm, R. Tütüncü, F.J. Fabozzi, 60 years of portfolio optimization: practical challenges and current trends. European Journal of Operational Research **234**(2), 356–371 (2014)
5. L. Ljung, *System identification* (Wiley, 1999)
6. R. Matai, S. Singh, and M. L. Mittal, "Traveling salesman problem: an overview of applications, formulations, and solution approaches," *in: Traveling Salesman Problem, Theory and Applications, IntechOpen*, 2010
7. O. Nelles, *Nonlinear system identification: from classical approaches to neural networks and fuzzy models* (Springer, 2013)
8. J. Nocedal, S.J. Wright, *Numerical optimization* (Springer, 2006)
9. M. Salazar, M. Tsao, I. Aguiar, M. Schiffer, and M. Pavone, "A congestion-aware routing scheme for autonomous mobility-on-demand systems," *European Control Conference*, 2019
10. D. Solow, "Linear and nonlinear programming," *in: Encyclopedia of Computer Science and Engineering, Wiley*, 2007
11. W. L. Winston, "Operations research: applications and algorithms," *Thomson/Brooks/Cole*, 2004

Part I
Model Predictive Control

Chapter 4
Linear Model Predictive Control

Abstract In the most basic and common linear model predictive control (LMPC) formulation, a deterministic linear model is used for a prediction of the system states along with a quadratic cost function and linear constraints. This chapter presents an overview of various aspects of LMPC. First, the unconstrained LMPC will be explained, including the solution of the optimization problem and the resulting control structure. Based on these results, the extension of the optimization problem by constraints is investigated. The constraints can apply for the actuated values or the system states, for instance. Two different ways to formulate the optimization problem are shown, namely the sparse and the dense formulations. It is shown that a quadratic program (QP) results for a constrained LMPC, which can be solved very efficiently. Additionally, the MPC scheme for linear time-variant (LTV) systems, called LTV MPC, is investigated. For the LTV MPC, a QP has to be solved in each time step. This is of particular interest, because with LTV MPC already appropriate control results can be achieved for slightly nonlinear systems. To show applications of the methods presented, the chapter concludes with numerical examples.

4.1 Linear Model Representation

A linear model is used for the prediction of the system states within LMPC. There are various possibilities to represent a linear model. The most common approach in LMPC is the use of a discrete-time state-space model.

The use of the state-space representation in LMPC offers several advantages compared to the other linear model representations. Multiple-input multiple-output (MIMO) systems can very naturally be handled and calculated. As the system states are explicitly given, the consideration of system states, e.g. for constraint handling is straightforward. Additionally, the use of a state observer provides the possibility

T. Albin Rajasingham, *Nonlinear Model Predictive Control of Combustion Engines*, Advances in Industrial Control,
https://doi.org/10.1007/978-3-030-68010-7_4

to incorporate additional information for the calculation of the system states, e.g. the covariance of the noise or the structure of the disturbance [16].

Besides the use of the state-space representation, the LMPC can also be based on other model representations [12]. Alternatives are the *Generalized Predictive Control (GPC)* which is an LMPC using discrete-time transfer functions. Another concept is the *Dynamic Matrix Control (DMC)* which is based on a discrete-time pulse-sequence model. Usually, limitations have to be considered for these alternative concepts, such as the restriction of DMC to stable plants.

In the following, the state-space model representation is introduced. First, an overview of the discrete-time state-space representation is given in Sect. 4.1.1. The discrete-time state-space representation can directly be used within the LMPC. Sometimes, instead of a discrete-time model, a continuous-time model is present. This is especially the case when the system model is governed by first principles. As discussed in Sect. 3.3 above, the consideration of a continuous-time model leads to an optimal control problem with an infinite number of optimization variables. Unfortunately, such a program cannot be solved easily, especially when additional constraints are considered. There are two possibilities to deal with a continuous-time model. The most common approach is to discretize the continuous-time linear model and then use the discretized model within the LMPC implementation. An overview of the discretization of continuous-time models is given in Sect. 4.1.2. An alternative way to deal with a continuous-time model within LMPC is the approximation of the trajectory of actuated values by a parametrized function. One exemplary approach is described in [13].

4.1.1 Discrete-Time State-Space Model for Linear Systems

A discrete-time state-space model is represented by the following equations.

$$\begin{aligned} \boldsymbol{x}(k+1) &= \boldsymbol{A}\boldsymbol{x}(k) + \boldsymbol{B}\boldsymbol{u}(k) \\ \boldsymbol{y}(k) &= \boldsymbol{C}\boldsymbol{x}(k) + \boldsymbol{D}\boldsymbol{u}(k) \end{aligned} \tag{4.1}$$

The system states are represented by $\boldsymbol{x} \in \mathbb{R}^n$, the actuated values by $\boldsymbol{u} \in \mathbb{R}^l$, and the output values by $\boldsymbol{y} \in \mathbb{R}^m$. Thus, the matrices have the dimensions $\boldsymbol{A} \in \mathbb{R}^{n\times n}$, $\boldsymbol{B} \in \mathbb{R}^{n\times l}$, $\boldsymbol{C} \in \mathbb{R}^{m\times n}$, and $\boldsymbol{D} \in \mathbb{R}^{m\times l}$. Without loss of generality, this book considers systems without feedthrough, i.e. $\boldsymbol{D} = \boldsymbol{0}$. This is reasonable for technical systems as they do not contain feedthrough. The state-space representation can also be visualized with the so-called functional diagram; see Fig. 4.1.

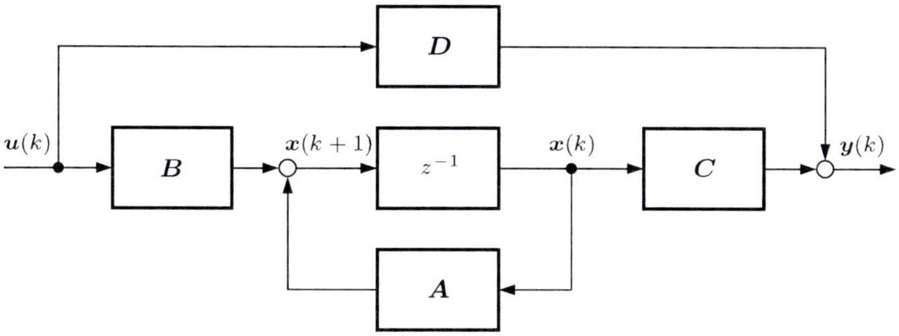

Fig. 4.1 Visualization of the state-space model by its functional diagram

4.1.2 Discretization of Linear Continuous-Time State-Space Models

A continuous-time representation with state-space models is given by (4.2).

$$\begin{aligned}\dot{\boldsymbol{x}}(t) &= \boldsymbol{A}\boldsymbol{x}(t) + \boldsymbol{B}\boldsymbol{u}(t)\\ \boldsymbol{y}(t) &= \boldsymbol{C}\boldsymbol{x}(t) + \boldsymbol{D}\boldsymbol{u}(t)\end{aligned} \tag{4.2}$$

In the following, the discretization of the continuous-time model is revisited. The discretized representation of the continuous-time model is denoted as

$$\begin{aligned}\boldsymbol{x}(k+1) &= \boldsymbol{A}_D\boldsymbol{x}(k) + \boldsymbol{B}_D\boldsymbol{u}(k)\\ \boldsymbol{y}(k) &= \boldsymbol{C}_D\boldsymbol{x}(k) + \boldsymbol{D}_D\boldsymbol{u}(k)\end{aligned} \tag{4.3}$$

The values $k, k+1, \ldots$ denote the discrete-time grid at the time instances $t, t + T_s, t + 2T_s, \ldots$ with T_s being the sampling time. When it is clear from the context that the system under consideration is a discrete-time system, the subscript D will usually be omitted. One way to discretize the continuous-time model is the use of the Euler discretization. The Euler approximation uses a constant derivative during the sampling time T_s:

$$\dot{\boldsymbol{x}}(t) = \frac{\boldsymbol{x}(t+T_s) - \boldsymbol{x}(t)}{T_s} \tag{4.4}$$

As a consequence, the discrete-time model is represented by

$$\boldsymbol{x}(k+1) = \boldsymbol{x}(k) + T_s\dot{\boldsymbol{x}}(k) \tag{4.5}$$

The following system matrices result for the state-space representation:

$$\begin{aligned} \boldsymbol{A}_D &= \boldsymbol{I} + T_s \boldsymbol{A} \\ \boldsymbol{B}_D &= T_s \boldsymbol{B} \\ \boldsymbol{C}_D &= \boldsymbol{C} \\ \boldsymbol{D}_D &= \boldsymbol{D} \end{aligned} \tag{4.6}$$

An alternative way to discretize the continuous-time model is the exact discretization. The discrete-time model calculates the same values for the states and outputs at the discrete-time grid as the continuous-time model as long as the actuated values u are constant during each sampling interval, i.e. zero-order hold. The discretization results from an analytic solution with the transition matrix [2].

$$\boldsymbol{x}(t) = e^{\boldsymbol{A}(t-t_0)}\boldsymbol{x}(t_0) + \int_{t_0}^{t} e^{\boldsymbol{A}(t-\tau)}\boldsymbol{B}\boldsymbol{u}(\tau)d\tau \tag{4.7}$$

With

$$\begin{aligned} t_0 &= kT_s, \\ t &= (k+1)T_s, \\ \boldsymbol{u}(\tau) &= \boldsymbol{u}(k) = \text{const.} \quad \forall\, kT_s \le \tau < (k+1)T_s \end{aligned}$$

the equation can be transformed to

$$\boldsymbol{x}(k+1) = \underbrace{e^{\boldsymbol{A}T_s}}_{\boldsymbol{A}_D}\boldsymbol{x}(k) + \underbrace{\int_{kT_s}^{(k+1)T_s} e^{\boldsymbol{A}(T_s-\tau')}\boldsymbol{B}d\tau'}_{\boldsymbol{B}_D}\boldsymbol{u}(k) \tag{4.8}$$

The corresponding system matrix $\boldsymbol{A}_D$ can be expressed as a series:

$$\boldsymbol{A}_D = e^{\boldsymbol{A}T_s} = \underbrace{\boldsymbol{I} + \frac{\boldsymbol{A}T_s}{1!}}_{\text{Euler disc.}} + \frac{\boldsymbol{A}^2T_s^2}{2!} + \cdots \tag{4.9}$$

If $\boldsymbol{A}$ is invertible, $\boldsymbol{B}_D$ can be calculated by the following series expansion:

$$\boldsymbol{B}_D = \boldsymbol{A}^{-1}\left(e^{\boldsymbol{A}T_s} - \boldsymbol{I}\right)\boldsymbol{B} = \underbrace{\frac{\boldsymbol{I}T_s}{1!}\boldsymbol{B}}_{\text{Euler disc.}} + \frac{\boldsymbol{A}T_s^2}{2!}\boldsymbol{B} + \cdots \tag{4.10}$$

The matrices $\boldsymbol{C}_D$ and $\boldsymbol{D}_D$ remain unchanged, similar to the Euler approximation.

$$\begin{aligned} \boldsymbol{C}_D &= \boldsymbol{C} \\ \boldsymbol{D}_D &= \boldsymbol{D}. \end{aligned} \tag{4.11}$$

4.2 Cost Function for Linear MPC

For MPC, the goals of the control task are mapped into a suitable optimization problem consisting of a cost function and constraints. In the following, the most conventional and basic structure of a cost function is discussed. For the standard reference tracking problem, two different goals have to be considered. The primary goal is that the controlled values should track the reference trajectory over the course of the prediction horizon as accurately as possible. At the same time, any change of actuated values should be penalized for several reasons. The change of actuated values is denoted by $\Delta \boldsymbol{u}(k) = \boldsymbol{u}(k) - \boldsymbol{u}(k-1)$. The penalization is needed for appropriate disturbance rejection. If a change of actuation was not penalized, there would be a strong change of actuated values for every arising disturbance. Based on classical control theory, an inspection of the sensitivity function reveals that there is a trade-off between the disturbance rejection and the reference tracking properties. The actuated values should also not be used with high dynamics for reasons of protection of the actuator. For instance, applying high dynamics to a valve reduces its lifetime. The penalization of a change of actuated values will help to ensure that the optimization problem is strictly convex as is to be shown below in Sect. 4.3.1. In order to set up the cost function, some notations are introduced in Fig. 4.2 and are listed in Table 4.1.

The cost function penalizes the deviation between the reference values during the prediction horizon $\boldsymbol{r}(\cdot|k)$ and the predicted values of the output $\boldsymbol{y}(\cdot|k)$. The current measurement data is represented by $\boldsymbol{y}(k)$ and $\boldsymbol{x}(k)$, while the first predicted state and output are denoted by $\boldsymbol{y}(k+1|k)$ and $\boldsymbol{x}(k+1|k)$. To account for the dead-time

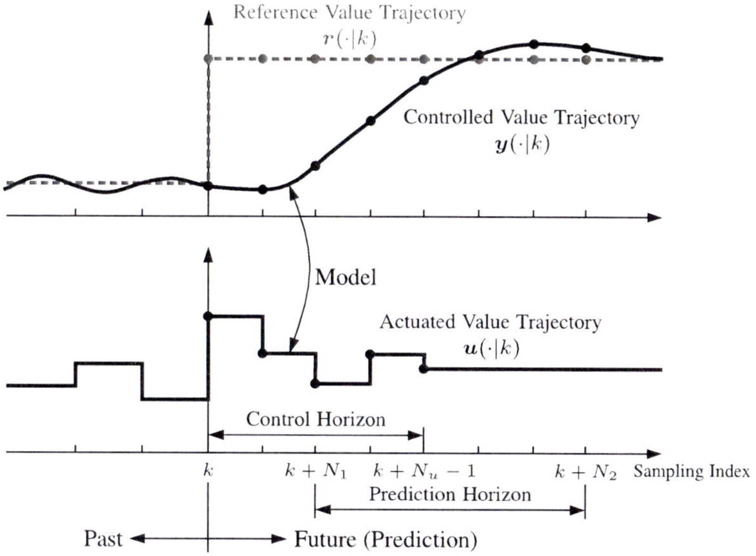

Fig. 4.2 Notation used for LMPC

Table 4.1 Nomenclature of LMPC

$\boldsymbol{a}(k+i\|k)$	Prediction of values a at time instance $k+i$ at the current time step k
$\boldsymbol{a}(\cdot\|k)$	Total trajectory of predicted values a in the horizon considered based on the current time step k
N_1	Lower prediction horizon
N_2	Upper prediction horizon
N_u	Control horizon

T_D, the lower prediction horizon N_1 is introduced. As the actuated values do not have an influence on the output before the time point $k + T_D + 1$, it is advantageous to not consider these time points in the optimization in order to reduce the size of the problem and thus also the computation time required. For a system without any dead-time, $N_1 = 1$ is chosen, else $N_1 = 1 + T_D$. The prediction is computed over a restricted prediction horizon from $N_1, \ldots, N_2$. When linear systems are considered, it is very often sufficient to account for a control horizon N_u that is smaller than the prediction horizon. The horizon N_u directly determines the number of optimization variables. As a consequence, the horizon is very critical for calculation times, and engineers are advised to choose it as low as possible. The actuated values $\boldsymbol{u}$ are held constant beyond the control horizon N_u for the prediction of the system states and outputs.

For the penalization of the two terms, namely the predicted change of the actuated values $\Delta\boldsymbol{u}(\cdot|k)$ and the predicted control error $\boldsymbol{y}(\cdot|k) - \boldsymbol{r}(\cdot|k)$, a suitable norm is needed. The most common choice is to quantify the penalization by a squared weighted 2-norm. The general squared weighted 2-norm of the vector $\boldsymbol{s} \in \mathbb{R}^n$ with the weighting matrix $\boldsymbol{A} \in \mathbb{R}^{n \times n}$ is given by (4.12).

$$||\boldsymbol{s}||_{\boldsymbol{A}}^2 := \boldsymbol{s}^\mathsf{T}\boldsymbol{A}\boldsymbol{s}, \quad \boldsymbol{s} \in \mathbb{R}^n, \boldsymbol{A} \in \mathbb{R}^{n \times n} \tag{4.12}$$

The main reason for this choice is that the optimization problem results in a QP, as is to be shown below, which can be solved efficiently. Moreover, for small changes in the optimization problem, e.g. due to disturbances, the minimizer, i.e. the actuated values, also change in a smooth manner. Additionally, large deviations lead to higher costs. An alternative is the use of the 1-norm in the cost function, which leads to a linear program (LP). The general 1-norm of the vector $\boldsymbol{s} \in \mathbb{R}^n$ is given by (4.13).

$$||\boldsymbol{s}||_1 := \sum_{i=1}^{n} |s_i| \tag{4.13}$$

In a few specific applications, the LP is a more "natural" way to describe the control goals; see [11] for an example. In general, the drawback is that for an LP, the minimizer always lies on an intersection of the linear constraints. As a consequence, for small changes, e.g. due to a disturbance, the minimizer and thus the actuated values change in a non-smooth manner which is often not desired in practical applications.

Using the squared weighted 2-norm, the entire cost function can be set up in a matrix representation:

$$J = \left\| \begin{bmatrix} \boldsymbol{y}(k+N_1|k) \\ \vdots \\ \boldsymbol{y}(k+N_2|k) \end{bmatrix} - \begin{bmatrix} \boldsymbol{r}(k+N_1|k) \\ \vdots \\ \boldsymbol{r}(k+N_2|k) \end{bmatrix} \right\|_{\boldsymbol{Q}}^2 + \left\| \begin{bmatrix} \Delta\boldsymbol{u}(k|k) \\ \vdots \\ \Delta\boldsymbol{u}(k+N_u-1|k) \end{bmatrix} \right\|_{\boldsymbol{R}}^2 \tag{4.14}$$

The weighting factors are given by $\boldsymbol{Q}$ and $\boldsymbol{R}$.

$$\boldsymbol{Q} = \begin{bmatrix} \boldsymbol{Q}_{N_1} & \boldsymbol{0} & \cdots \boldsymbol{0} \\ \boldsymbol{0} & \boldsymbol{Q}_{N_1+1} & \cdots \boldsymbol{0} \\ \vdots & \vdots & \ddots \vdots \\ \boldsymbol{0} & \boldsymbol{0} & \cdots \boldsymbol{Q}_{N_2} \end{bmatrix} \qquad \boldsymbol{R} = \begin{bmatrix} \boldsymbol{R}_0 & \boldsymbol{0} & \cdots \boldsymbol{0} \\ \boldsymbol{0} & \boldsymbol{R}_1 & \cdots \boldsymbol{0} \\ \vdots & \vdots & \ddots \vdots \\ \boldsymbol{0} & \boldsymbol{0} & \cdots \boldsymbol{R}_{N_u-1} \end{bmatrix}. \tag{4.15}$$

4.3 Unconstrained Linear MPC

The cost function resulting for the unconstrained LMPC investigated can be summarized in a compact form:

$$J = \sum_{i=N_1}^{N_2} ||\boldsymbol{y}(k+i|k) - \boldsymbol{r}(k+i|k)||_{\boldsymbol{Q}_i}^2 + \sum_{i=0}^{N_u-1} ||\Delta\boldsymbol{u}(k+i|k)||_{\boldsymbol{R}_i}^2 \tag{4.16}$$

In the following, the solution of this optimization problem is described. It is detailed that the resulting optimization problem can actually be solved analytically. Based on the solution, it can even be shown that a linear state-feedback controller results for an unconstrained LMPC.

4.3.1 Analytic Solution of the Optimization Problem

In order to derive the solution of the optimization problem, the linear model is incorporated into the optimization problem. Within the optimization problem, the change of actuated values over the horizon $\Delta\boldsymbol{u}(\cdot|k)$ are chosen as optimization variables.

Thus, the sequence of actuated values is reformulated to

$$
\begin{aligned}
\boldsymbol{u}(k|k) &= \boldsymbol{u}(k-1) + \Delta\boldsymbol{u}(k|k) \\
\boldsymbol{u}(k+1|k) &= \boldsymbol{u}(k|k) + \Delta\boldsymbol{u}(k+1|k) \\
&= \boldsymbol{u}(k-1) + \Delta\boldsymbol{u}(k|k) + \Delta\boldsymbol{u}(k+1|k) \\
&\ \ \vdots \\
\boldsymbol{u}(k+N_u-1|k) &= \boldsymbol{u}(k+N_u-2|k) + \Delta\boldsymbol{u}(k+N_u-1|k) \\
\boldsymbol{u}(k+N_u|k) &= \boldsymbol{u}(k+N_u-1|k) \\
\boldsymbol{u}(k+N_u+1|k) &= \boldsymbol{u}(k+N_u-1|k) \\
&\ \ \vdots \\
\boldsymbol{u}(k+N_2-1|k) &= \boldsymbol{u}(k+N_u-1|k)
\end{aligned}
\tag{4.17}
$$

The control sequence takes into account that the actuated values are set constant after the control horizon which corresponds to $\Delta\boldsymbol{u}(k+i|k) = 0, i = N_u, \ldots, N_2 - 1$. This definition of the sequence of actuated values allows to reformulate the general system equations:

$$
\begin{aligned}
\boldsymbol{x}(k+1|k) &= \boldsymbol{A}\boldsymbol{x}(k) + \boldsymbol{B}\boldsymbol{u}(k-1) + \boldsymbol{B}\Delta\boldsymbol{u}(k|k) \\
\boldsymbol{y}(k+1|k) &= \boldsymbol{C}(\boldsymbol{A}\boldsymbol{x}(k) + \boldsymbol{B}\boldsymbol{u}(k-1) + \boldsymbol{B}\Delta\boldsymbol{u}(k|k))
\end{aligned}
\tag{4.18}
$$

The initial system states, obtained by measurement or estimation, are denoted by $\boldsymbol{x}(k)$. The last actuated values applied are given by $\boldsymbol{u}(k-1)$. Both $\boldsymbol{x}(k)$ and $\boldsymbol{u}(k-1)$ cannot be adjusted. The variables $\Delta\boldsymbol{u}(\cdot|k)$ are adjustable and serve as degrees of freedom for the optimization. From here on and for the duration of this derivation, the optimization variables are underlined as a visual aid where necessary. The prediction of the system states $\boldsymbol{x}(\cdot|k)$ can be calculated in dependence of the optimization variables:

$$
\begin{aligned}
\boldsymbol{x}(k+1|k) &= \boldsymbol{A}\boldsymbol{x}(k) + \boldsymbol{B}\boldsymbol{u}(k-1) + \boldsymbol{B}\underline{\Delta\boldsymbol{u}(k|k)} \\
\boldsymbol{x}(k+2|k) &= \boldsymbol{A}\boldsymbol{x}(k+1|k) + \boldsymbol{B}\boldsymbol{u}(k|k) + \boldsymbol{B}\underline{\Delta\boldsymbol{u}(k+1|k)} \\
\boldsymbol{x}(k+3|k) &= \boldsymbol{A}\boldsymbol{x}(k+2|k) + \boldsymbol{B}\boldsymbol{u}(k+1|k) + \boldsymbol{B}\underline{\Delta\boldsymbol{u}(k+2|k)} \\
&\ \ \vdots \\
\boldsymbol{x}(k+N_u|k) &= \boldsymbol{A}\boldsymbol{x}(k+N_u-1|k) + \boldsymbol{B}\boldsymbol{u}(k+N_u-2|k) + \boldsymbol{B}\underline{\Delta\boldsymbol{u}(k+N_u-1|k)} \\
\Rightarrow \boldsymbol{x}(k+j|k) &= \boldsymbol{A}^j\boldsymbol{x}(k) + \sum_{i=1}^{j}\left(\boldsymbol{A}^{i-1}\boldsymbol{B}\boldsymbol{u}(k-1)\right) + \sum_{m=0}^{j-1}\left(\sum_{i=1}^{j-m}\left(\boldsymbol{A}^{i-1}\boldsymbol{B}\right)\underline{\Delta\boldsymbol{u}(k+m|k)}\right)
\end{aligned}
\tag{4.19}
$$

After the control horizon,

$$
\begin{aligned}
\boldsymbol{x}(k+N_u+1|k) &= \boldsymbol{A}\boldsymbol{x}(k+N_u|k) + \boldsymbol{B}\boldsymbol{u}(k+N_u-1|k) + \overbrace{\boldsymbol{B}\Delta\boldsymbol{u}(k+N_u|k)}^{=0 \text{ as } \boldsymbol{u} \text{ is const. after } N_u \text{ steps}} \\
&\vdots \\
\boldsymbol{x}(k+N_2|k) &= \boldsymbol{A}\boldsymbol{x}(k+N_2-1|k) + \boldsymbol{B}\boldsymbol{u}(k+N_u-1|k) + \underbrace{\boldsymbol{B}\Delta\boldsymbol{u}(k+N_u|k)}_{=0 \text{ as } \boldsymbol{u} \text{ is const. after } N_u \text{ steps}}
\end{aligned}
\tag{4.20}
$$

All equations are substituted such that only the initial values $\boldsymbol{x}(k)$, $\boldsymbol{u}(k-1)$, and the optimization variables $\underline{\Delta\boldsymbol{u}(\cdot|k)}$ remain.

$$
\begin{aligned}
\boldsymbol{x}(\cdot|k) = \begin{bmatrix} \boldsymbol{x}(k+N_1|k) \\ \boldsymbol{x}(k+N_1+1|k) \\ \vdots \\ \boldsymbol{x}(k+N_2|k) \end{bmatrix} &= \underbrace{\begin{bmatrix} \boldsymbol{A}^{N_1} \\ \boldsymbol{A}^{N_1+1} \\ \vdots \\ \boldsymbol{A}^{N_2} \end{bmatrix}}_{=:\,\boldsymbol{\Theta} \in \mathbb{R}^{n\cdot(N_2-N_1+1)\times n}} \boldsymbol{x}(k) + \underbrace{\begin{bmatrix} \sum_{i=1}^{N_1} \boldsymbol{A}^{i-1}\boldsymbol{B} \\ \sum_{i=1}^{N_1+1} \boldsymbol{A}^{i-1}\boldsymbol{B} \\ \vdots \\ \sum_{i=1}^{N_2} \boldsymbol{A}^{i-1}\boldsymbol{B} \end{bmatrix}}_{=:\,\boldsymbol{\Pi} \in \mathbb{R}^{n\cdot(N_2-N_1+1)\times l}} \boldsymbol{u}(k-1) \\
&+ \underbrace{\begin{bmatrix} \boldsymbol{\Lambda}(N_1) & \boldsymbol{\Lambda}(N_1-1) & \cdots & \boldsymbol{\Lambda}(N_1-N_u+1) \\ \boldsymbol{\Lambda}(N_1+1) & \boldsymbol{\Lambda}(N_1) & \cdots & \boldsymbol{\Lambda}(N_1-N_u+2) \\ \vdots & \vdots & \ddots & \vdots \\ \boldsymbol{\Lambda}(N_2) & \boldsymbol{\Lambda}(N_2-1) & \cdots & \boldsymbol{\Lambda}(N_2-N_u+1) \end{bmatrix}}_{=:\,\boldsymbol{\Upsilon} \in \mathbb{R}^{n\cdot(N_2-N_1+1)\times l\cdot N_u}} \underbrace{\begin{bmatrix} \Delta\boldsymbol{u}(k|k) \\ \Delta\boldsymbol{u}(k+1|k) \\ \vdots \\ \Delta\boldsymbol{u}(k+N_u-1|k) \end{bmatrix}}_{=\,\underline{\Delta\boldsymbol{u}(\cdot|k)}}
\end{aligned}
\tag{4.21}
$$

$$
\text{with} \quad \mathbb{R}^{n\times l} \ni \boldsymbol{\Lambda}(i) = \begin{cases} \sum_{j=1}^{i} \boldsymbol{A}^{j-1}\boldsymbol{B}, & i \geq 1 \\ 0, & i < 1 \end{cases} \tag{4.22}
$$

The prediction of the system outputs is given by the simple relation:

$$
\boldsymbol{y}(k+i|k) = \boldsymbol{C}\boldsymbol{x}(k+i|k) \tag{4.23}
$$

The prediction of the system states and outputs over the entire horizon can be rewritten in a compact matrix representation:

$$
\begin{aligned}
\boldsymbol{x}(\cdot|k) &= \boldsymbol{\Theta}\boldsymbol{x}(k) + \boldsymbol{\Pi}\boldsymbol{u}(k-1) + \boldsymbol{\Upsilon}\Delta\boldsymbol{u}(\cdot|k) \\
\boldsymbol{y}(\cdot|k) &= \boldsymbol{\Gamma}\boldsymbol{x}(\cdot|k) \\
\text{with} \quad \boldsymbol{\Gamma} &= \operatorname{diag}(\boldsymbol{C},\ldots,\boldsymbol{C}) \in \mathbb{R}^{m\cdot(N_2-N_1+1)\times n\cdot(N_2-N_1+1)}
\end{aligned}
\tag{4.24}
$$

Some new variables are introduced that are helpful for interpretation. The free response $\boldsymbol{f}$ is the vector of the output prediction for the case that the actuated values are kept constant over the entire prediction horizon, i.e. $\Delta\boldsymbol{u}(\cdot|k) = 0$.

$$
\boldsymbol{f}(\cdot|k) = \boldsymbol{\Gamma}\big(\boldsymbol{\Theta}\boldsymbol{x}(k) + \boldsymbol{\Pi}\boldsymbol{u}(k-1)\big) \tag{4.25}
$$

The free error $\boldsymbol{e}(\cdot|k)$ is the resulting predicted error for the constant actuated values $\underline{\Delta \boldsymbol{u}(\cdot|k)} = 0$, with $\boldsymbol{r}(\cdot|k)$ being the reference.

$$\boldsymbol{e}(\cdot|k) = \boldsymbol{r}(\cdot|k) - \boldsymbol{f}(\cdot|k) \tag{4.26}$$

Hence, the controller has to compensate for the free error by an adjustment of the optimization variables $\underline{\Delta \boldsymbol{u}(\cdot|k)}$. As a linear system model is considered, the overall predicted output $\boldsymbol{y}(\cdot|k)$ results from a superposition:

$$\boldsymbol{y}(\cdot|k) = \boldsymbol{f}(\cdot|k) + \boldsymbol{\Gamma}\boldsymbol{\Upsilon}\underline{\Delta \boldsymbol{u}(\cdot|k)} \tag{4.27}$$

The cost function can be reformulated using the defined variables:

$$\begin{aligned} J &= ||\boldsymbol{y}(\cdot|k) - \boldsymbol{r}(\cdot|k)||_{\boldsymbol{Q}}^2 + ||\underline{\Delta \boldsymbol{u}(\cdot|k)}||_{\boldsymbol{R}}^2 \\ &= ||\boldsymbol{\Gamma}\boldsymbol{\Upsilon}\underline{\Delta \boldsymbol{u}(\cdot|k)} - \boldsymbol{e}(\cdot|k)||_{\boldsymbol{Q}}^2 + ||\underline{\Delta \boldsymbol{u}(\cdot|k)}||_{\boldsymbol{R}}^2 \\ &= \left(\underline{\Delta \boldsymbol{u}(\cdot|k)}^\mathsf{T}\boldsymbol{\Upsilon}^\mathsf{T}\boldsymbol{\Gamma}^\mathsf{T} - \boldsymbol{e}^\mathsf{T}(\cdot|k)\right)\boldsymbol{Q}\left(\boldsymbol{\Gamma}\boldsymbol{\Upsilon}\underline{\Delta \boldsymbol{u}(\cdot|k)} - \boldsymbol{e}(\cdot|k)\right) + \underline{\Delta \boldsymbol{u}(\cdot|k)}^\mathsf{T}\boldsymbol{R}\underline{\Delta \boldsymbol{u}(\cdot|k)} \\ &= \underline{\Delta \boldsymbol{u}(\cdot|k)}^\mathsf{T}\left(\boldsymbol{\Upsilon}^\mathsf{T}\boldsymbol{\Gamma}^\mathsf{T}\boldsymbol{Q}\boldsymbol{\Gamma}\boldsymbol{\Upsilon} + \boldsymbol{R}\right)\underline{\Delta \boldsymbol{u}(\cdot|k)} - 2\underline{\Delta \boldsymbol{u}(\cdot|k)}^\mathsf{T}\boldsymbol{\Upsilon}^\mathsf{T}\boldsymbol{\Gamma}^\mathsf{T}\boldsymbol{Q}\boldsymbol{e}(\cdot|k) + \boldsymbol{e}(\cdot|k)^\mathsf{T}\boldsymbol{Q}\boldsymbol{e}(\cdot|k) \end{aligned} \tag{4.28}$$

The cost function can be rearranged to show the quadratic function in the optimization variables $\underline{\Delta \boldsymbol{u}(\cdot|k)}$.

$$\begin{aligned} J =2\Big(&\frac{1}{2}\underline{\Delta \boldsymbol{u}(\cdot|k)}^\mathsf{T}\left(\boldsymbol{\Upsilon}^\mathsf{T}\boldsymbol{\Gamma}^\mathsf{T}\boldsymbol{Q}\boldsymbol{\Gamma}\boldsymbol{\Upsilon} + \boldsymbol{R}\right)\underline{\Delta \boldsymbol{u}(\cdot|k)} && \text{Quadratic term} \\ &+ \underline{\Delta \boldsymbol{u}(\cdot|k)}^\mathsf{T}\left(-\boldsymbol{\Upsilon}^\mathsf{T}\boldsymbol{\Gamma}^\mathsf{T}\boldsymbol{Q}\boldsymbol{e}(\cdot|k)\right) && \text{Linear term} \\ &+ \frac{1}{2}\boldsymbol{e}(\cdot|k)^\mathsf{T}\boldsymbol{Q}\boldsymbol{e}(\cdot|k)\Big) && \text{Constant term} \end{aligned}$$

The constant term is invariant with respect to $\underline{\Delta \boldsymbol{u}(\cdot|k)}$. As it has no influence on the minimizer, i.e. the actuated values, it thus can be neglected for the purpose of optimization. Only the non-constant terms have to be considered for the optimization. The same holds for the constant factor two, it can also be neglected.

$$\min_{\underline{\Delta \boldsymbol{u}(\cdot|k)}} \quad \frac{1}{2}\underline{\Delta \boldsymbol{u}(\cdot|k)}^\mathsf{T}\underbrace{\left(\boldsymbol{\Upsilon}^\mathsf{T}\boldsymbol{\Gamma}^\mathsf{T}\boldsymbol{Q}\boldsymbol{\Gamma}\boldsymbol{\Upsilon} + \boldsymbol{R}\right)}_{=:\boldsymbol{H}}\underline{\Delta \boldsymbol{u}(\cdot|k)} + \underline{\Delta \boldsymbol{u}(\cdot|k)}^\mathsf{T}\underbrace{\left(-\boldsymbol{\Upsilon}^\mathsf{T}\boldsymbol{\Gamma}^\mathsf{T}\boldsymbol{Q}\boldsymbol{e}(\cdot|k)\right)}_{=:\boldsymbol{q}} \tag{4.29}$$

This optimization problem can be solved by the means discussed in Sect. 3.4.1 above. The first-order necessary condition (FONC) for the minimizer $\underline{\Delta \boldsymbol{u}^*(\cdot|k)}$ can be set up as

$$\nabla_{\underline{\Delta \boldsymbol{u}(\cdot|k)}} J(\underline{\Delta \boldsymbol{u}^*(\cdot|k)}) = \boldsymbol{0} \tag{4.30}$$

The calculation of the gradient yields

$$\nabla_{\Delta \boldsymbol{u}(\cdot|k)} J(\Delta \boldsymbol{u}^*(\cdot|k)) = \boldsymbol{H} \Delta \boldsymbol{u}^*(\cdot|k) + \boldsymbol{q} \tag{4.31}$$

Thus, the FONC can be reformulated as in the following equation:

$$\boldsymbol{H} \Delta \boldsymbol{u}^*(\cdot|k) + \boldsymbol{q} = 0 \tag{4.32}$$

One way to solve this equation for $\Delta \boldsymbol{u}^*(\cdot|k)$ is given by the use of the matrix inversion. For real-time implementations, it is computationally more efficient and robust to solve this equation as a least-squares problem. Still, the matrix inversion is denoted for ease of notation:

$$\begin{aligned} \Delta \boldsymbol{u}^*(\cdot|k) &= \boldsymbol{H}^{-1} \boldsymbol{q} \\ &= \underbrace{\left(\boldsymbol{\Upsilon}^{\mathsf{T}} \boldsymbol{\Gamma}^{\mathsf{T}} \boldsymbol{Q} \boldsymbol{\Gamma} \boldsymbol{\Upsilon} + \boldsymbol{R}\right)^{-1}}_{\text{Matrix Inversion}} \boldsymbol{\Upsilon}^{\mathsf{T}} \boldsymbol{\Gamma}^{\mathsf{T}} \boldsymbol{Q} \boldsymbol{e}(\cdot|k) \end{aligned} \tag{4.33}$$

In order to check if the candidate $\Delta \boldsymbol{u}^*(\cdot|k)$ is a minimum, the second-order sufficient condition (SOSC) can be used, as follows:

$$\nabla^2_{\Delta \boldsymbol{u}(\cdot|k)} J(\Delta \boldsymbol{u}^*(\cdot|k)) \succ \boldsymbol{0} \tag{4.34}$$

The Hessian of J needs to be positive definite. The SOSC can be denoted by the following condition for the minimizer:

$$\boldsymbol{\Upsilon}^{\mathsf{T}} \boldsymbol{\Gamma}^{\mathsf{T}} \boldsymbol{Q} \boldsymbol{\Gamma} \boldsymbol{\Upsilon} + \boldsymbol{R} \succ \boldsymbol{0} \tag{4.35}$$

The positive definiteness of the Hessian can be ensured by an appropriate choice of the weighting matrices $\boldsymbol{Q}$ and $\boldsymbol{R}$. The conditions on $\boldsymbol{Q}$ and $\boldsymbol{R}$ are as follows:

$$\boldsymbol{R} \succ \boldsymbol{0} \quad \text{and} \quad \boldsymbol{Q} \succeq \boldsymbol{0} \tag{4.36}$$

With a positive-definite Hessian, the QP also becomes strictly convex. As a consequence, if there exists a minimizer it is the unique global minimizer.

4.3.2 Resulting Control Law

Thus far, the text showed that there exists an analytic solution for the optimization problem arising in an unconstrained LMPC. The optimal trajectory over the entire control horizon can be calculated with (4.33). According to the receding-horizon principle, only the first values of $\Delta \boldsymbol{u}^*(\cdot|k)$, i.e. $\Delta \boldsymbol{u}^*(k|k)$ are applied to the plant, while the remaining values are discarded. The current actuated values $\boldsymbol{u}(k)$ are given by

$$\begin{aligned}\Delta\boldsymbol{u}^*(k|k) &= \boldsymbol{K}_{mpc}\boldsymbol{e}(\cdot|k)\\ \boldsymbol{K}_{mpc} &= \left[\boldsymbol{I}\ \boldsymbol{0}\ \cdots\ \boldsymbol{0}\right]\left(\left(\boldsymbol{\Upsilon}^{\mathsf{T}}\boldsymbol{\Gamma}^{\mathsf{T}}\boldsymbol{Q}\boldsymbol{\Gamma}\boldsymbol{\Upsilon}+\boldsymbol{R}\right)^{-1}\boldsymbol{\Upsilon}^{\mathsf{T}}\boldsymbol{\Gamma}^{\mathsf{T}}\boldsymbol{Q}\right)\\ \boldsymbol{u}(k) &= \boldsymbol{u}(k-1)+\Delta\boldsymbol{u}^*(k|k)\end{aligned} \tag{4.37}$$

The matrix $\boldsymbol{K}_{mpc}$ is linear time-invariant. It depends solely on the linear system model and the weighting matrices. The free error $\boldsymbol{e}(\cdot|k)$ depends linearly on the current state measurement $\boldsymbol{x}(k)$ and the reference values $\boldsymbol{r}(\cdot|k)$. Thus, the error typically changes at every time step. As a consequence, the unconstrained LMPC is actually a linear state-feedback controller.

As a linear controller results for an LMPC, classical properties such as pole-zero locations of the closed-loop system can be calculated. It is expectable that a linear state-feedback controller results when the concept of an unconstrained LMPC is compared to the linear quadratic regulator (LQR). The concepts are very similar, with the only difference being that the LQR conducts an infinite-horizon prediction rather than a finite-time prediction for an unconstrained LMPC. With the LQR concept, a linear time-invariant state-feedback controller results, as well [5].

4.4 Constrained Linear MPC

One of the major advantages of MPC is the ability it offers to systematically consider constraints. The constraints can arise on the actuated values $\boldsymbol{u}(\cdot|k)$, the change of the actuated values $\Delta\boldsymbol{u}(\cdot|k)$, on the system states $\boldsymbol{x}(\cdot|k)$, and on the system output $\boldsymbol{y}(\cdot|k)$. Examples are the maximum opening position of a valve, the maximum rate of change of an electric motor, the maximum allowed temperature for safety reasons, and the prevention of overshoots on an output due to performance demands. In the following, two approaches are demonstrated to set up the optimization problem for a constrained LMPC. The first approach is the so-called dense formulation where the model equations are substituted into the cost function as done already in the unconstrained case. The second approach is the so-called sparse formulation where the model equations are added to the optimization problem as equality constraints.

4.4.1 Dense Formulation

The goal is to formulate the constrained LMPC problem in such a way that it results in the typical constrained QP formulation. The general structure of a constrained QP with optimization variables $\boldsymbol{z} \in \mathbb{R}^l$ is given by (4.38). To define p equality constraints, the matrices $\boldsymbol{A}_{eq} \in \mathbb{R}^{p\times l}$, $\boldsymbol{b}_{eq} \in \mathbb{R}^p$ are used. Additionally, q inequality constraints are defined by $\boldsymbol{A}_{in} \in \mathbb{R}^{q\times l}$, $\boldsymbol{b}_{in} \in \mathbb{R}^q$.

$$\begin{aligned}
\min_{z \in \mathbb{R}^l} \quad & \frac{1}{2} z^{\mathsf{T}} \boldsymbol{H} z + z^{\mathsf{T}} \boldsymbol{q} \\
\text{s.t.} \quad & \boldsymbol{A}_{eq} z - \boldsymbol{b}_{eq} = \boldsymbol{0}, \quad \boldsymbol{A}_{eq} \in \mathbb{R}^{p \times l}, \ \boldsymbol{b}_{eq} \in \mathbb{R}^{p}, \\
& \boldsymbol{A}_{in} z - \boldsymbol{b}_{in} \leq \boldsymbol{0}, \quad \boldsymbol{A}_{in} \in \mathbb{R}^{q \times l}, \ \boldsymbol{b}_{in} \in \mathbb{R}^{q}
\end{aligned} \tag{4.38}$$

For a constrained, LMPC the same cost function (4.28) is considered as in the case of an unconstrained LMPC. Within the dense formulation, the model equations for the system dynamics are substituted into the cost function. The entire cost function is formulated in terms of the optimization variables, i.e. $\Delta \boldsymbol{u}(\cdot|k)$. In the case of a constrained LMPC, the constraints on the system output $\boldsymbol{y}(\cdot|k)$, the system states $\boldsymbol{x}(\cdot|k)$, the change of the actuated values $\Delta \boldsymbol{u}(\cdot|k)$, and the absolute actuated values $\boldsymbol{u}(\cdot|k)$ are added. The transformation to the standard QP formulation, as shown in (4.38), is conducted exemplarily for box constraints, i.e. inequality-constrained variables with upper and lower bounds.

$$\begin{aligned}
\min_{\Delta \boldsymbol{u}(\cdot|k)} \quad & \frac{1}{2} \Delta \boldsymbol{u}(\cdot|k)^{\mathsf{T}} \overbrace{\left(\boldsymbol{\Upsilon}^{\mathsf{T}} \boldsymbol{\Gamma}^{\mathsf{T}} \boldsymbol{Q} \boldsymbol{\Gamma} \boldsymbol{\Upsilon} + \boldsymbol{R}\right)}^{=:\boldsymbol{H}} \Delta \boldsymbol{u}(\cdot|k) + \Delta \boldsymbol{u}(\cdot|k)^{\mathsf{T}} \overbrace{\left(-\boldsymbol{\Upsilon}^{\mathsf{T}} \boldsymbol{\Gamma}^{\mathsf{T}} \boldsymbol{Q} \boldsymbol{e}(\cdot|k)\right)}^{=:\boldsymbol{q}} \\
\text{s.t.} \quad & \Delta \boldsymbol{u}_{min} \leq \Delta \boldsymbol{u}(\cdot|k) \leq \Delta \boldsymbol{u}_{max}, \\
& \boldsymbol{u}_{min} \leq \boldsymbol{u}(\cdot|k) \leq \boldsymbol{u}_{max}, \\
& \boldsymbol{x}_{min} \leq \boldsymbol{x}(\cdot|k) \leq \boldsymbol{x}_{max}, \\
& \boldsymbol{y}_{min} \leq \boldsymbol{y}(\cdot|k) \leq \boldsymbol{y}_{max}
\end{aligned} \tag{4.39}$$

In order to formulate the LMPC optimization problem in a QP setting, all constraints have to be expressed in dependence of the optimization variables $\Delta \boldsymbol{u}(\cdot|k)$. Using the prediction matrices derived above, the lower and upper bounds have to be reformulated to match the standard form given by $\boldsymbol{A}_{in} \Delta \boldsymbol{u}(\cdot|k) \leq \boldsymbol{b}_{in}$.

First, the transformation of the bounds on $\Delta \boldsymbol{u}(\cdot|k)$ is discussed.

$$\Delta \boldsymbol{u}_{min} \leq \Delta \boldsymbol{u}(\cdot|k) \leq \Delta \boldsymbol{u}_{max} \tag{4.40}$$

Expressing lower and upper bounds as *less than or equal to* inequalities yields

$$\begin{aligned}
\Delta \boldsymbol{u}(k|k) &\leq \Delta \boldsymbol{u}_{\max} \\
&\vdots \\
\Delta \boldsymbol{u}(k + N_u - 1|k) &\leq \Delta \boldsymbol{u}_{\max} \\
-\Delta \boldsymbol{u}(k|k) &\leq -\Delta \boldsymbol{u}_{\min} \\
&\vdots \\
-\Delta \boldsymbol{u}(k + N_u - 1|k) &\leq -\Delta \boldsymbol{u}_{\min}
\end{aligned} \tag{4.41}$$

The changes on the actuated values have to be considered until the time point $N_u - 1$, as once the control horizon is reached, the actuated values are fixed. This can be rewritten in a matrix-vector representation:

$$\underbrace{\begin{bmatrix} \boldsymbol{I} & \boldsymbol{0} & \dots & \boldsymbol{0} \\ \boldsymbol{0} & \boldsymbol{I} & \dots & \boldsymbol{0} \\ \vdots & \vdots & \ddots & \vdots \\ \boldsymbol{0} & \boldsymbol{0} & \dots & \boldsymbol{I} \\ -\boldsymbol{I} & \boldsymbol{0} & \dots & \boldsymbol{0} \\ \boldsymbol{0} & -\boldsymbol{I} & \dots & \boldsymbol{0} \\ \vdots & \vdots & \ddots & \vdots \\ \boldsymbol{0} & \boldsymbol{0} & \dots & -\boldsymbol{I} \end{bmatrix}}_{=:\, \boldsymbol{A}_{in,\,\Delta u} \in \mathbb{R}^{2lN_u \times lN_u}} \underbrace{\begin{bmatrix} \Delta\boldsymbol{u}(k|k) \\ \vdots \\ \Delta\boldsymbol{u}(k+N_u-1|k) \end{bmatrix}}_{= \Delta\boldsymbol{u}(\cdot|k)} \leq \underbrace{\begin{bmatrix} \Delta\boldsymbol{u}_{\max} \\ \vdots \\ \Delta\boldsymbol{u}_{\max} \\ -\Delta\boldsymbol{u}_{\min} \\ \vdots \\ -\Delta\boldsymbol{u}_{\min} \end{bmatrix}}_{=:\, \boldsymbol{b}_{in,\,\Delta u} \in \mathbb{R}^{2lN_u}} \tag{4.42}$$

Analogously, the bounds on $\boldsymbol{u}(\cdot|k)$ are attained by the following *less than or equal to* inequalities.

$$\begin{aligned} \boldsymbol{u}(k|k) &\leq \boldsymbol{u}_{\max} \\ &\vdots \\ \boldsymbol{u}(k+N_u-1|k) &\leq \boldsymbol{u}_{\max} \\ -\boldsymbol{u}(k|k) &\leq -\boldsymbol{u}_{\min} \\ &\vdots \\ -\boldsymbol{u}(k+N_u-1|k) &\leq -\boldsymbol{u}_{\min} \end{aligned} \tag{4.43}$$

Expressed in terms of the optimization variables $\Delta\boldsymbol{u}(\cdot|k)$, these bounds may be rewritten as the following inequalities:

$$\begin{aligned} \boldsymbol{u}(k-1) + \Delta\boldsymbol{u}(k|k) &\leq \boldsymbol{u}_{\max} \\ \boldsymbol{u}(k-1) + \Delta\boldsymbol{u}(k|k) + \Delta\boldsymbol{u}(k+1|k) &\leq \boldsymbol{u}_{\max} \\ &\vdots \\ \boldsymbol{u}(k-1) + \Delta\boldsymbol{u}(k|k) + \dots + \Delta\boldsymbol{u}(k+N_u-1|k) &\leq \boldsymbol{u}_{\max} \\ -\boldsymbol{u}(k-1) - \Delta\boldsymbol{u}(k|k) &\leq -\boldsymbol{u}_{\min} \\ -\boldsymbol{u}(k+1) - \Delta\boldsymbol{u}(k|k) - \Delta\boldsymbol{u}(k+1|k) &\leq -\boldsymbol{u}_{\min} \\ &\vdots \\ -\boldsymbol{u}(k-1) - \Delta\boldsymbol{u}(k|k) - \dots - \Delta\boldsymbol{u}(k+N_u-1|k) &\leq -\boldsymbol{u}_{\min} \end{aligned} \tag{4.44}$$

The corresponding matrix-vector notation reads as follows:

$$\underbrace{\begin{bmatrix} \boldsymbol{I} & \boldsymbol{0} & \dots & \boldsymbol{0} \\ \vdots & \vdots & \ddots & \vdots \\ \boldsymbol{I} & \boldsymbol{I} & \dots & \boldsymbol{I} \\ -\boldsymbol{I} & \boldsymbol{0} & \dots & \boldsymbol{0} \\ \vdots & \vdots & \ddots & \vdots \\ -\boldsymbol{I} & -\boldsymbol{I} & \dots & -\boldsymbol{I} \end{bmatrix}}_{=:\boldsymbol{A}_{in,u} \in \mathbb{R}^{2lN_u \times lN_u}} \Delta \boldsymbol{u}(\cdot|k) \leq \underbrace{\begin{bmatrix} \boldsymbol{u}_{\max} \\ \vdots \\ \boldsymbol{u}_{\max} \\ -\boldsymbol{u}_{\min} \\ \vdots \\ -\boldsymbol{u}_{\min} \end{bmatrix} - \begin{bmatrix} \boldsymbol{u}(k-1) \\ \vdots \\ \boldsymbol{u}(k-1) \\ -\boldsymbol{u}(k-1) \\ \vdots \\ -\boldsymbol{u}(k-1) \end{bmatrix}}_{=:\boldsymbol{b}_{in,u} \in \mathbb{R}^{2lN_u}} \tag{4.45}$$

The bounds on the system states $\boldsymbol{x}(\cdot|k)$ are defined by the following equations:

$$\begin{aligned} \boldsymbol{x}(k+N_1|k) &\leq \boldsymbol{x}_{\max} \\ &\vdots \\ \boldsymbol{x}(k+N_2|k) &\leq \boldsymbol{x}_{\max} \\ -\boldsymbol{x}(k+N_1|k) &\leq -\boldsymbol{x}_{\min} \\ &\vdots \\ -\boldsymbol{x}(k+N_2|k) &\leq -\boldsymbol{x}_{\min} \end{aligned} \tag{4.46}$$

As shown in Sect. 4.3.1, the system states predicted can be expressed by $\boldsymbol{\Theta}$, $\boldsymbol{\Pi}$, $\boldsymbol{\Upsilon}$ in accordance to the definitions in (4.21).

$$\boldsymbol{x}(\cdot|k) = \boldsymbol{\Theta}\boldsymbol{x}(k) + \boldsymbol{\Pi}\boldsymbol{u}(k-1) + \boldsymbol{\Upsilon}\Delta\boldsymbol{u}(\cdot|k) \tag{4.47}$$

As a consequence, the inequalities can be expressed by the following matrix-vector notation:

$$\begin{bmatrix} \boldsymbol{x}(k+N_1|k) \\ \vdots \\ \boldsymbol{x}(k+N_2|k) \\ -\boldsymbol{x}(k+N_1|k) \\ \vdots \\ -\boldsymbol{x}(k+N_2|k) \end{bmatrix} = \begin{bmatrix} \boldsymbol{\Theta} \\ -\boldsymbol{\Theta} \end{bmatrix} \boldsymbol{x}(k) + \begin{bmatrix} \boldsymbol{\Pi} \\ -\boldsymbol{\Pi} \end{bmatrix} \boldsymbol{u}(k-1) + \begin{bmatrix} \boldsymbol{\Upsilon} \\ -\boldsymbol{\Upsilon} \end{bmatrix} \begin{bmatrix} \Delta\boldsymbol{u}(k|k) \\ \vdots \\ \Delta\boldsymbol{u}(k+N_u-1|k) \end{bmatrix} \leq \begin{bmatrix} \boldsymbol{x}_{max} \\ \vdots \\ \boldsymbol{x}_{max} \\ -\boldsymbol{x}_{min} \\ \vdots \\ -\boldsymbol{x}_{min} \end{bmatrix} \tag{4.48}$$

The transformation into standard form yields

$$\underbrace{\begin{bmatrix} \boldsymbol{\Upsilon} \\ -\boldsymbol{\Upsilon} \end{bmatrix}}_{\substack{=:\boldsymbol{A}_{in,x} \\ \in \mathbb{R}^{2n(N_2-N_1+1)\times lN_u}}} \Delta\boldsymbol{u}(\cdot|k) \leq \underbrace{\begin{bmatrix} \boldsymbol{x}_{\max} \\ \vdots \\ \boldsymbol{x}_{\max} \\ -\boldsymbol{x}_{\min} \\ \vdots \\ -\boldsymbol{x}_{\min} \end{bmatrix} - \begin{bmatrix} \boldsymbol{\Theta} \\ -\boldsymbol{\Theta} \end{bmatrix} \boldsymbol{x}(k) - \begin{bmatrix} \boldsymbol{\Pi} \\ -\boldsymbol{\Pi} \end{bmatrix} \boldsymbol{u}(k-1)}_{=:\boldsymbol{b}_{in,x} \in \mathbb{R}^{2n(N_2-N_1+1)}} \tag{4.49}$$

In a similar fashion, the constraints on the outputs $\boldsymbol{y}(\cdot|k)$ can be defined as

$$\underbrace{\begin{bmatrix} \boldsymbol{\Gamma\Upsilon} \\ -\boldsymbol{\Gamma\Upsilon} \end{bmatrix}}_{=:\,\boldsymbol{A}_{in,y}\,\in\,\mathbb{R}^{2m(N_2-N_1+1)\times lN_u}} \Delta\boldsymbol{u}(\cdot|k) \leq \underbrace{\begin{bmatrix} \boldsymbol{y}_{\max} \\ \vdots \\ \boldsymbol{y}_{\max} \\ -\boldsymbol{y}_{\min} \\ \vdots \\ -\boldsymbol{y}_{\min} \end{bmatrix} - \begin{bmatrix} \boldsymbol{\Gamma\Theta} \\ -\boldsymbol{\Gamma\Theta} \end{bmatrix} \boldsymbol{x}(k) - \begin{bmatrix} \boldsymbol{\Gamma\Pi} \\ -\boldsymbol{\Gamma\Pi} \end{bmatrix} \boldsymbol{u}(k-1)}_{=:\,\boldsymbol{b}_{in,y}\,\in\,\mathbb{R}^{2m(N_2-N_1+1)}} \tag{4.50}$$

By these transformations, the constrained LMPC can be brought into the standard QP formulation. The transformations were presented here for inequalities which are given as box constraints. However, the general framework allows more flexibility. For instance, the linear combination of various variables or equality constraints can be considered as well. The transformations have to be conducted in a similar way. A more general formulation of a constrained LMPC with a dense formulation is given by the following optimization problem, where the system dynamics are incorporated into the cost function via the prediction matrices:

$$\begin{aligned} \min_{\Delta\boldsymbol{u}(\cdot|k)} \quad & \frac{1}{2}\Delta\boldsymbol{u}(\cdot|k)^{\mathsf{T}}\boldsymbol{H}\Delta\boldsymbol{u}(\cdot|k) + \Delta\boldsymbol{u}(\cdot|k)^{\mathsf{T}}\boldsymbol{q} \\ \text{s.t.} \quad & \boldsymbol{A}_{eq}\Delta\boldsymbol{u}(\cdot|k) - \boldsymbol{b}_{eq} = 0, \quad \boldsymbol{A}_{eq} \in \mathbb{R}^{p\times lN_u},\ \boldsymbol{b}_{eq} \in \mathbb{R}^{p}, \\ & \boldsymbol{A}_{in}\Delta\boldsymbol{u}(\cdot|k) - \boldsymbol{b}_{in} \leq 0, \quad \boldsymbol{A}_{in} \in \mathbb{R}^{q\times lN_u},\ \boldsymbol{b}_{in} \in \mathbb{R}^{q} \end{aligned} \tag{4.51}$$

$$\text{with} \quad \Delta\boldsymbol{u}(\cdot|k) = \begin{bmatrix} \Delta\boldsymbol{u}(k|k) \\ \vdots \\ \Delta\boldsymbol{u}(k+N_u-1|k) \end{bmatrix} \in \mathbb{R}^{lN_u}.$$

4.4.2 Sparse Formulation

An alternative approach for setting up the QP problem for the constrained LMPC is given by the sparse problem formulation. In this approach, the state equations of the linear model are directly incorporated as equality constraints in the optimization problem. As optimization variables, the system states $\boldsymbol{x}(\cdot|k)$ and the actuated values $\boldsymbol{u}(\cdot|k)$ are used additionally to the changes of the actuated values $\Delta\boldsymbol{u}(\cdot|k)$. Compared to the dense formulation, the sparse problem formulation changes the size of the optimization problem as well as the sparsity pattern of the resulting matrices of the QP problem. The sparsity pattern shows where non-zero entries are structurally possible in the matrix. The resulting sparsity pattern allows to apply computationally efficient methods. As a result, the sparse formulation is computationally faster than the dense formulation in specific cases, as detailed below in Sect. 4.4.3. The optimization problem can be formulated as follows:

$$
\begin{aligned}
&\min_{\boldsymbol{x}(\cdot|k),\,\boldsymbol{u}(\cdot|k),\,\Delta\boldsymbol{u}(\cdot|k)} \quad ||\underbrace{\boldsymbol{\Gamma x}(\cdot|k)}_{=\boldsymbol{y}(\cdot|k)} - \boldsymbol{r}(\cdot|k)||^2_{\boldsymbol{Q}} + ||\Delta\boldsymbol{u}(\cdot|k)||^2_{\boldsymbol{R}} \\
&\text{s.t.} \\
&\left.\begin{aligned}
\boldsymbol{x}(k+i+1|k) &= \boldsymbol{A x}(k+i|k) + \boldsymbol{B u}(k+i|k) \quad i = 0,\dots,N_2-1 \\
\boldsymbol{x}(k|k) &= \boldsymbol{x}(k) \\
\boldsymbol{u}(k+i|k) &= \boldsymbol{u}(k-1+i|k) + \Delta\boldsymbol{u}(k+i|k) \quad i = 1,\dots,N_u-1 \\
\boldsymbol{u}(k+i|k) &= \boldsymbol{u}(k+N_u-1) \qquad i = N_u,\dots,N_2-1 \\
\boldsymbol{u}(k|k) &= \boldsymbol{u}(k-1) + \Delta\boldsymbol{u}(k|k)
\end{aligned}\right\} \text{system dynamics,} \\
&\left.\begin{aligned}
\Delta\boldsymbol{u}_{min} \le\ &\Delta\boldsymbol{u}(\cdot|k) \le \Delta\boldsymbol{u}_{max} \\
\boldsymbol{u}_{min} \le\ &\boldsymbol{u}(\cdot|k) \le \boldsymbol{u}_{max} \\
\boldsymbol{x}_{min} \le\ &\boldsymbol{x}(\cdot|k) \le \boldsymbol{x}_{max} \\
\boldsymbol{y}_{min} \le\ &\boldsymbol{\Gamma x}(\cdot|k) \le \boldsymbol{y}_{max}
\end{aligned}\right\} \text{constraints}
\end{aligned}
\tag{4.52}
$$

For simplicity, the resulting matrices will be derived for the confined case of $N_u = N_2 = N$ and $N_1 = 1$. Additionally, inequalities on $\boldsymbol{u}(\cdot|k)$, $\Delta\boldsymbol{u}(\cdot|k)$, $\boldsymbol{x}(\cdot|k)$, and $\boldsymbol{y}(\cdot|k)$ are considered, which are given as box constraints. Other cases can be handled analogously. All optimization variables are merged in the vector $\boldsymbol{\xi}(\cdot|k)$.

$$
\boldsymbol{\xi}(\cdot|k) := \begin{bmatrix} \Delta\boldsymbol{u}(k|k) \\ \boldsymbol{u}(k|k) \\ \boldsymbol{x}(k+1|k) \\ \vdots \\ \Delta\boldsymbol{u}(k+N-1|k) \\ \boldsymbol{u}(k+N-1|k) \\ \boldsymbol{x}(k+N|k) \end{bmatrix} \in \mathbb{R}^{(2l+n)N} \tag{4.53}
$$

The linear system model has to be considered in the optimization problem as a linear equality constraint in dependence of $\boldsymbol{\xi}(\cdot|k)$ by the form $\boldsymbol{A}_{eq,s}\boldsymbol{\xi}(\cdot|k) = \boldsymbol{b}_{eq,s}$. This results in a *banded sparse structure*.

$$
\underbrace{\begin{bmatrix}
\boldsymbol{I} & -\boldsymbol{I} & \boldsymbol{0} & & & & & & & & & & \boldsymbol{0} \\
\boldsymbol{0} & \boldsymbol{B} & -\boldsymbol{I} & \boldsymbol{0} & & & & & & & & & \\
\boldsymbol{0} & \boldsymbol{I} & \boldsymbol{0} & \boldsymbol{I} & -\boldsymbol{I} & \boldsymbol{0} & & & & & & & \\
& \boldsymbol{0} & \boldsymbol{A} & \boldsymbol{0} & \boldsymbol{B} & -\boldsymbol{I} & \boldsymbol{0} & & & & & & \\
& & & \boldsymbol{0} & \boldsymbol{I} & \boldsymbol{0} & \boldsymbol{I} & -\boldsymbol{I} & \boldsymbol{0} & & & & \\
& & & & \boldsymbol{0} & \boldsymbol{A} & \boldsymbol{0} & \boldsymbol{B} & -\boldsymbol{I} & \boldsymbol{0} & & & \\
& & & & & \ddots & \ddots & \ddots & \ddots & \ddots & \ddots & & \\
& & & & & & \boldsymbol{0} & \boldsymbol{I} & \boldsymbol{0} & \boldsymbol{I} & -\boldsymbol{I} & \boldsymbol{0} & \\
\boldsymbol{0} & & & & & & & \boldsymbol{0} & \boldsymbol{A} & \boldsymbol{0} & \boldsymbol{B} & -\boldsymbol{I} &
\end{bmatrix}}_{=:\, \boldsymbol{A}_{eq,s} \in \mathbb{R}^{(l+n)N \times (2l+n)N}}
\begin{bmatrix}
\Delta \boldsymbol{u}(k|k) \\ \boldsymbol{u}(k|k) \\ \boldsymbol{x}(k+1|k) \\ \Delta \boldsymbol{u}(k+1|k) \\ \boldsymbol{u}(k+1|k) \\ \boldsymbol{x}(k+2|k) \\ \vdots \\ \Delta \boldsymbol{u}(k+N-1|k) \\ \boldsymbol{u}(k+N-1|k) \\ \boldsymbol{x}(k+N|k)
\end{bmatrix}
$$

$$
= \underbrace{\begin{bmatrix} \boldsymbol{0} \\ -\boldsymbol{A} \\ \boldsymbol{0} \\ \vdots \\ \boldsymbol{0} \end{bmatrix} \boldsymbol{x}(k) + \begin{bmatrix} -\boldsymbol{I} \\ \boldsymbol{0} \\ \boldsymbol{0} \\ \vdots \\ \boldsymbol{0} \end{bmatrix} \boldsymbol{u}(k-1)}_{=:\, \boldsymbol{b}_{eq,s} \in \mathbb{R}^{(l+n)N}} \tag{4.54}
$$

The cost function and the box constraints on the system states, outputs, and actuated values have to be formulated in dependence of $\boldsymbol{\xi}(\cdot|k)$ as well. This can be done by appropriate linear mappings, as denoted for the example of $\boldsymbol{x}(\cdot|k)$.

$$
\boldsymbol{x}(\cdot|k) = \begin{bmatrix} \boldsymbol{x}(k+1|k) \\ \vdots \\ \boldsymbol{x}(k+N|k) \end{bmatrix} = \underbrace{\begin{bmatrix}
\boldsymbol{0}\ \boldsymbol{0}\ \boldsymbol{I}\ \boldsymbol{0} & \cdots & & \boldsymbol{0} \\
\boldsymbol{0} \cdots \boldsymbol{0}\ \boldsymbol{0} & \boldsymbol{I} & \boldsymbol{0} \cdots & \boldsymbol{0} \\
\vdots & \ddots & \ddots\ \ddots & \vdots \\
\boldsymbol{0} & \cdots & \boldsymbol{0}\ \ \boldsymbol{0} & \boldsymbol{I}
\end{bmatrix}}_{=:\, \boldsymbol{I}_x} \boldsymbol{\xi}(\cdot|k) \tag{4.55}
$$

With linear mappings such as $\boldsymbol{I}_x$, the cost function can be expressed in terms of $\boldsymbol{\xi}(\cdot|k)$. The Hessian matrix $\boldsymbol{H}_s$ again has a sparse structure. Additionally, $\boldsymbol{q}_s$ does not depend any more on the initial states, instead it only depends on the reference trajectory $\boldsymbol{r}(\cdot|k)$.

$$
\begin{aligned}
J &= (\boldsymbol{\Gamma x}(\cdot|k) - \boldsymbol{r}(\cdot|k))^{\mathsf{T}} \boldsymbol{Q}\, (\boldsymbol{\Gamma x}(\cdot|k) - \boldsymbol{r}(\cdot|k)) + \Delta \boldsymbol{u}(\cdot|k)^{\mathsf{T}} \boldsymbol{R} \Delta \boldsymbol{u}(\cdot|k) \\
&= (\boldsymbol{\Gamma I}_x \boldsymbol{\xi}(\cdot|k) - \boldsymbol{r}(\cdot|k))^{\mathsf{T}} \boldsymbol{Q}\, (\boldsymbol{\Gamma I}_x \boldsymbol{\xi}(\cdot|k) - \boldsymbol{r}(\cdot|k)) + (\boldsymbol{I}_{\Delta u} \boldsymbol{\xi}(\cdot|k))^{\mathsf{T}} \boldsymbol{R}\, (\boldsymbol{I}_{\Delta u} \boldsymbol{\xi}(\cdot|k)) \\
&= \boldsymbol{\xi}(\cdot|k)^{\mathsf{T}} \underbrace{(\boldsymbol{I}_x^{\mathsf{T}} \boldsymbol{\Gamma}^{\mathsf{T}} \boldsymbol{Q \Gamma I}_x + \boldsymbol{I}_{\Delta u}^{\mathsf{T}} \boldsymbol{R I}_{\Delta u})}_{=:\, \boldsymbol{H}_s \in \mathbb{R}^{(2l+n)N \times (2l+n)N}} \boldsymbol{\xi}(\cdot|k) - 2\boldsymbol{\xi}(\cdot|k)^{\mathsf{T}} \underbrace{(\boldsymbol{I}_x^{\mathsf{T}} \boldsymbol{\Gamma}^{\mathsf{T}} \boldsymbol{Q r}(\cdot|k))}_{=:\, -\boldsymbol{q}_s \in \mathbb{R}^{(2l+n)N}} + \text{const.}
\end{aligned} \tag{4.56}
$$

The inequality constraints are similarly expressed by these linear mappings. All matrices generated for the handling of the constraints have sparse structures and reordering the rows results in sparse banded structures. The constraints on the state equations, for example, can be formulated by the following matrix-vector notation.

$$\underbrace{\begin{bmatrix} \mathbf{0}\;\mathbf{0}\;\boldsymbol{I}\;\mathbf{0} & & \cdots & & \mathbf{0} \\ \mathbf{0} & \cdots & \mathbf{0}\;\mathbf{0}\;\boldsymbol{I} & \mathbf{0} \cdots & \mathbf{0} \\ \vdots & & \ddots\;\ddots & \ddots & \vdots \\ \mathbf{0} & \cdots & & \mathbf{0}\;\;\mathbf{0} & \boldsymbol{I} \\ \mathbf{0}\;\mathbf{0}\;-\boldsymbol{I}\;\mathbf{0} & & \cdots & & \mathbf{0} \\ \mathbf{0} & \cdots & \mathbf{0}\;\mathbf{0}\;-\boldsymbol{I} & \mathbf{0} \cdots & \mathbf{0} \\ \vdots & & \ddots\;\ddots & \ddots & \vdots \\ \mathbf{0} & \cdots & & \mathbf{0}\;\;\mathbf{0} & -\boldsymbol{I} \end{bmatrix}}_{=:\,\boldsymbol{A}_{in,s,x}\,\in\,\mathbb{R}^{2nN\times(2l+n)N}} \boldsymbol{\xi}(\cdot|k) \leq \underbrace{\begin{bmatrix} \boldsymbol{x}_{\max} \\ \vdots \\ \boldsymbol{x}_{\max} \\ -\boldsymbol{x}_{\min} \\ \vdots \\ -\boldsymbol{x}_{\min} \end{bmatrix}}_{\substack{=:\,\boldsymbol{b}_{in,s,x} \\ \in\,\mathbb{R}^{2nN}}} \tag{4.57}$$

The transformations mentioned above allow the constrained LMPC to be brought into the standard QP formulation. To sum up, the constrained LMPC results overall in the following QP. Other cases, such as $N_u < N_2$ or the constraint of a linear combination of various variables, can be handled analogously.

$$\begin{aligned} \min_{\boldsymbol{\xi}(\cdot|k)} \quad & \frac{1}{2}\boldsymbol{\xi}(\cdot|k)^{\mathsf{T}}\boldsymbol{H}_s\boldsymbol{\xi}(\cdot|k) + \boldsymbol{\xi}(\cdot|k)^{\mathsf{T}}\boldsymbol{q}_s + \text{const.} \\ \text{s.t.} \quad & \boldsymbol{A}_{eq,s}\boldsymbol{\xi}(\cdot|k) = \boldsymbol{b}_{eq,s}, \\ & \underbrace{\begin{bmatrix} \boldsymbol{A}_{in,s,\Delta u} \\ \boldsymbol{A}_{in,s,u} \\ \boldsymbol{A}_{in,s,x} \\ \boldsymbol{A}_{in,s,y} \end{bmatrix}}_{=:\,\boldsymbol{A}_{in,s}} \boldsymbol{\xi}(\cdot|k) \leq \underbrace{\begin{bmatrix} \boldsymbol{b}_{in,s,\Delta u} \\ \boldsymbol{b}_{in,s,u} \\ \boldsymbol{b}_{in,s,x} \\ \boldsymbol{b}_{in,s,y} \end{bmatrix}}_{=:\,\boldsymbol{b}_{in}} \end{aligned} \tag{4.58}$$

$$\text{with} \quad \boldsymbol{\xi}(\cdot|k) = \begin{bmatrix} \Delta\boldsymbol{u}(k|k) \\ \boldsymbol{u}(k|k) \\ \boldsymbol{x}(k+1|k) \\ \vdots \\ \Delta\boldsymbol{u}(k+N-1|k) \\ \boldsymbol{u}(k+N-1|k) \\ \boldsymbol{x}(k+N|k) \end{bmatrix} \tag{4.59}$$

The matrices have the following dimensions:

$$\begin{aligned} \boldsymbol{A}_{eq,s} &\in \mathbb{R}^{(l+n)N\times(2l+n)N} \\ \boldsymbol{b}_{eq,s} &\in \mathbb{R}^{(l+n)N} \\ \boldsymbol{A}_{in,s} &\in \mathbb{R}^{(m+n+2l)2N\times(2l+n)N} \\ \boldsymbol{b}_{in,s} &\in \mathbb{R}^{(m+n+2l)2N} \\ \boldsymbol{\xi}(\cdot|k) &\in \mathbb{R}^{(2l+n)N}. \end{aligned} \tag{4.60}$$

4.4.3 Comparison of Dense and Sparse Formulations

The resulting structures of the dense and the sparse formulations are summarized in Fig. 4.3. As the dense formulation has fewer optimization variables, fewer constraints, and as the resulting matrices are much smaller, one would expect the computation time to be lower than that of the sparse formulation. However, due to its dense structure, the decomposition of the matrices is numerically expensive. As a consequence, solving the QP in the sparse formulation might still be faster. The sparse banded structure of the Hessian, the equality, and the inequality constraint matrices result in an overall sparse structure of the resulting QP. This can be exploited by using tailored numerics of the QP solver, e.g. for an efficient lower–upper (LU) decomposition [15]. Additionally, the sparse formulation results in a better conditioning for unstable systems, as the matrix potentiation is avoided. In the case of a linear time-variant MPC, see Sect. 4.5, the constraints and system matrices change at each time step. With the sparse structure, less time is needed for constructing the necessary matrices.

Sparse Formulation

$$\min_{\boldsymbol{\xi}=[\Delta\boldsymbol{u},\boldsymbol{u},\boldsymbol{x}]^T} \frac{1}{2}\boldsymbol{\xi}^T\boldsymbol{H}_s\boldsymbol{\xi} + \boldsymbol{\xi}^T\boldsymbol{q}_s$$

s.t.

$$\boldsymbol{A}_{eq,s}\boldsymbol{\xi} = \boldsymbol{b}_{eq,s}$$
$$\boldsymbol{A}_{in,s}\boldsymbol{\xi} \leq \boldsymbol{b}_{in,s}$$

Sparse structured matrices
$\boldsymbol{H}_s, \boldsymbol{A}_{eq,s}, \boldsymbol{A}_{in,s}$

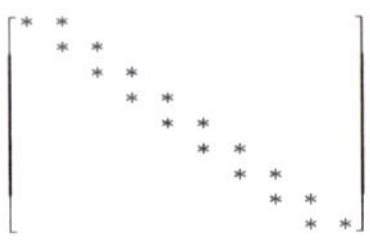

Dense Formulation

$$\min_{\Delta\boldsymbol{u}} \frac{1}{2}\Delta\boldsymbol{u}^T\boldsymbol{H}_d\Delta\boldsymbol{u} + \Delta\boldsymbol{u}^T\boldsymbol{q}_d$$

s.t.

$$\boldsymbol{A}_{in,d}\Delta\boldsymbol{u} \leq \boldsymbol{b}_{in,d}$$

Dense structured matrices
$\boldsymbol{H}_d, \boldsymbol{A}_{in,d}$

Fig. 4.3 Comparison of the sparse and dense formulations for LMPC

In general, the dense formulation is especially advantageous for smaller problem sizes and for cases in which $N_u << N_2$. For the sparse formulation, drawbacks arise for the case where $N_u << N_2$ as many additional optimization variables have to be added that are not adjustable. The sparse formulation is better suited for larger problems and in the case of $N_u \approx N_2$.

4.4.4 Control Structure of Constrained Linear MPC

From the discussions thus far, it is possible to conclude that a linear control law results if only equality constraints are considered within the constrained LMPC. The equality constraints can be substituted in the cost function, such that a new unconstrained optimization problem with reduced degrees of freedom results. A linear controller also results if there is one fixed active set that is not changing, containing equality constraints and some active inequalities. As soon as inequality constraints are present that are both active and inactive such that the active set is changing, a nonlinear control behavior results. Thus, although the control concept is called "Linear MPC", it is actually a nonlinear controller. In the following, some remarks are added concerning the structure of the resulting nonlinear controller.

The constrained LMPC is represented by the following QP. For the sake of simplicity, no equality constraints are considered and the dense formulation is investigated. Additionally, a positive-definite Hessian matrix is assumed.

$$\begin{aligned} \min_{\boldsymbol{\Delta u}} \quad & \frac{1}{2}\boldsymbol{\Delta u}^{\mathsf{T}}\boldsymbol{H}\boldsymbol{\Delta u}+\boldsymbol{\Delta u}^{\mathsf{T}}\boldsymbol{q} \qquad \text{with } \boldsymbol{H} \succ \boldsymbol{0} \\ \text{s.t.} \quad & \boldsymbol{A}_{in,d}\boldsymbol{\Delta u} - \boldsymbol{b}_{in,d} \leq \boldsymbol{0} \end{aligned} \tag{4.61}$$

To detail the structure of the resulting controller, it is assumed for now that a certain set of inequalities are active. The active set is given by $\mathcal{A}$. As soon as the inequalities are known, the optimization problem turns into one with only equality constraints. The inactive inequalities can be neglected for the calculation of the minimizer. The active inequality constraints are represented by $\boldsymbol{A}_{\mathcal{A}}\boldsymbol{\Delta u} - \boldsymbol{b}_{\mathcal{A}} = \boldsymbol{0}$. The new optimization problem can be presented as follows:

$$\begin{aligned} \min_{\boldsymbol{\Delta u}} \quad & \frac{1}{2}\boldsymbol{\Delta u}^{\mathsf{T}}\boldsymbol{H}\boldsymbol{\Delta u}+\boldsymbol{\Delta u}^{\mathsf{T}}\boldsymbol{q} \qquad \text{with } \boldsymbol{H} \succ \boldsymbol{0} \\ \text{s.t.} \quad & \boldsymbol{A}_{\mathcal{A}}\boldsymbol{\Delta u} - \boldsymbol{b}_{\mathcal{A}} = \boldsymbol{0} \end{aligned} \tag{4.62}$$

If the Hessian is positive definite, i.e. $\boldsymbol{H} \succ \boldsymbol{0}$ and linear independence constraint qualification (LICQ) holds, there exists a unique minimizer $\boldsymbol{\Delta u}^*$ that can be directly calculated by the optimality conditions. The corresponding FONC are given by a system of linear equations with $\nabla L = 0$.

$$\nabla L(\Delta \boldsymbol{u}^*, \boldsymbol{\lambda}^*) = \begin{bmatrix} \boldsymbol{H} & \boldsymbol{A}_{\mathcal{A}}{}^{\mathsf{T}} \\ \boldsymbol{A}_{\mathcal{A}} & \boldsymbol{0} \end{bmatrix} \begin{bmatrix} \Delta \boldsymbol{u}^* \\ \boldsymbol{\lambda}^* \end{bmatrix} - \begin{bmatrix} -\boldsymbol{q} \\ \boldsymbol{b}_{\mathcal{A}} \end{bmatrix} = 0 \tag{4.63}$$

The minimizer $\Delta \boldsymbol{u}^*$ can be calculated as follows:

$$\begin{bmatrix} \Delta \boldsymbol{u}^* \\ \boldsymbol{\lambda}^* \end{bmatrix} = \begin{bmatrix} \boldsymbol{H} & \boldsymbol{A}_{\mathcal{A}}{}^{\mathsf{T}} \\ \boldsymbol{A}_{\mathcal{A}} & \boldsymbol{0} \end{bmatrix}^{-1} \begin{bmatrix} -\boldsymbol{q} \\ \boldsymbol{b}_{\mathcal{A}} \end{bmatrix} \tag{4.64}$$

Based on the definitions of the various matrices, the matrix $\boldsymbol{H}$ turns out to be constant, i.e. $\boldsymbol{H} = const.$, as it depends solely on the weighting matrices and on the system matrices. For a fixed active set, the matrix $\boldsymbol{A}_{\mathcal{A}} = const.$ is constant. However, with a different active set, the matrix changes. For a fixed active set, the following holds:

$$\begin{bmatrix} \boldsymbol{H} & \boldsymbol{A}_{\mathcal{A}}{}^{\mathsf{T}} \\ \boldsymbol{A}_{\mathcal{A}} & \boldsymbol{0} \end{bmatrix}^{-1} = const. \tag{4.65}$$

The vector $\boldsymbol{q}$ is a linear function of $\boldsymbol{x}(k)$, $\boldsymbol{u}(k-1)$, and the reference values $\boldsymbol{r}(\cdot|k)$. If state or output constraints are considered, the vector $\boldsymbol{b}_{\mathcal{A}}$ is an affine function of $\boldsymbol{x}(k)$ and $\boldsymbol{u}(k-1)$. The structure of the resulting optimal actuated values $\boldsymbol{u}^*(\cdot|k)$ thus becomes

$$\boldsymbol{u}^*(\cdot|k) = \boldsymbol{K}_x \boldsymbol{x}(k) + \boldsymbol{K}_u \boldsymbol{u}(k-1) + \boldsymbol{K}_r \boldsymbol{r}(k) + \boldsymbol{K}_c \tag{4.66}$$

The matrices $\boldsymbol{K}_x$, $\boldsymbol{K}_u$, $\boldsymbol{K}_r$, and $\boldsymbol{K}_c$ are all constant. Hence, for a fixed active set, an *affine control law* in the feedback variable $\boldsymbol{x}(k)$ results. When the active set changes, the parameters of $\boldsymbol{K}_x$, $\boldsymbol{K}_u$, $\boldsymbol{K}_r$, and $\boldsymbol{K}_c$ change. Thus, for each possible set of active constraints, a corresponding affine control law can be computed. Overall, a piecewise affine (PWA) control law results. The authors of [3] demonstrate that the resulting PWA control law is continuous in the output $\boldsymbol{u}^*(\cdot|k)$. All in all, the control structure for a constrained LMPC results to be nonlinear and more specifically, a PWA control law results.

The fact that a PWA control law results is exploited for the use within explicit MPC [1]. For explicit MPC, this control law is precomputed offline using multi-parametric programming. Subsequently, by evaluation of the function, the PWA control law can be used online to calculate the optimal actuated values. However, it is only feasible for small systems and low amounts of possible active set changes. In the worst case, for an amount of q constraints, 2^q different active sets can result. Most engine MPC applications are too large for the use of explicit MPC. Instead, the solution of the optimization problem is found by the use of numerical solution methods. These allow the solution of the optimization problem to be found in an iterative manner during the runtime of the process. An example of such a method is described in the next section.

4.4.5 Numerical Solution of the Resulting Quadratic Program

The above sections have shown that the LMPC with constraints leads to a QP that is typically convex. A variety of approaches exist to solve a convex QP in a fast and reliable manner [14]. A number of powerful numerical tools can be used to solve QPs, even for demanding problems in real time. The possibility to efficiently solve QPs is one of the reasons why LMPC is in widespread use, even in industrial applications.

Prominent approaches are the active set method, which is implemented, for example, in qpOASES [8], the interior point (IP) method implemented, for example, in FORCESPRO [7], and the gradient method implemented, for example, in FiOrdOs [10]. Some of the software tools originate from academic research, such as qpOASES [8], and others are commercially available and professionally maintained, such as FORCESPRO [7]. The algorithms differ in terms of properties such as their convergence rate, numerical robustness, and capability to include an initial guess, which is also called warm-starting. To show the fundamental idea behind one of these techniques, the basic idea behind the IP method is briefly introduced.

For constrained LMPC in every time step, related QP problems are solved, i.e. they have the same number of constraints and optimization variables. The instances differ, as for example the initial state or the reference is changing. For such kinds of related problems, IP methods show relatively constant computation times. Additionally, IP methods are well suited for exploiting the sparsity pattern. Thus, it is advantageous to use IP methods along with a sparse formulation for large problems. The dense formulation is often used in combination with active set methods for small problems.

The convex QP investigated has the following form. For simplicity, a problem with q inequality constraints is investigated. The additional handling of equality constraints does not pose a challenge.

$$\begin{aligned} \min_{z} \quad & \frac{1}{2} z^{\mathsf{T}} \boldsymbol{H} z + z^{\mathsf{T}} \boldsymbol{q}, \qquad \text{with } \boldsymbol{H} \succ \boldsymbol{0} \\ \text{s.t.} \quad & \boldsymbol{A}_{in} z - \boldsymbol{b}_{in} \leq \boldsymbol{0} \end{aligned} \tag{4.67}$$

The Lagrange function of the convex QP is defined by $L(z, \boldsymbol{\mu})$.

$$L(z, \boldsymbol{\mu}) = \frac{1}{2} z^{\mathsf{T}} \boldsymbol{H} z + z^{\mathsf{T}} \boldsymbol{q} + \boldsymbol{\mu}^{\mathsf{T}} (\boldsymbol{A}_{in} z - \boldsymbol{b}_{in}) \tag{4.68}$$

If z^* is a minimizer of the constrained optimization problem and LICQ holds, then the following conditions are satisfied:

$$\begin{aligned} \nabla_z L(z^*, \boldsymbol{\mu}^*) = \boldsymbol{H} z^* + \boldsymbol{q} + \boldsymbol{A}_{in}{}^{\mathsf{T}} \boldsymbol{\mu}^* &= 0 \\ \boldsymbol{A}_{in} \boldsymbol{z}^* - \boldsymbol{b}_{in} &\leq 0 \\ \boldsymbol{\mu}^* &\geq 0 \\ \boldsymbol{\mu}^{*\mathsf{T}} (\boldsymbol{A}_{in} \boldsymbol{z}^* - \boldsymbol{b}_{in}) &= 0 \end{aligned} \tag{4.69}$$

As discussed in Sect. 3.4.3, the inequality constraints make the optimization task demanding. They lead to the complementary slackness conditions which are non-smooth. The idea behind the IP methods is to reformulate the optimization problem such that the inequality constraints are avoided. The inequality constraints are replaced by an additional term in the cost function which penalizes any constraint violation. For the penalty term, various functions can be used; a common choice is the application of logarithmic functions. The addition of logarithmic barrier functions yields a new unconstrained NLP. For unconstrained NLPs, many solution techniques exist, such as the Newton's method, which is to be described in Sect. 5.2. The augmented cost function, with γ being a scalar positive value, is represented by the following equations:

$$\begin{aligned} J_{aug}(z,\gamma) &= J(z) + \frac{1}{\gamma} J_{log}(z) \\ J_{log}(z) &= -\sum_{i=1}^{q} \log\Big(-(\boldsymbol{A}_{in,i} z - \boldsymbol{b}_{in,i})\Big) \end{aligned} \tag{4.70}$$

As the Hessian is supposed to be positive-definite $\boldsymbol{H} \succ \boldsymbol{0}$, the original cost function $J(z)$ is convex. The sum of two convex functions is a convex function again [6]. Thus, the augmented cost function is a convex function also. Additionally, the augmented cost function is smooth in its domain. The function J_{log} yields costs of infinity as the argument goes to zero. This occurs when the original limits described by the inequality constraints are approached. Therefore, the penalty term forces the optimization variables to lie inside the feasible set. Indeed, every minimizer z_γ^* of the augmented cost function $z_\gamma^* = \arg\min J_{aug}(z,\gamma)$ satisfies all constraints. As the original cost function is augmented by additional terms, the minimizer of the augmented cost function z_γ^* differs from the minimizer of the original cost function z^*. However, in the case of $\gamma \to \infty$, the minimizer of the augmented cost function converges to the solution of the original problem $z_\gamma^* \to z^*$. The point $z_0^* = \arg\min J_{log}(z)$ is called the analytic center of constraints. The central path is defined as the trajectory given by $z_\gamma^* = \arg\min J_{aug}(z,\gamma)$ from the analytic center z_0^* to the original minimizer z^* with positive values of γ.

The numerical solution of the original problem can be calculated by following the central path. First, an initial point inside the feasible set is needed. This point serves as an initial guess for the unconstrained optimization of the augmented cost function $J_{aug}(z,\gamma)$ with a small value of γ. In every iteration, a different instance of the augmented unconstrained optimization problem is then solved. The value of γ is continuously increased in every iteration. The initial guess of every iteration is the solution of the previous iterate. This procedure allows finding very efficiently a minimizer of the augmented cost function, as a good initial guess is present. The solutions of the iterations lie on the central path and the algorithm converges to the original minimizer z^*. Among the various versions of IP methods that exist, especially the primal-dual method is used for online solutions within MPC [17].

4.5 Linear Time-Variant (LTV) Model Predictive Control

In the following, LTV problems are investigated. A plant with LTV system dynamics can be denoted by the following discrete-time representation:

$$\begin{aligned} \boldsymbol{x}(k+1) &= \boldsymbol{A}_k \boldsymbol{x}(k) + \boldsymbol{B}_k \boldsymbol{u}(k) \\ \boldsymbol{y}(k) &= \boldsymbol{C}_k \boldsymbol{x}(k) \end{aligned} \tag{4.71}$$

The time-variant system matrices $\boldsymbol{A}_k, \boldsymbol{B}_k, \boldsymbol{C}_k$ are valid for the sampling instant k. For an analysis of MPC based on an LTV system, the prediction of the system is considered. The optimization variables $\underline{\Delta \boldsymbol{u}}$ are underlined.

$$\begin{aligned} \boldsymbol{x}(k+1|k) &= \boldsymbol{A}_k \boldsymbol{x}(k) + \boldsymbol{B}_k \boldsymbol{u}(k|k) \\ &= \boldsymbol{A}_k \boldsymbol{x}(k) + \boldsymbol{B}_k \boldsymbol{u}(k-1) + \boldsymbol{B}_k \underline{\Delta \boldsymbol{u}(k|k)} \\ \boldsymbol{x}(k+2|k) &= \boldsymbol{A}_{k+1} \boldsymbol{x}(k+1|k) + \boldsymbol{B}_{k+1} \boldsymbol{u}(k+1|k) \\ &= \boldsymbol{A}_{k+1} \Big(\boldsymbol{A}_k \boldsymbol{x}(k) + \boldsymbol{B}_k \boldsymbol{u}(k-1) + \boldsymbol{B}_k \underline{\Delta \boldsymbol{u}(k|k)} \Big) + \\ &\quad \boldsymbol{B}_{k+1} \Big(\boldsymbol{u}(k-1) + \underline{\Delta \boldsymbol{u}(k|k)} + \underline{\Delta \boldsymbol{u}(k+1|k)} \Big) \\ &= \boldsymbol{A}_{k+1} \boldsymbol{A}_k \boldsymbol{x}(k) + \Big(\boldsymbol{A}_{k+1} \boldsymbol{B}_k + \boldsymbol{B}_{k+1} \Big) \boldsymbol{u}(k-1) + \\ &\quad \Big(\boldsymbol{A}_{k+1} \boldsymbol{B}_k + \boldsymbol{B}_{k+1} \Big) \underline{\Delta \boldsymbol{u}(k|k)} + \boldsymbol{B}_{k+1} \underline{\Delta \boldsymbol{u}(k+1|k)} \\ &\;\;\vdots \end{aligned} \tag{4.72}$$

These equations can be summarized in matrix form as functions of the initial values $(\boldsymbol{x}(k), \boldsymbol{u}(k-1))$ and the optimization variables $(\underline{\Delta \boldsymbol{u}(\cdot|k)})$:

$$\begin{aligned} \boldsymbol{x}(\cdot|k) = \begin{bmatrix} \boldsymbol{x}(k+N_1|k) \\ \boldsymbol{x}(k+N_1+1|k) \\ \vdots \\ \boldsymbol{x}(k+N_2|k) \end{bmatrix} &= \underbrace{\begin{bmatrix} \prod_{p=1}^{N_1} \boldsymbol{A}_{k+N_1-p} \\ \prod_{p=1}^{N_1+1} \boldsymbol{A}_{k+N_1+1-p} \\ \vdots \\ \prod_{p=1}^{N_2} \boldsymbol{A}_{k+N_2-p} \end{bmatrix}}_{\boldsymbol{\Theta} \in \mathbb{R}^{n(N_2-N_1+1)\times n}} \boldsymbol{x}(k) + \underbrace{\begin{bmatrix} \boldsymbol{\Lambda}(N_1,1) \\ \boldsymbol{\Lambda}(N_1+1,1) \\ \vdots \\ \boldsymbol{\Lambda}(N_2,1) \end{bmatrix}}_{\boldsymbol{\Pi} \in \mathbb{R}^{n(N_2-N_1+1)\times l}} \boldsymbol{u}(k-1) \\ &+ \underbrace{\begin{bmatrix} \boldsymbol{\Lambda}(N_1,1) & \dots & \boldsymbol{\Lambda}(N_1,N_u) \\ \vdots & \ddots & \vdots \\ \boldsymbol{\Lambda}(N_2,1) & \dots & \boldsymbol{\Lambda}(N_2,N_u) \end{bmatrix}}_{\boldsymbol{\Upsilon} \in \mathbb{R}^{n(N_2-N_1+1)\times lN_u}} \underbrace{\begin{bmatrix} \Delta \boldsymbol{u}(k|k) \\ \Delta \boldsymbol{u}(k+1|k) \\ \dots \\ \Delta \boldsymbol{u}(k+N_u-1|k) \end{bmatrix}}_{\underline{\Delta \boldsymbol{u}(\cdot|k)}} \end{aligned} \tag{4.73}$$

with $\boldsymbol{\Lambda}(i, r)$ being defined recursively as

$$\mathbb{R}^{n\times l} \ni \boldsymbol{A}(i,r) = \begin{cases} \boldsymbol{A}_{k+i-1}\boldsymbol{A}(i-1,r) + \boldsymbol{B}_{k+i-1}, & i > r \\ \boldsymbol{B}_{k+i-1}, & i = r \\ \boldsymbol{0}, & i < r \end{cases} \tag{4.74}$$

The output equation is defined by

$$\boldsymbol{y}(\cdot|k) = \boldsymbol{\Gamma}\boldsymbol{x}(\cdot|k) \quad \text{with} \quad \boldsymbol{\Gamma} = \text{diag}(\boldsymbol{C}_{N_1},\ldots,\boldsymbol{C}_{N_2}) \in \mathbb{R}^{m(N_2-N_1+1)\times n(N_2-N_1+1)} \tag{4.75}$$

The findings from (4.73) to (4.75) show that the prediction still depends linearly on $\boldsymbol{x}(k)$, $\boldsymbol{u}(k-1)$ and $\Delta\boldsymbol{u}(\cdot|k)$. Therefore, the same solution procedure as for the linear time-invariant (LTI) case is applicable. The LTV MPC formulation thus results in a QP of the same form as that of the LTI case. For the dense formulation, the QP is defined by the following optimization problem:

$$\begin{aligned} \min_{\Delta\boldsymbol{u}(\cdot|k)} \quad & \frac{1}{2}\Delta\boldsymbol{u}(\cdot|k)^{\mathsf{T}}\boldsymbol{H}\Delta\boldsymbol{u}(\cdot|k) + \Delta\boldsymbol{u}(\cdot|k)^{\mathsf{T}}\boldsymbol{q} + \text{const.} \\ \text{s.t.} \quad & \boldsymbol{A}_{in,d}\Delta\boldsymbol{u}(\cdot|k) - \boldsymbol{b}_{in,d} \le 0, \quad \boldsymbol{A}_{in,d} \in \mathbb{R}^{q\times lN_u},\ \boldsymbol{b}_{in} \in \mathbb{R}^{q} \end{aligned} \tag{4.76}$$

However, there exists one difference between the LTI and the LTV cases. The matrices describing the system dynamics are changing in every time step for the LTV case. As a result, the matrices $\boldsymbol{H}$ and $\boldsymbol{q}$ have to be recalculated in every time step, while for the LTI case $\boldsymbol{H}$ and $\boldsymbol{q}$ are constant. Hence, even for the unconstrained LTV case, a nonlinear control law results, since the matrix $\boldsymbol{K}_{mpc}$ has to be recalculated in every time step.

Why is the LTV MPC interesting at all? First of all, the original system model might be described by an LTV system. Secondly, the LTV MPC can be used to approximately take into account the nonlinear system behavior. The linearization of a nonlinear system model $\dot{\boldsymbol{x}} = \boldsymbol{f}(\boldsymbol{x},\boldsymbol{u})$ can be calculated by the Taylor series expansion around the operating point (OP) given by $\boldsymbol{x}_{op}$, $\boldsymbol{u}_{op}$. The linearized continuous-time state-space model at the current operating point has the following form:

$$\dot{\boldsymbol{x}} = \boldsymbol{f}(\boldsymbol{x},\boldsymbol{u}) \approx \boldsymbol{f}(\boldsymbol{x}_{op},\boldsymbol{u}_{op}) + \underbrace{\left.\frac{\partial \boldsymbol{f}}{\partial \boldsymbol{x}}\right|_{\boldsymbol{x}_{op},\boldsymbol{u}_{op}}}_{=:\boldsymbol{A}}(\boldsymbol{x}-\boldsymbol{x}_{op}) + \underbrace{\left.\frac{\partial \boldsymbol{f}}{\partial \boldsymbol{u}}\right|_{\boldsymbol{x}_{op},\boldsymbol{u}_{op}}}_{=:\boldsymbol{B}}(\boldsymbol{u}-\boldsymbol{u}_{op}) \tag{4.77}$$

For a stationary operating point, the expression $\boldsymbol{f}(\boldsymbol{x}_{op},\boldsymbol{u}_{op}) = \boldsymbol{0}$ holds. Otherwise, the non-stationary term has to be considered in the state prediction. The overall approach is depicted in Fig. 4.4. The following steps are conducted:

1. Obtain a new measurement or estimation for $\boldsymbol{x}(k)$.
2. Build one linear model around the measured or estimated value $\boldsymbol{x}(k)$.
3. Use this model for an LTI prediction over the entire prediction horizon.
4. Repeat this procedure in every sampling step.

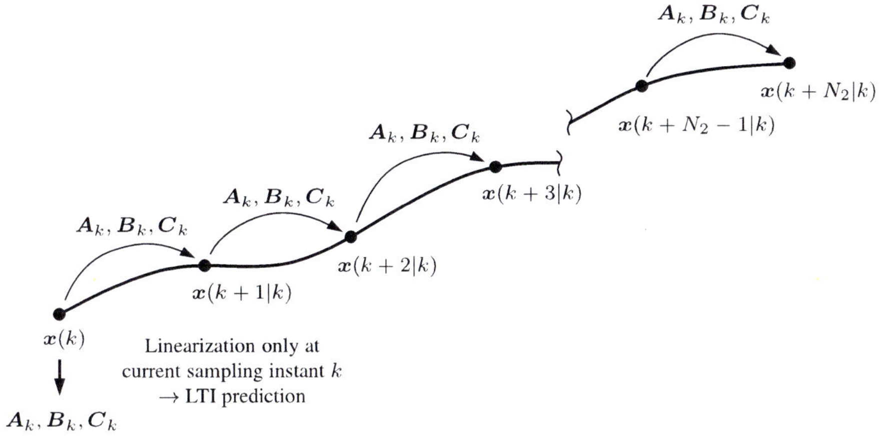

Fig. 4.4 LTV MPC by linearizations of current measured or estimated states

In each time step, one new linearized LTI state-space model results, which is used within MPC for a prediction.

$$x(k+i|k) = A_k x(k+i-1|k) + B_k u(k+i-1|k), \quad \forall i = 1, \ldots, N_2 \quad (4.78)$$

The LTV approach presented so far uses the linearized information around the current measured or estimated system state $x(k)$ for handling a nonlinear system. This results in one QP which has to be solved in each time step. Compared to the LTI MPC approach, the computational load is increased as the linearization has to be calculated as well.

An alternative approach to handle a nonlinear system is to incorporate additionally the information of the system around future states in the LTV model. For instance, the optimal actuated values from the last time step $u^*(\cdot|k-1)$ could be used along with the recent measured or estimated state $x(k)$ for a nonlinear simulation of the system states $x_{u^*}(\cdot|k)$. Subsequently, linear system models can be derived around the measured or estimated value $x(k)$ and the predicted states $x_{u^*}(\cdot|k)$. A new state-space model with time-varying matrices is governed in each time step for the prediction within MPC. The approach is depicted in Fig. 4.5. The computational load for one time step results from the nonlinear simulation for calculating the future states, the determination of the linearized models for the various states and solving one QP. However, there exist also other possibilities for incorporating the nonlinear information of the system while having comparable computational loads. One example is the so-called real-time iteration (RTI) scheme, where similarly one QP is solved in each time step. It is introduced in Chap. 5. The RTI scheme offers various algorithmic choices. For specific choices, the RTI scheme is the same as the LTV with time-varying matrices. In general, due to the algorithmic flexibility, the RTI scheme is usually superior to the LTV approach.

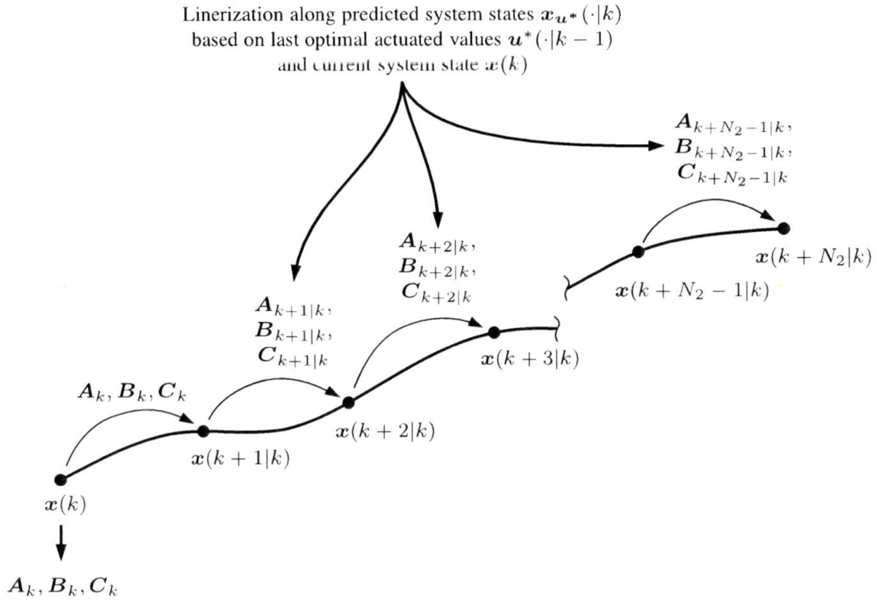

Fig. 4.5 LTV MPC by linearization along current measured or estimated states and the predicted states' trajectory

The LTV MPC approach can also be used to handle time-variant constraints. For example, non-convex feasible sets can be approximated by time-variant linear constraints that define time-varying convex sets. An application example is described in [4].

4.6 Numerical Examples for Linear MPC

In the following, some numerical examples are given to demonstrate the capabilities of LMPC.

Example 4.1 (*Unconstrained LMPC for unstable non-minimum phase system*) This example demonstrates that LMPC is able to handle complex system dynamics. A plant model is used that is unstable and is a non-minimum phase system. The continuous-time transfer function is given by $G(s)$:

$$G(s) = \frac{s - 1.3}{s^2 + 1.1s - 2} \tag{4.79}$$

The poles are located at $\pi_1 \approx 0.97$ and $\pi_2 \approx -2.1$, while the zero is set at $\zeta_1 = 1.3$. Due to the system characteristics present, the control with conventional linear

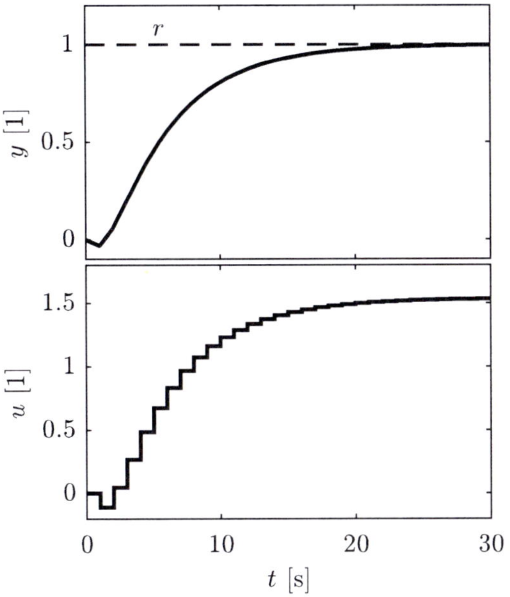

Fig. 4.6 Closed-loop control result for an unconstrained LMPC with sparse problem formulation for Example 4.1

controllers is quite complex. With the use of an LMPC approach, it turns out to be quite straightforward. A conventional cost function J is used.

$$J = \sum_{i=N_1}^{N_2} ||\boldsymbol{y}(k+i|k) - \boldsymbol{r}||_{\boldsymbol{Q}}^2 + \sum_{i=0}^{N_u-1} ||\Delta \boldsymbol{u}(k+i|k)||_{\boldsymbol{R}}^2 \tag{4.80}$$

The parameters used for the MPC are $N_1 = 1$, $N_2 = N_u = 45$, $\boldsymbol{R} = 10\boldsymbol{I}$, $\boldsymbol{Q} = \boldsymbol{I}$, and the sampling time is chosen to be $T_s = 1$ s. A constant reference signal of $r = 1$ is used. As the system is unstable, the sparse formulation is used for the implementation of the LMPC. Figure 4.6 shows that with this conventional cost function and without detailed tuning of the weighting matrices, already very good closed-loop control results can be achieved.

Example 4.2 (*Comparing the dense and the sparse formulations*) In this example, the differences between the dense formulation as introduced in Sect. 4.4.1 and the sparse formulation as introduced in Sect. 4.4.2 are demonstrated. For this purpose, the system described in Example 4.1 is investigated again. In Example 4.1, the closed-loop control was achieved by the use of the sparse formulation. It might seem counter-intuitive to use the sparse formulation instead of the dense formulation, as the optimization problem is blown up. This example demonstrates that there exist certain settings where the sparse formulation is advantageous due to the numerical differences arising. As mentioned in Sect. 4.4.2, the performance of the sparse formulation is superior to that of the dense formulation for large problems and settings

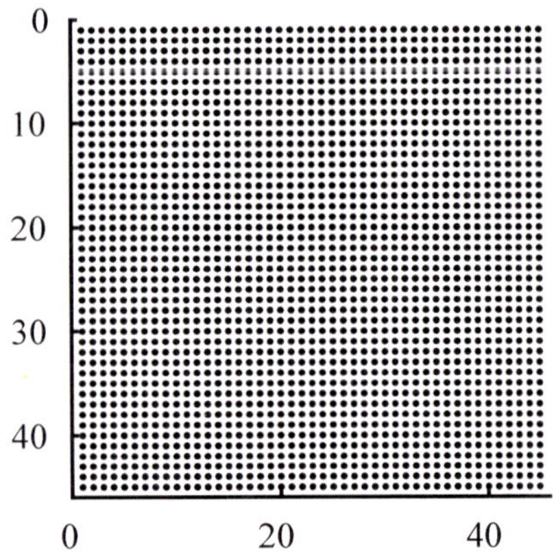

Fig. 4.7 Sparsity pattern of the Hessian of the dense MPC formulation

with $N_u \approx N_2$. The parameters chosen in the example are $N_1 = 1, N_2 = N_u = 45$ such that both criteria are fulfilled.

To illustrate the effect, the sparsity pattern of the Hessian of the cost function is investigated. The dimension of the Hessian for the dense formulation of the optimization problem is

$$N_u \times N_u = 45 \times 45 \tag{4.81}$$

The dense character of the Hessian is shown in Fig. 4.7. All 2,025 entries of the Hessian are structurally non-zero.

For the sparse formulation, the Hessian results have 32,400 entries. The dimensions are

$$(2lN_u + nN_2) \times (2lN_u + nN_2) = 180 \times 180 \tag{4.82}$$

The sparsity pattern in Fig. 4.8 shows that only 225 entries are non-zero. Additionally, it is a so-called tridiagonal matrix, the 225 entries are located exclusively on the main, super-, and sub-diagonals. In conclusion, the matrix for the Hessian of the sparse formulation is much larger than that of the dense formulation, but nevertheless, much fewer entries are structurally non-zero. If a solver is used that exploits the sparsity pattern, advantages for the computation time can be realized.

The second advantage of the sparse formulation can be observed in terms of the numerical conditioning of the resulting matrices, which is illustrated next. Exactly the same plant model and MPC settings as those described in Example 4.1 are investigated again. The only difference is that instead of the sparse formulation the dense formulation is considered. Figure 4.9 shows the control result obtained with the dense formulation.

The closed-loop system is unstable. In contrast, the sparse formulation resulted in a stable closed-loop system, although all the settings were the same. For the dense formulation, matrix potentiations are used for setting up the relevant matrices, such

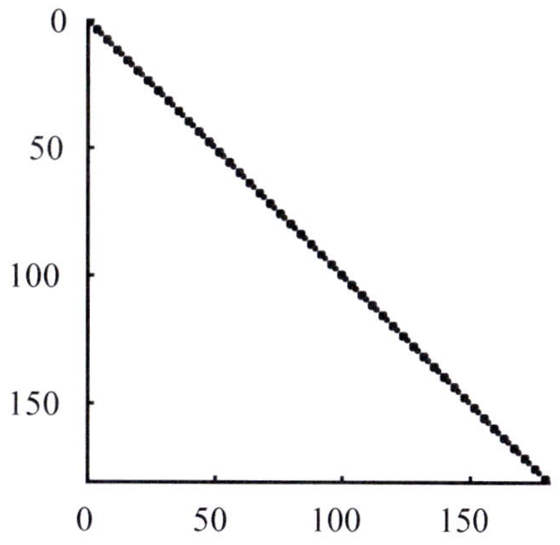

Fig. 4.8 Sparsity pattern of the Hessian of the sparse MPC formulation

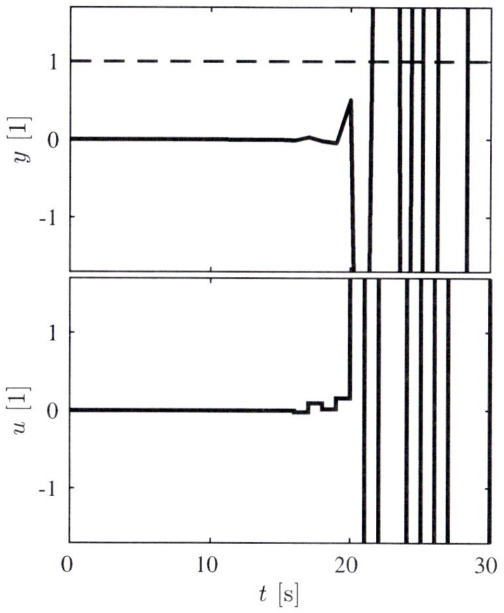

Fig. 4.9 Closed-loop control result for unconstrained LMPC (dense problem formulation) – Example 4.2

as the Hessian. In the case of an unstable system, which is predicted over many sampling steps, huge numbers result as entries in the matrix. Thus, small errors, which may occur due to the restricted precision of digital computing, can lead to big differences in the values of the matrix. As the sparse formulation does not use matrix potentiations, it is less sensitive to these kinds of errors. In the worst case, these propagated errors lead to an unstable closed-loop behavior, as given in this example.

Example 4.3 (*Unconstrained LMPC with anticipatory action*) The MPC algorithm allows future reference values to be considered, as shown in this numerical example for the unconstrained LMPC. The effect is demonstrated on the following simple example system. It is given by the continuous state-space model with matrices $\boldsymbol{A}$, $\boldsymbol{B}$, $\boldsymbol{C}$, and $\boldsymbol{D}$.

$$\boldsymbol{A} = \begin{bmatrix} 4.8 & -4.8 \\ 5.1 & -5.1 \end{bmatrix} \quad \boldsymbol{B} = \begin{bmatrix} 1 \\ -1 \end{bmatrix} \quad \boldsymbol{C} = \begin{bmatrix} 0.04 & 0.04 \end{bmatrix} \quad D = 0 \tag{4.83}$$

A standard cost function J with the possibility to consider future reference values is chosen.

$$J = \sum_{i=N_1}^{N_2} ||\boldsymbol{y}(k+i|k) - \boldsymbol{r}(k+i|k)||_{\boldsymbol{Q}}^2 + \sum_{i=0}^{N_u-1} ||\Delta\boldsymbol{u}(k+i|k)||_{\boldsymbol{R}}^2 \tag{4.84}$$

For the weighting matrices $\boldsymbol{Q}$ and $\boldsymbol{R}$, the identity matrix is chosen. The horizons are $N_1 = 1$ (no dead-time is considered), $N_2 = N_u = 10$, and the sampling time is $T_s = 0.2$ s. For the reference, a step signal is applied.

Two different cases are investigated. In the first scenario, only the current reference value is known. In each time step, the reference is assumed to be constant over the entire prediction horizon $\boldsymbol{r}(k+i|k) = \boldsymbol{r}(k|k)$ for $i \in \{N_1, \ldots, N_2\}$. In the second scenario, the entire reference trajectory is known to the controller in advance. In this case, the MPC is able to anticipate the reference and act already in advance, such that the reference is reached faster. The response of the system for both scenarios is displayed in Fig. 4.10. For the anticipatory action, the actuated values are not only shifted forward in time, instead an entirely different trajectory results.

Example 4.4 (*Constrained LMPC*) Example 4.3 is revisited. The system dynamics, the cost function, and also its weighting are unchanged. However, this time, additional constraints on the actuated values are considered. Specifically, the constraints are given by $u_{min} = -0.2$ and $u_{max} = 0.2$. The resulting closed-loop step response is shown in Fig. 4.11. The LMPC is able to systematically deal with the constraints on the actuated values. A high-performance reference tracking can be achieved. The resulting closed-loop controller is nonlinear due to the consideration of constraints. Compared to conventional PID-based controllers, this performance can be achieved without any additional measures such as anti-windup schemes.

Example 4.5 (*Control of a nonlinear system with LMPC*) In the following, a numerical example is detailed which shows the ability of the LMPC algorithm to control a nonlinear system. The plant model is based on the so-called Brusselator which was formulated first in the 1950s by Prigogine and collaborators at the Universite Libre de Bruxelles [9]. The Brusselator describes the chemical reactions from two educts A, B to two products C, D with two intermediate species X_1, X_2. Specifically, four different chemical reactions are considered.

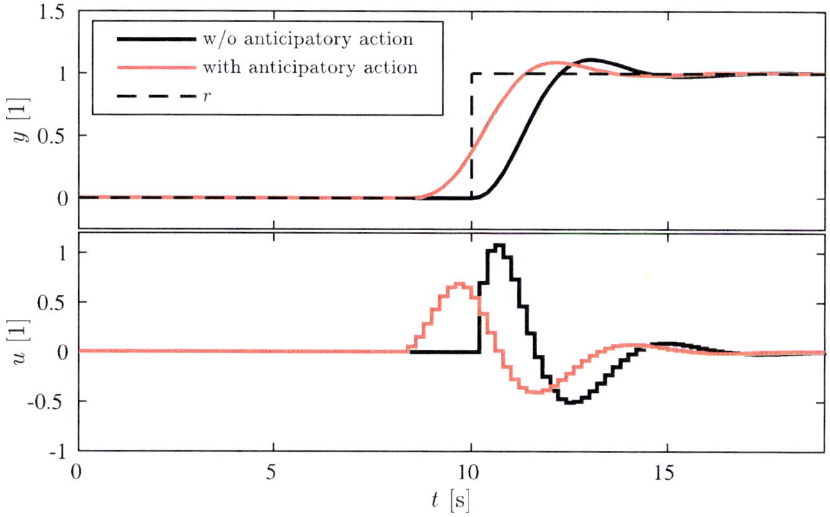

Fig. 4.10 Influence of anticipatory action on closed-loop control for Example 4.3

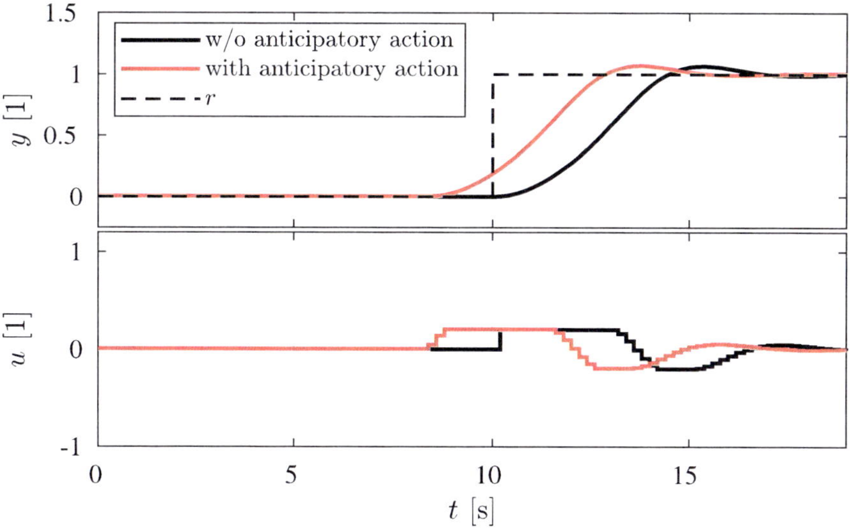

Fig. 4.11 Influence of constraints on actuated value on closed-loop control – Example 4.4

$$\begin{aligned}
\mathrm{A} &\xrightarrow{k_1} \mathrm{X_1} \\
\mathrm{B} + \mathrm{X_1} &\xrightarrow{k_2} \mathrm{X_2} + \mathrm{C} \\
2\mathrm{X_1} + \mathrm{X_2} &\xrightarrow{k_3} 3\mathrm{X_1} \\
\mathrm{X_1} &\xrightarrow{k_4} \mathrm{D}
\end{aligned}$$

The global reaction is given by:

$$\mathrm{A} + \mathrm{B} \longrightarrow \mathrm{C} + \mathrm{D}$$

The time trajectory of the intermediate species can be described by the following ODEs:

$$\dot{x}_1 = k_1 a - k_2 b x_1 + 3k_3 {x_1}^2 x_2 - k_4 x_1 \tag{4.85}$$

$$\dot{x}_2 = k_2 b x_1 - 3k_3 x_1^2 x_2 \tag{4.86}$$

The variables a, b, c, d, x_1, and x_2 denote the concentrations of the various species. The constants $k_1, \ldots, k_4$ are related to the reaction rates. The conventional Brusselator assumes constant concentrations a, b. However, in order to allow for closed-loop control, the system is modified in that the concentration a can be adjusted during operation. Thus, it serves as actuated value u for closed-loop control. The concentration b is assumed constant, i.e. the species B is continuously fed into the system to keep the concentration constant. The task of the closed-loop controller shall be the reference-tracking of both intermediate species concentrations x_1, x_2. By using the parameters $k_1 = 1$, $k_2 = 2$, $k_3 = 1$, $k_4 = 4$, and $b = 2.475$, the state-space system becomes

$$\dot{x}_1 = -8.95x_1 + 3x_1^2 x_2 + u \tag{4.87}$$

$$\dot{x}_2 = 4.95x_1 - 3x_1^2 x_2 \tag{4.88}$$

The Brusselator is characterized by nonlinear dynamics and oscillatory behavior. A distinct system property is the fact that the system enters a limit cycle for a constant input of $u = 1.7$, for all initial conditions except of $\boldsymbol{x}_0 = [0.425\ 3.882]^\mathrm{T}$. This point is actually the only stationary point. Figure 4.12 depicts the system evolution in the phase portrait for various initial conditions $\boldsymbol{x}_{0,1}, \ldots, \boldsymbol{x}_{0,4}$. Clearly, the trajectories $\boldsymbol{x}_{0,1}, \ldots, \boldsymbol{x}_{0,3}$ converge to the limit cycle regardless of whether they start inside or outside of it. The only trajectory that does not enter the limit cycle is the one starting at the stationary point $\boldsymbol{x}_{0,4} = [0.425\ 3.882]^\mathrm{T}$. It remains stationary at this point.

Since the stationary point is unstable, a challenging control task is to force the Brusselator to this point and then stabilize it there. For this task, the performance of

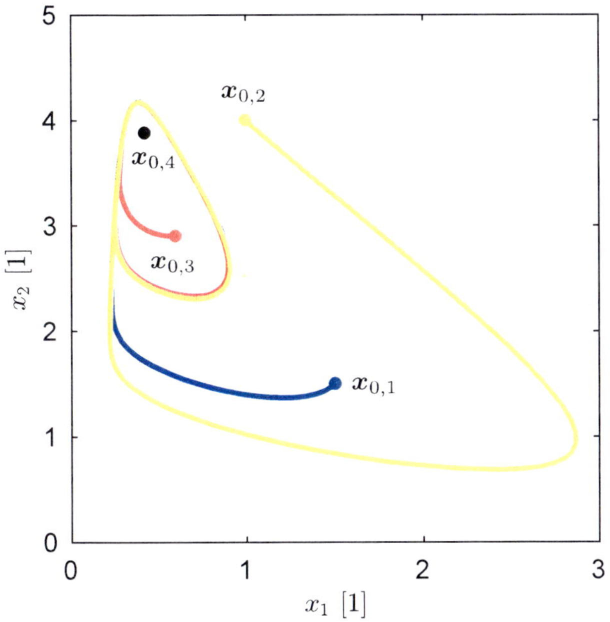

Fig. 4.12 Phase portrait of the open-loop Brusselator for various initial conditions

the LMPC approach using a linearized model of the system is investigated. For the optimization, the cost function J is used.

$$J = \sum_{i=1}^{N} \left|\left|\boldsymbol{x}(k+i|k) - \boldsymbol{x}_{ref}\right|\right|_{\boldsymbol{Q}}^{2} + \sum_{i=0}^{N-1} \left|\left|u(k+i|k) - u_{ref}\right|\right|_{R}^{2} \tag{4.89}$$

Within the cost function J, the deviation of the actuated values and the system states to the references are penalized. The reference values are given by the stationary point $\boldsymbol{x}_{ref} = [0.425\ 3.882]^{\mathsf{T}}$ and $u_{ref} = 1.7$. The weighting matrices are chosen to be $\boldsymbol{Q} = \boldsymbol{I}$ and $R = 1$. The initial condition is set to $\boldsymbol{x}_0 = [1\ 2]^{\mathsf{T}}$, the prediction and control horizons to $N = 50$, and the lower control horizon to $N_1 = 1$ as no dead-time is considered. The sampling time is chosen to be $T_s = 0.2s$. Following constraints on the actuated value, the change of actuated value and on the system states are considered:

$$0 \leq u(\cdot|k) \leq 5 \quad \text{and} \quad -2 \leq \Delta u(\cdot|k) \leq 2 \tag{4.90}$$

$$0 \leq x_1(\cdot|k) \leq 2 \quad \text{and} \quad 0 \leq x_2(\cdot|k) \leq 5 \tag{4.91}$$

Within LMPC, an LTI model is used. In order to derive this model, the nonlinear model equation is linearized around a single operating point defined by $\boldsymbol{x}_{op}$, u_{op}. All predictions are performed using the resulting LTI model. The linearization is conducted around the stationary point that is to be reached.

$$\boldsymbol{x}_{op} = \begin{bmatrix} 0.425 \\ 3.882 \end{bmatrix} \quad \text{and} \quad u_{op} = 1.7 \tag{4.92}$$

Using (4.77), the linearized continuous-time state-space model is obtained:

$$\dot{\boldsymbol{x}}(t) = \begin{bmatrix} 0.950 & 0.542 \\ -4.950 & -0.542 \end{bmatrix} \boldsymbol{x} + \begin{bmatrix} 1 \\ 0 \end{bmatrix} u \tag{4.93}$$

With the poles of that system being at $\pi_{1,2} = 0.204 \pm 1.458i$, the steady-state point is an unstable equilibrium of the nonlinear system. Therefore, the sparse formulation is used. An exact discretization with the sampling time T_s leads to the following discrete-time state-space model:

$$\boldsymbol{x}(k+1) = \begin{bmatrix} 1.151 & 0.111 \\ -1.017 & 0.844 \end{bmatrix} \boldsymbol{x}(k) + \begin{bmatrix} 0.216 \\ -0.101 \end{bmatrix} u(k) \tag{4.94}$$

Figure 4.13 shows the closed-loop control results with LTI MPC. It depicts the evolution of the state trajectories over time. Clearly, the LTI MPC is able to bring the states to the steady-state point and stabilize them there. The system can be stabilized

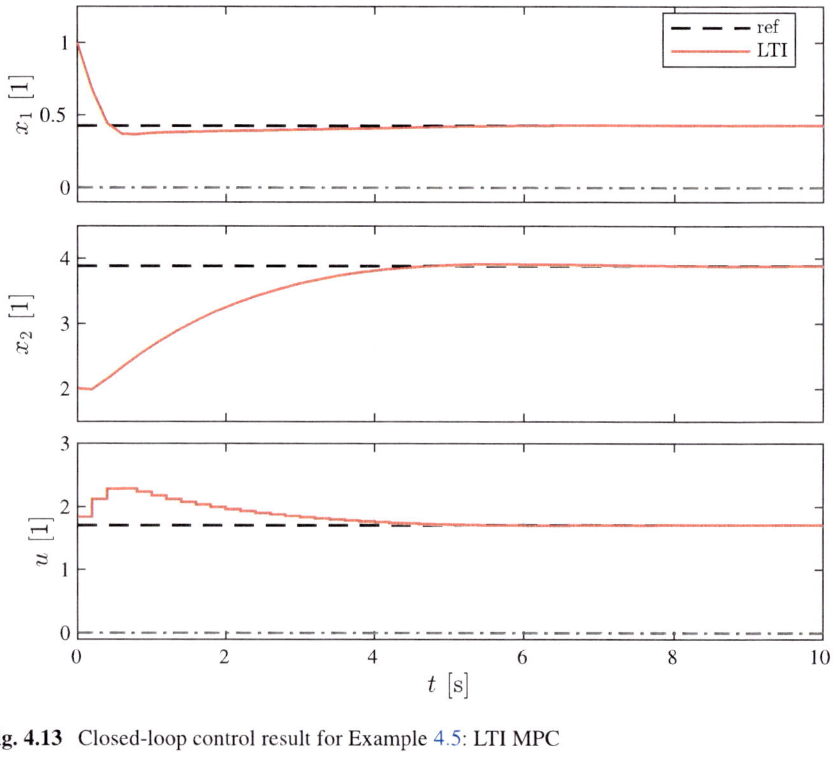

Fig. 4.13 Closed-loop control result for Example 4.5: LTI MPC

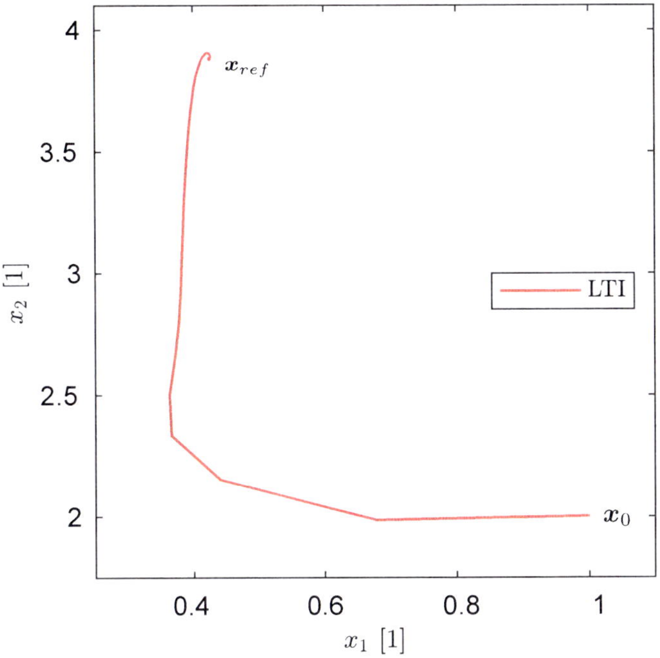

Fig. 4.14 Closed-loop control result in the phase portrait for Example 4.5: LTI MPC

although it is nonlinear. The result shows that robustness is inherently present in MPC. The state evolution is also visualized by the phase portrait in Fig. 4.14.

References

1. A. Alessio, A. Bemporad, A survey on explicit model predictive control, in *Nonlinear Model Predictive Control* (Springer, Berlin, 2009), pp. 345–369
2. K.J. Aström, R.M. Murray, *Feedback Systems: An Introduction for Scientists and Engineers* (Princeton University Press, Princeton, 2010)
3. A. Bemporad, M. Morari, V. Dua, E.N. Pistikopoulos, The explicit linear quadratic regulator for constrained systems. Automatica **38**(1), 3–20 (2002)
4. A. Bemporad, C. Rocchi, Decentralized linear time-varying model predictive control of a formation of unmanned aerial vehicles, in *IEEE Conference on Decision and Control* (2011), pp. 7488–7493
5. F. Borrelli, A. Bemporad, M. Morari, *Predictive Control for Linear and Hybrid Systems* (Cambridge University Press, Cambridge, 2017)
6. S. Boyd, L. Vandenberghe, *Convex Optimization* (Cambridge University Press, Cambridge, 2004)
7. A. Domahidi, A.U. Zgraggen, M.N. Zeilinger, M. Morari, C.N. Jones, Efficient interior point methods for multistage problems arising in receding horizon control, in *IEEE Conference on Decision and Control* (2012), pp. 668–674

8. H.J. Ferreau, C. Kirches, A. Potschka, H.G. Bock, M. Diehl, qpOASES: a parametric active-set algorithm for quadratic programming. Math. Progr. Comput. **6**(4), 327–363 (2014)
9. J.A. Gallas, Periodic oscillations of the forced brusselator. Mod. Phys. Lett. B **29**, 1530018 (2015)
10. C.N. Jones, A. Domahidi, M. Morari, S. Richter, F. Ullmann, M. Zeilinger, Fast predictive control: real-time computation and certification. IFAC Proc. Vol. **45**(17), 94–98 (2012)
11. J. Ma, S.J. Qin, B. Li, T. Salsbury, Economic model predictive control for building energy systems, in *IEEE PES Innovative Smart Grid Technologies* (2011)
12. J.M. Maciejowski, *Predictive Control: with Constraints* (Pearson Education, London, 2002)
13. M. Muehlebach, R. D'Andrea, Parametrized infinite-horizon model predictive control for linear time-invariant systems with input and state constraints, in *American Control Conference* (2016), pp. 2669–2674
14. J. Nocedal, S.J. Wright, *Numerical Optimization* (Springer, Berlin, 2006)
15. W.H. Press, S.A. Teukolsky, W.T. Vetterling, B.P. Flannery, *Numerical Recipes: The Art of Scientific Computing* (Cambridge University Press, Cambridge, 2007)
16. D. Simon, *Optimal State Estimation: Kalman, H-infinity, and Nonlinear Approaches* (Wiley, New York, 2006)
17. Y. Wang, S. Boyd, Fast model predictive control using online optimization. IEEE Trans. Control Syst. Technol. **18**(2), 267–278 (2009)

Chapter 5
Nonlinear Model Predictive Control

Abstract The nonlinear model predictive control (NMPC) approach allows more challenging control problems to be handled than those dealt with by the linear model predictive control (LMPC) method. For engine control problems, the consideration of their nonlinear process behavior is particularly important as their treatment as a linearized system is often not sufficient. In this chapter, the fundamentals of NMPC are presented. A special focus is set on the real-time feasibility of the computation algorithm, as this aspect poses a major challenge for engine control. The numerical solution of the resulting nonlinear program (NLP) is discussed. As a suitable solution technique, the sequential quadratic programming (SQP) approach is explained. Additionally, multiple shooting versus single shooting techniques are detailed, which allows optimal control problems (OCP) to be discretized efficiently. Numerical examples are investigated to show the capabilities of the algorithms described.

5.1 Introduction to Nonlinear MPC

Most engine control tasks can be classified as reference tracking problems. Hence, the MPC approaches are usually characterized by a quadratic cost function accounting for deviations of the predicted states to the reference states. Additionally, affine constraints on the actuated values and on the system states are considered. The system model within LMPC is restricted to being a linear or a linearized model. However, the use of a linear model is often insufficient for representing the system behavior. In these cases, the NMPC approach allows for a better control performance than the LMPC. The term NMPC is used typically when the resulting optimization problem is neither a QP nor an LP. An NLP optimization problem results for MPC when a nonlinear system model, nonlinear constraints or a non-convex cost function is considered. For engine control tasks, the NMPC algorithm is especially used for consideration of a nonlinear system model.

T. Albin Rajasingham, *Nonlinear Model Predictive Control of Combustion Engines*, Advances in Industrial Control,
https://doi.org/10.1007/978-3-030-68010-7_5

Typically, nonlinear models are present as continuous-time models, e.g. because they are derived from physical modeling. The consideration of this continuous-time nonlinear models in an optimization-based control algorithm results in an OCP. An example of an OCP is depicted by the following optimization problem with cost function J_{ocp}.

$$J_{ocp}(\boldsymbol{x}(t), \boldsymbol{u}(t)) = l_f(\boldsymbol{x}(t_f)) + \int_{t_0}^{t_f} l_s(\boldsymbol{x}(t), \boldsymbol{u}(t))\mathrm{d}t \tag{5.1}$$

The cost function J_{ocp} in this example is subject to the system dynamics in (5.2b), the initial state of the system at $t = t_0$ in (5.2c), the equality constraints on the final states in (5.2d), and the path constraints which are present as inequality constraints in (5.2e).

$$\min_{\boldsymbol{x}(t), \boldsymbol{u}(t)} \quad J_{ocp}(\boldsymbol{x}(t), \boldsymbol{u}(t)) \tag{5.2a}$$

$$\text{s.t.} \quad \dot{\boldsymbol{x}}(t) = \boldsymbol{f}(\boldsymbol{x}(t), \boldsymbol{u}(t), t), \forall t \in [t_0, t_f], \tag{5.2b}$$

$$\boldsymbol{x}(t_0) = \boldsymbol{x}_0, \tag{5.2c}$$

$$\boldsymbol{h}(\boldsymbol{x}(t_f)) = \boldsymbol{0}, \tag{5.2d}$$

$$\boldsymbol{g}(\boldsymbol{x}(t), \boldsymbol{u}(t)) \leq \boldsymbol{0}, \forall t \in [t_0, t_f] \tag{5.2e}$$

Due to the infinite number of optimization variables, OCPs usually cannot be solved in real time, especially when a nonlinear model is considered. Thus, the OCP cannot be directly used for practical NMPC applications. However, in order to obtain a tractable optimization problem, the OCP can be approximated by an NLP. The transformation from the OCP to the NLP is conducted by a suitable discretization. The process of the discretization is detailed in Sect. 5.3. An NLP also results if the system model, cost function, and the constraints are directly given in discrete time. For data-based modeling, usually a discrete-time system model results. The general NLP is usually of the following form:

$$\begin{aligned} \min_{z \in \mathbb{R}^n} \quad & J(z) \\ \text{s.t.} \quad & \boldsymbol{h}(z) = \boldsymbol{0}, \\ & \boldsymbol{g}(z) \leq \boldsymbol{0} \end{aligned} \tag{5.3}$$

The optimization variables are denoted by $z \in \mathbb{R}^n$, while $\boldsymbol{h}$ defines the equality constraints and $\boldsymbol{g}$ the inequality constraints. A typical NMPC optimization problem for engine control problems is presented in (5.4). To account for the reference tracking problem, a quadratic cost function is used where the difference between predicted states and reference is penalized along with the deviation of the actuated values to the steady-state values. For nonlinear systems, characteristics such as superposition no longer exist. Thus, in contrast to LMPC, for NMPC no difference is made between the horizons for control actuation and prediction of the states. While in LMPC problems, the actuated values can be left constant after the control horizon, in NMPC problems this approach is inappropriate, even for stable systems. To account for the system dynamics, a time-discrete nonlinear system model of the form $\boldsymbol{x}(k+1) = \boldsymbol{f}_{dis}(\boldsymbol{x}(k+i|k), \boldsymbol{u}(k+i|k)$ is considered along with the initial states. Additionally, box constraints on the actuated values and on the system states are present in the optimization problem.

$$
\begin{aligned}
\min_{\boldsymbol{x}(\cdot|k),\, \boldsymbol{u}(\cdot|k)} \quad & \sum_{i=1}^{N} ||\boldsymbol{x}(k+i|k) - \boldsymbol{r}(k)||^2_{\boldsymbol{Q}_i} + \sum_{i=0}^{N-1} ||\boldsymbol{u}(k+i|k) - \boldsymbol{u}_{ss}(k)||^2_{\boldsymbol{R}_i} \\
\text{s.t.} \quad & \boldsymbol{x}(k+i+1|k) = \boldsymbol{f}_{dis}(\boldsymbol{x}(k+i|k), \boldsymbol{u}(k+i|k)), \quad i = 0, \ldots, N-1, \\
& \boldsymbol{x}(k|k) = \boldsymbol{x}(k), \\
& \boldsymbol{u}_{min} \leq \boldsymbol{u}(k+i|k) \leq \boldsymbol{u}_{max}, \quad i = 0, \ldots, N-1, \\
& \boldsymbol{x}_{min} \leq \boldsymbol{x}(k+i|k) \leq \boldsymbol{x}_{max}, \quad i = 1, \ldots, N
\end{aligned}
\tag{5.4}
$$

One of the main challenges in applying NMPC for engine control is the calculation of a suitable solution of the arising NLP in real time. Since in the case of engine control problems, both the system dynamics and the constraint functions can be modeled as continuously differentiable functions, gradient-based numerical methods can be applied. There exist various gradient-based solution methods that allow for the calculation of the minimizer within small timescales. They rely on first-order derivative information and some algorithms additionally on second-order derivative information. For real-time applications, these are typically calculated by automatic differentiation (AD) which is efficient concerning the code size, fast and accurate, and thus better suited than numerical differentiation and symbolic differentiation. Software packages such as CasADi [2] can be used for AD.

The SQP method is a gradient-based optimization method with special advantages for optimization problems comprising structures as arising in engine control problems (5.4). Thus, it is introduced as a powerful numerical solution method for engine control problems in Sect. 5.2. The best solution method depends on the specific real-time decision-making problem. Usually, the interior point (IP) methods are superior to the SQP methods for motion planning problems, e.g. for autonomous driving and robotics. In these problems, typically no reference exists and there are nonlinear inequality constraints.

While the local minimizers can be calculated with the SQP method, the results do not necessarily represent global minimizers. In the case of non-convex optimization in the context of an online algorithm for NMPC, one is satisfied with local minimizers [7]. Ideally, the local minimizer of the optimization problem is obtained within each sampling step. For the engine control tasks, very small sampling times are used, e.g. in the order of milliseconds. In this case, the fully converged solution might not be obtainable. Alternatively, the so-called real-time iteration (RTI) scheme can be used [9]. Here, in every time step, a suboptimal solution is calculated and applied to the plant, as explained in Sect. 5.3.5.

In Fig. 5.1, various MPC schemes are compared along the two dimensions of computational complexity and capability of the resulting controller. Clearly, there is a direct correlation between these two characteristics. Depending on the complexity

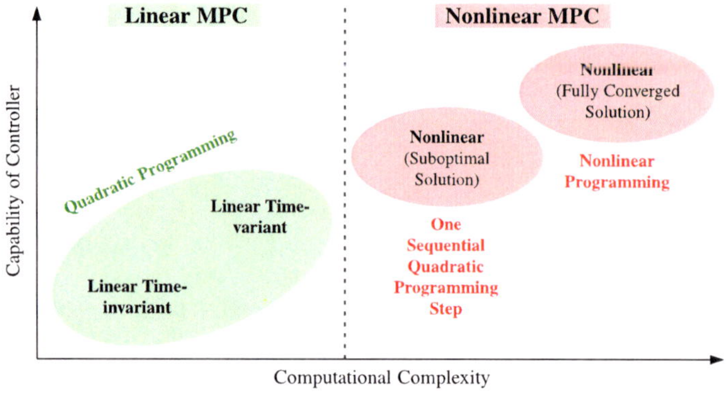

Fig. 5.1 Classification of various MPC schemes

of the control task, the available computation time, and hardware power, a suitable algorithm can be chosen. The order of the complexity is given by LMPC, LTV MPC, NMPC with a suboptimal solution, and NMPC with a fully converged solution.

5.2 Numerical Solution of the Resulting Nonlinear Program

For real-time NMPC, the arising NLP is solved during the runtime of the process. Typically, the optimization problem cannot be solved analytically, instead the solution has to be calculated numerically. In the following, solution methods are presented for the unconstrained and constrained NLPs that are well established for the application within NMPC. A general treatment of various solution methods for NLPs can be found in [3, 4, 10].

5.2.1 *Solving the Unconstrained NLP*

First, unconstrained NLP problems are investigated. In this case, the optimization problem is given by the minimization of the cost function $J(z)$.

$$\min_{z \in \mathbb{R}^n} \quad J(z) \tag{5.5}$$

In most cases, it is not possible to solve the optimality condition $\nabla J(z^*) = \mathbf{0}$ analytically. Instead, numerical methods are used to find the minimizers. Most of the numerical methods rely on a reduction of the cost function in every iteration step $J(z_{k+1}) < J(z_k)$ starting from an initial guess z_0. For finding the new iterate z_{k+1}, there exist two fundamental approaches. These are the *line-search* algorithms and the *trust-region* algorithms. Compared to trust-region methods, the line-search algorithms are characterized by first determining the search direction and subsequently the step length. In the following, the line-search algorithm is detailed. Its general approach is as follows:

1. Set the index of iteration $k = 0$.
2. Choose the initial guess z_0.
3. Determine the search direction $\boldsymbol{p}_k$.
4. Determine the step size $\alpha_k > 0$.
5. Set the new iterate $z_{k+1} = z_k + \alpha_k \boldsymbol{p}_k$.
6. Check stop criterion, e.g. $||\nabla J(z_{k+1})|| \leq \epsilon$.
7. Start new iteration with $k := k + 1$ and continue with step 3.

The various line-search algorithms differ by the approach for the determination of the search direction $\boldsymbol{p}_k$ and for the step size α_k. A suitable search direction has to ensure that a reduction of the cost function is possible for a sufficiently small step size $\alpha > 0$, i.e. it is a descent direction. This can be translated into a sufficient condition for the search direction [10] with $\cdot$ being the dot product:

$$\boldsymbol{p}_k \cdot \nabla J(z_k) = \boldsymbol{p}_k^\mathsf{T} \nabla J(z_k) < 0 \tag{5.6}$$

If this condition is fulfilled, there exists an α_{max} such that for any α with $0 < \alpha < \alpha_{max}$, the following inequality must hold:

$$J(z_k + \alpha \boldsymbol{p}_k) < J(z_k) \tag{5.7}$$

Two suitable search directions $\boldsymbol{p}_k$ are outlined: the steepest descent algorithm which is based on gradient information and the Newton method which is based on gradient and curvature information.

The Steepest Descent Method

The *steepest descent* method is the simplest method for the calculation of a descent direction [10]. The search direction $\boldsymbol{p}_k$ is set to

$$\boldsymbol{p}_k = -\nabla J(z_k) \tag{5.8}$$

The vector $-\nabla J(z_k)$ corresponds to the direction of the steepest descent. This method is easy to implement as only first-order and no second-order derivative information are needed. Additionally, it guarantees a decrease in J for a sufficiently small step size α_k, as the condition (5.6) is fulfilled. The steepest descent method converges linearly. It converges quickly into an area around the minimum. However, it has a slow convergence around the minimum. Especially for ill-conditioned optimization problems, small step sizes might be necessary which leads to slow overall convergence. In this case, scaling of the optimization problem is a possible remedy [11].

The Newton Method

The search direction $\boldsymbol{p}_k$ for the *Newton method* can be derived by two different approaches that lead to the same result. The search direction can be derived from the minimization of the second-order Taylor approximation. The second approach, from where also its name originates, applies the Newton method for root-finding to the optimality condition [4].

The first approach minimizes the second-order Taylor approximation $q(\boldsymbol{p}_k)$ of J at each iterate z_k. The minimization is denoted by

$$\min_{\boldsymbol{p}_k} \quad q(\boldsymbol{p}_k) \tag{5.9}$$

with

$$q(\boldsymbol{p}_k) := J(z_k) + \boldsymbol{p}_k^\mathsf{T} \nabla J(z_k) + \frac{1}{2} \boldsymbol{p}_k^\mathsf{T} \nabla^2 J(z_k) \boldsymbol{p}_k$$

The stationary point for this optimization problem is given by

$$\nabla q(\boldsymbol{p}_k^*) = \nabla J(z_k) + \nabla^2 J(z_k) \boldsymbol{p}_k^* = 0 \tag{5.10}$$

The resulting solution of this optimization problem is denoted by $\boldsymbol{p}_k^*$. The search direction $\boldsymbol{p}_k$ for the Newton method uses the solution of the optimization problem as search direction within the line-search approach:

$$\boldsymbol{p}_k = -\left(\nabla^2 J(z_k)\right)^{-1} \nabla J(z_k) \tag{5.11}$$

The same search direction can also be derived by applying the Newton method to the optimality condition $\nabla J(z) = \mathbf{0}$. In general, the Newton method can be utilized to solve nonlinear root-finding problems. Assume the function $\boldsymbol{f}(z)$ is present and the following equation is to be solved:

$$\boldsymbol{f}(z) = \mathbf{0} \tag{5.12}$$

In order to iteratively determine a solution z which satisfies the equation, it is approximated as a linear function in each iteration step.

$$\boldsymbol{f}(z_k + \boldsymbol{p}_k) \approx \boldsymbol{f}(z_k) + \boldsymbol{p}_k^\mathsf{T} \nabla \boldsymbol{f}(z_k) = \mathbf{0} \tag{5.13}$$

For the linear approximation, the root can be directly calculated as

$$\boldsymbol{p}_k = -\left(\nabla \boldsymbol{f}(z_k)^\mathsf{T}\right)^{-1} \boldsymbol{f}(z_k) \tag{5.14}$$

By the solution of the linear approximation, the new iterate z_{k+1} can be calculated as

$$z_{k+1} = z_k + \boldsymbol{p}_k \tag{5.15}$$

This Newton method can be applied to the first-order necessary optimality condition $\nabla J(z) = \boldsymbol{0}$. In this case, the search direction in each iteration is given by

$$\boldsymbol{p}_k = -\left(\nabla^2 J(z_k)\right)^{-1} \nabla J(z_k) \tag{5.16}$$

This is the same search direction as the one obtained with the second-order Taylor approximation. The search direction can be used within the line-search algorithm. Following update equation is used in every iteration:

$$z_{k+1} = z_k - \alpha_k \left(\nabla^2 J(z_k)\right)^{-1} \nabla J(z_k) \tag{5.17}$$

The Newton method shows a fast local convergence rate. Its convergence rate is quadratic rather than linear, as obtained by the steepest descent method [10]. The natural step size of the Newton method is $\alpha_k = 1$. If the initial points are chosen sufficiently close to the local minimizer, the Newton method is mathematically guaranteed to converge to this local minimizer. From remote initial points, z_0 convergence cannot be guaranteed. Globalization strategies are applied to improve convergence also from remote initial points to the local minimizer. Globalization strategies concern especially the step size and the search direction.

Applying the full step length can cause non-convergence, i.e. oscillation, or even divergence. To overcome this problem, the step size $\alpha_k \in (0, 1]$ must be chosen accordingly to guarantee a decrease in the cost function. In practice, an initial step size α_k is chosen, which is iteratively reduced until some criteria on progress are met. There exist various methods, such as the Armijo rule and the Wolfe conditions, to evaluate whether sufficient progress has been made [10]. The Newton method with reduced step size is also called the damped Newton method.

For remote initial points, besides the step size, the search direction can be inappropriate as well. The Newton method results in a descent direction as long as the Hessian matrix is positive definite. It can fail when the Hessian matrix turns indefinite or negative definite. If such ill-conditioned Hessians occur, this problem can be overcome by substituting the Hessian with a positive-definite approximation. Actually, any positive-definite approximation will lead to a descent direction. One possibility to obtain a positive-definite approximation is the use of regularization. A simple approach is adding a positive curvature until the modified Hessian matrix $\boldsymbol{H}'$ is positive definite again.

$$\boldsymbol{H}' = \boldsymbol{H} + \beta \boldsymbol{I} \tag{5.18}$$

However, completely excluding the Hessian matrix, e.g. by using high values for β, respectively, the identity matrix for $\boldsymbol{H}'$ leads to a simple steepest-descent algorithm. This often features inferior convergence properties. There exist also more advanced approaches for regularization to still exploit curvature information as much as possible [10].

5.2.2 *Solving the Constrained NLP via Sequential Quadratic Programming*

Various solution algorithms exist to solve the constrained NLP, such as the IP NLP method, the augmented Lagrangian method, and the SQP method [3]. In the following, SQP methods are investigated, since they perform very fast and have some favorable properties for the reference-tracking NMPC problems that arise in engine control.

The SQP method can be used within a line-search approach. In this context, the SQP method is used to find an appropriate search direction $\boldsymbol{p}_k$. In every iteration step, the search direction results from a solution of a quadratic program (QP). The resulting QP can be solved, e.g. by using an IP QP method or an active set strategy. After the calculation of the search direction, the step size α_k can be determined. Only by an appropriate choice of α_k the convergence of the numerical solution method can be ensured, as explained for the unconstrained NLP case. Similarly, the Hessian matrix must be positive definite.

For the line search, a measure is needed to evaluate the progress toward the KKT point. In the unconstrained case, the cost function J can be used to evaluate the progress in each iteration. Within the constrained case, a measure is needed that takes into account the progress in terms of the cost as well as in terms of the constraint compliance. These functions are called merit functions [10]. One example is the use of the so-called l_1-penalty function $m(z_k)$. It is based on the 1-norm denoted by $||a||_1$. The factor σ weighs the constraint compliance against the reduction of cost. The weight σ has to be chosen bigger than the maximum Lagrange multiplier associated with the minimizer z^* to ensure convergence within a line-search method [10].

$$m(z_k) = J(z_k) + \sigma \sum_{i=1}^{p} ||h_i(z_k)||_1 + \sigma \sum_{i=1}^{q} \max\{0, g_i(z_k)\} \tag{5.19}$$

The calculation of the search direction $\boldsymbol{p}_k$ via the SQP method is introduced for equality and inequality constrained problems in the following sections.

5.2.2.1 NLP with Equality Constraints

The numerical solution of the equality-constrained NLP with the SQP method is investigated. Two different viewpoints are sketched to derive the resulting formulas, namely, the application of the Newton method for root-finding on the Karush–Kuhn–Tucker (KKT) conditions and the quadratic approximation of the Lagrange function. In both cases, the following equality-constrained NLP is investigated:

$$\begin{aligned} \min_{z \in \mathbb{R}^n} \quad & J(z) \\ \text{s.t.} \quad & \boldsymbol{h}(z) = \boldsymbol{0} \end{aligned} \tag{5.20}$$

Applying the Newton Method on the KKT Conditions

The derivation with the first viewpoint examines the Lagrange function $L(z, \boldsymbol{\lambda})$. It is given by

$$L(z, \boldsymbol{\lambda}) = J(z) + \boldsymbol{\lambda}^\mathsf{T} \boldsymbol{h}(z) \tag{5.21}$$

In order to find the local minimizers z^* of the equality-constrained problem, the KKT conditions can be applied, which are denoted by

$$\nabla L(z^*, \boldsymbol{\lambda}^*) = \begin{bmatrix} \nabla_z L(z^*, \boldsymbol{\lambda}^*) \\ \nabla_\lambda L(z^*, \boldsymbol{\lambda}^*) \end{bmatrix} = \begin{bmatrix} \nabla J(z^*) + \nabla \boldsymbol{h}(z^*)\boldsymbol{\lambda}^* \\ \boldsymbol{h}(z^*) \end{bmatrix} = \mathbf{0} \tag{5.22}$$

Here $\nabla \boldsymbol{h}(z)$, i.e. the transposed Jacobian matrix, results from the derivatives of the constraints $\nabla \boldsymbol{h}(z) = [\nabla h_1(z), \nabla h_2(z), \ldots, \nabla h_p(z)]$. If $\nabla \boldsymbol{h}(z)^\mathsf{T}$ has full row rank, the linear independence constraint qualification (LICQ) is fulfilled, and there exist unique Lagrange multipliers $\boldsymbol{\lambda}^*$ such that every minimizer z^* solves $\nabla L(z^*, \boldsymbol{\lambda}^*) = \mathbf{0}$.

The KKT conditions are a system of nonlinear equations that consist of the same number of unknowns as the number of equations. The main idea to solve this problem is to use the Newton method for solving nonlinear root-finding problems, analogous to the unconstrained case. This time, the Newton method is applied to root-finding of the equation $\nabla L(z, \boldsymbol{\lambda}) = \mathbf{0}$. The search direction $\boldsymbol{p}_k$ for the Newton iteration follows as

$$\nabla L(z_k, \boldsymbol{\lambda}_k) + \boldsymbol{p}_k^\mathsf{T} \nabla^2 L(z_k, \boldsymbol{\lambda}_k) = \mathbf{0} \tag{5.23}$$

The KKT matrix is given by

$$\nabla^2 L(z_k, \boldsymbol{\lambda}_k) = \begin{bmatrix} \nabla_{zz}^2 L(z_k, \boldsymbol{\lambda}_k) & \nabla \boldsymbol{h}(z_k) \\ \nabla \boldsymbol{h}(z_k)^\mathsf{T} & \mathbf{0} \end{bmatrix} \tag{5.24}$$

The problem can be expanded to

$$\begin{bmatrix} \nabla_{zz}^2 L(z_k, \boldsymbol{\lambda}_k) & \nabla \boldsymbol{h}(z_k) \\ \nabla \boldsymbol{h}(z_k)^\mathsf{T} & \mathbf{0} \end{bmatrix} \begin{bmatrix} \boldsymbol{p}_{z_k} \\ \boldsymbol{p}_{\lambda_k} \end{bmatrix} = - \begin{bmatrix} \nabla J(z_k) + \nabla \boldsymbol{h}(z_k)\boldsymbol{\lambda}_k \\ \boldsymbol{h}(z_k) \end{bmatrix} \tag{5.25}$$

The update for the iterate can be calculated by

$$\begin{bmatrix} z_{k+1} \\ \boldsymbol{\lambda}_{k+1} \end{bmatrix} = \begin{bmatrix} z_k \\ \boldsymbol{\lambda}_k \end{bmatrix} + \begin{bmatrix} \boldsymbol{p}_{z_k}^* \\ \boldsymbol{p}_{\lambda_k}^* \end{bmatrix} \tag{5.26}$$

Moving the term $\nabla \boldsymbol{h}(z_k)\boldsymbol{\lambda}_k$ to the other side yields

$$\begin{bmatrix} \nabla_{zz}^2 L(z_k, \boldsymbol{\lambda}_k) & \nabla \boldsymbol{h}(z_k) \\ \nabla \boldsymbol{h}(z_k)^\mathsf{T} & 0 \end{bmatrix} \begin{bmatrix} \boldsymbol{p}_{z_k} \\ \boldsymbol{\lambda}_{k+1} \end{bmatrix} = - \begin{bmatrix} \nabla J(z_k) \\ \boldsymbol{h}(z_k) \end{bmatrix} \tag{5.27}$$

To calculate the update for the variables, the linear system of equations needs to be solved. Solving (5.27) directly yields $\boldsymbol{\lambda}_{k+1}^*$ and $\boldsymbol{p}_{z_k}^*$ with which the update on z_{k+1}

and $\boldsymbol{\lambda}_{k+1}$ can be calculated. Instead of using the full step also a damped step can be applied. The standard equation for a line-search update can be used to finally calculate the new iterate:

$$\begin{aligned} z_{k+1} &= z_k + \alpha_k \boldsymbol{p}_{z_k}^* \\ \boldsymbol{\lambda}_{k+1} &= (1-\alpha_k)\boldsymbol{\lambda}_k + \alpha_k \boldsymbol{\lambda}_{k+1}^* \end{aligned} \tag{5.28}$$

Second-Order Taylor Approximation of the Lagrange Function

The same search direction can be derived by an alternative approach. The alternative approach uses a second-order Taylor approximation of the Lagrange function. The Lagrange function is given by

$$L(z, \boldsymbol{\lambda}) = J(z) + \boldsymbol{\lambda}^\mathsf{T} \boldsymbol{h}(z) \tag{5.29}$$

It can be approximated by a quadratic function around z_k, $\boldsymbol{\lambda}_k$ with

$$L_{QP,k}(\boldsymbol{p}_{z_k}, \boldsymbol{p}_{\lambda_k}) = L(z_k, \boldsymbol{\lambda}_k) + \left[\boldsymbol{p}_{z_k}^\mathsf{T}\ \boldsymbol{p}_{\lambda_k}^\mathsf{T}\right] \nabla L(z_k, \boldsymbol{\lambda}_k) + \frac{1}{2}\left[\boldsymbol{p}_{z_k}^\mathsf{T}\ \boldsymbol{p}_{\lambda_k}^\mathsf{T}\right] \nabla^2 L(z_k, \boldsymbol{\lambda}_k) \begin{bmatrix} \boldsymbol{p}_{z_k} \\ \boldsymbol{p}_{\lambda_k} \end{bmatrix} \tag{5.30}$$

As conducted above, the variables can be exchanged, e.g. $\boldsymbol{p}_{\lambda_k} = \boldsymbol{\lambda}_{k+1} - \boldsymbol{\lambda}_k$. Using this substitution and expanding the equation leads to

$$\begin{aligned} L_{QP,k}(\boldsymbol{p}_{z_k}, \boldsymbol{\lambda}_{k+1}) =& \frac{1}{2} \boldsymbol{p}_{z_k}^\mathsf{T} \nabla_{zz}^2 L(z_k, \boldsymbol{\lambda}_k) \boldsymbol{p}_{z_k} + \boldsymbol{p}_{z_k}^\mathsf{T} \nabla J(z_k) \\ &+ \boldsymbol{\lambda}_{k+1}^\mathsf{T} (\nabla \boldsymbol{h}(z_k)^\mathsf{T} \boldsymbol{p}_{z_k} + \boldsymbol{h}(z_k)) - \boldsymbol{\lambda}_k^\mathsf{T} \boldsymbol{h}(z_k) + L(z_k, \boldsymbol{\lambda}_k) \end{aligned} \tag{5.31}$$

Neglecting the constant terms, the approximated Lagrange function $L_{QP,k}(\boldsymbol{p}_{z_k}, \boldsymbol{\lambda}_{k+1})$ corresponds to the following equality-constrained QP [3]:

$$\begin{aligned} \min_{\boldsymbol{p}_{z_k}} \quad & \frac{1}{2} \boldsymbol{p}_{z_k}^\mathsf{T} \nabla_{zz}^2 L(z_k, \boldsymbol{\lambda}_k) \boldsymbol{p}_{z_k} + \boldsymbol{p}_{z_k}^\mathsf{T} \nabla J(z_k) \\ \text{s.t.} \quad & \nabla \boldsymbol{h}(z_k)^\mathsf{T} \boldsymbol{p}_{z_k} + \boldsymbol{h}(z_k) = \boldsymbol{0} \end{aligned} \tag{5.32}$$

The optimal Lagrange multipliers of this QP directly yield $\boldsymbol{\lambda}_{k+1}^*$. In this QP, the term $\nabla_{zz}^2 L(z_k, \boldsymbol{\lambda}_k)$ contains local curvature information of both J and $\boldsymbol{h}$, while the term $\nabla \boldsymbol{h}(z_k)^\mathsf{T} \boldsymbol{p}_{z_k} + \boldsymbol{h}(z_k)$ is the linearization of the equality constraints $\boldsymbol{h}(z) = \boldsymbol{0}$ at z_k. The first-order necessary optimality conditions of this QP result in

$$\begin{aligned} \nabla L_{QP,k}(\boldsymbol{p}_{z_k}^*, \boldsymbol{\lambda}_{k+1}^*) &= \begin{bmatrix} \nabla_{\boldsymbol{p}_{z_k}} L_{QP,k}(\boldsymbol{p}_{z_k}^*, \boldsymbol{\lambda}_{k+1}^*) \\ \nabla_{\boldsymbol{\lambda}_{k+1}} L_{QP,k}(\boldsymbol{p}_{z_k}^*, \boldsymbol{\lambda}_{k+1}^*) \end{bmatrix} \\ &= \begin{bmatrix} \nabla_{zz}^2 L(z_k, \boldsymbol{\lambda}_k) \boldsymbol{p}_{z_k}^* + \nabla J(z_k) + \nabla \boldsymbol{h}(z_k) \boldsymbol{\lambda}_{k+1}^* \\ \nabla \boldsymbol{h}(z_k)^\mathsf{T} \boldsymbol{p}_{z_k}^* + \boldsymbol{h}(z_k) \end{bmatrix} = \boldsymbol{0} \end{aligned} \tag{5.33}$$

This equation can be reformulated to

$$\begin{bmatrix} \nabla^2_{zz} L(z_k, \boldsymbol{\lambda}_k) & \nabla \boldsymbol{h}(z_k) \\ \nabla \boldsymbol{h}(z_k)^\top & \mathbf{0} \end{bmatrix} \begin{bmatrix} \boldsymbol{p}^*_{z_k} \\ \boldsymbol{\lambda}^*_{k+1} \end{bmatrix} = - \begin{bmatrix} \nabla J(z_k) \\ \boldsymbol{h}(z_k) \end{bmatrix} \tag{5.34}$$

This update equation is equal to the one derived by the application of the Newton method on the KKT condition. The second derivation leads to a quadratic program (QP) that is solved in each iterate. This is where the name Sequential Quadratic Program originates.

In conclusion, for an equality-constrained optimization problem, the Newton method applied on the KKT condition is the same as the SQP with the characteristic of exact Hessian. The general approach for the SQP method within a line-search framework is as follows:

1. Set the index of iteration $k = 0$.
2. Choose the initial guess $z_0, \boldsymbol{\lambda}_0$.
3. Compute necessary values for QP approximation, i.e. $\nabla^2_{zz} L(z_k, \boldsymbol{\lambda}_k)$, $\nabla J(z_k), \nabla \boldsymbol{h}(z_k)^\top, \boldsymbol{h}(z_k)$.
4. Determine search direction $\boldsymbol{p}^*_{z_k}$ and $\boldsymbol{\lambda}^*_{k+1}$ by solving the QP (5.32).
5. Determine step size $\alpha_k > 0$.
6. Set the new iterate $z_{k+1} = z_k + \alpha_k \boldsymbol{p}^*_{z_k}$, $\boldsymbol{\lambda}_{k+1} = (1 - \alpha_k)\boldsymbol{\lambda}_k + \alpha_k \boldsymbol{\lambda}^*_{k+1}$.
7. Check stop criterion, e.g. $||\nabla_z L(z_{k+1}, \boldsymbol{\lambda}_{k+1})|| \leq \epsilon_L$ and $||\boldsymbol{h}(z_{k+1})|| \leq \epsilon_h$.
8. Start new iteration, $k := k + 1$, and continue with step 3.

Example 5.1 (*Sequential Quadratic Programming*) For demonstrating the working principle of the SQP algorithm, the following optimization problem is investigated. It consists of a quadratic cost function $J(z)$ with the optimization variable $z = [z_1, z_2]^\top$.

$$J(z) = 1.5z_1^2 + 2z_2^2 - 1.5z_1z_2 + 2z_1 - z_2 \tag{5.35}$$

For the optimization problem, a quadratic constraint function $h(z)$ is considered.

$$h(z) = z_1^2 + z_2^2 - 0.5z_1 - z_2 + 0.1 \tag{5.36}$$

The overall optimization problem is given by

$$\begin{aligned} \min_{z_1, z_2} \quad & J(z) \\ \text{s.t.} \quad & h(z) = 0 \end{aligned} \tag{5.37}$$

Figure 5.2 shows the path starting from three initial guesses z_0 when using the SQP algorithm with exact Hessian calculation, i.e. no additional regularization is used. The initial guess for the Lagrange multipliers is always set to $\lambda_0 = 0$, and the step size is chosen constant as $\alpha = 0.5$. From all three initial guesses, the algorithm converges to the solution of the optimization problem. All initial guesses are in the

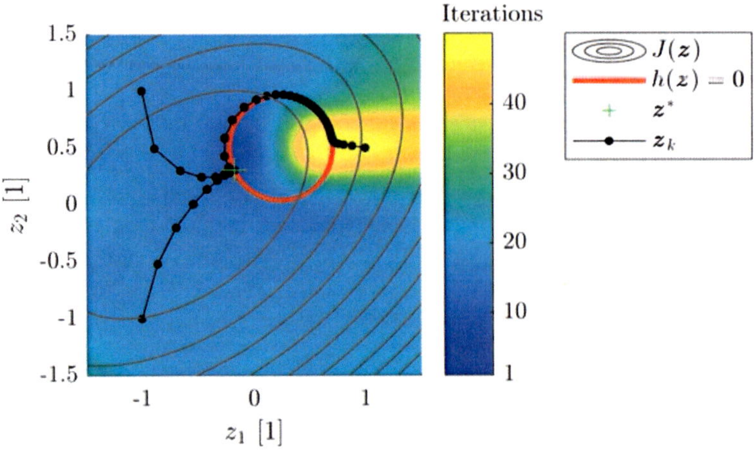

Fig. 5.2 Graphical representation of the iterative solution of Example 5.1 with the SQP method

infeasible region. According to the function principle of the SQP algorithm, the iterates move toward the KKT point, while the intermediate iterates are often in the infeasible region of the original optimization problem.

Besides the paths from the three starting points, also the speed of convergence depending on the initial point is shown in the figure. It is quantified by the number of iterations needed until reaching the KKT point. The number of iterations strongly depends on the initial guess. If the initial guess is close to the solution, only very few iterations are needed. In contrast, from remote initial points, much more iterations are necessary.

5.2.2.2 NLP with Equality and Inequality Constraints

The additional treatment of inequality constraints is discussed now. The following general NLP optimization problem is investigated:

$$\begin{aligned} \min_{z \in \mathbb{R}^n} \quad & J(z) \\ \text{s.t.} \quad & h_i(z) = \mathbf{0}, \quad i = 1, \ldots, p, \\ & \boldsymbol{g}_i(z) \leq \mathbf{0}, \quad i = 1, \ldots, q \end{aligned} \tag{5.38}$$

The corresponding Lagrange function is denoted by $L(z, \boldsymbol{\lambda}, \boldsymbol{\mu})$.

$$\begin{aligned} L(z, \boldsymbol{\lambda}, \boldsymbol{\mu}) &= J(z) + \sum_{i=1}^{p} \lambda_i h_i(z) + \sum_{i=1}^{q} \mu_i \boldsymbol{g}_i(z) \\ &= J(z) + \boldsymbol{\lambda}^{\mathsf{T}} \boldsymbol{h}(z) + \boldsymbol{\mu}^{\mathsf{T}} \boldsymbol{g}(z) \end{aligned} \tag{5.39}$$

For this optimization problem, the KKT conditions can be set up as follows:

$$\begin{aligned}\nabla_z L(z^*, \boldsymbol{\lambda}^*, \boldsymbol{\mu}^*) &= 0 \\ h_i(z^*) &= 0, \ \forall i = 1, \dots, p \\ g_i(z^*) &\leq 0, \ \forall i = 1, \dots, q \\ \mu_i^* &\geq 0, \ \forall i = 1, \dots, q \\ \mu_i^* g_i(z^*) &= 0, \ \forall i = 1, \dots, q\end{aligned} \tag{5.40}$$

The complementary slackness conditions imply a non-smooth solution manifold; as for $g_i(z^*) < 0$, the corresponding Lagrange multiplier has to be $\mu_i^* = 0$ and for $g_i(z^*) = 0$, the Lagrange multiplier can take values $\mu_i^* \geq 0$. Due to this non-smoothness, the Newton method for root-finding cannot be applied to the KKT conditions. However, a quadratic approximation of the Lagrange function can be set up as in the case of the equality constrained NLP. This approximation can be transformed into a corresponding QP again [10]. It is defined by

$$\begin{aligned}\min_{\boldsymbol{p}_{z_k}} \quad & \frac{1}{2}\boldsymbol{p}_{z_k}^\mathsf{T} \nabla_{zz}^2 L(z_k, \boldsymbol{\lambda}_k, \boldsymbol{\mu}_k)\boldsymbol{p}_{z_k} + \boldsymbol{p}_{z_k}^\mathsf{T} \nabla J(z_k) \\ \text{s.t.} \quad & \nabla \boldsymbol{h}(z_k)^\mathsf{T} \boldsymbol{p}_{z_k} + \boldsymbol{h}(z_k) = \boldsymbol{0}, \\ & \nabla \boldsymbol{g}(z_k)^\mathsf{T} \boldsymbol{p}_{z_k} + \boldsymbol{g}(z_k) \leq \boldsymbol{0}\end{aligned} \tag{5.41}$$

The solution of this QP results in the search direction $\boldsymbol{p}_{z_k}^*$. The optimal Lagrange multipliers of this QP correspond to $\boldsymbol{\lambda}_{k+1}^*$ and $\boldsymbol{\mu}_{k+1}^*$. The cost function term $\nabla_{zz}^2 L(z_k, \boldsymbol{\lambda}_k)$ contains local curvature information of J, $\boldsymbol{h}$ and $\boldsymbol{g}$. The constraints consider the linearization of the equality and the inequaltiy constraints at z_k. With the values $\boldsymbol{p}_{z_k}^*$, $\boldsymbol{\lambda}_{k+1}^*$, and $\boldsymbol{\mu}_{k+1}^*$, the next iterate can be calculated as

$$\begin{aligned}z_{k+1} &= z_k + \alpha_k \boldsymbol{p}_{z_k}^* \\ \boldsymbol{\lambda}_{k+1} &= (1 - \alpha_k)\boldsymbol{\lambda}_k + \alpha_k \boldsymbol{\lambda}_{k+1}^* \\ \boldsymbol{\mu}_{k+1} &= (1 - \alpha_k)\boldsymbol{\mu}_k + \alpha_k \boldsymbol{\mu}_{k+1}^*\end{aligned} \tag{5.42}$$

For a full step, the step size can be set to $\alpha_k = 1$. The overall procedure can be summarized as follows:

1. Set the index of iteration $k = 0$.
2. Choose the initial guess z_0, $\boldsymbol{\lambda}_0$, $\boldsymbol{\mu}_0$.
3. Compute necessary values for QP approximation, i.e. $\nabla_{zz}^2 L(z_k, \boldsymbol{\lambda}_k, \boldsymbol{\mu}_k)$, $\nabla J(z_k)$, $\nabla \boldsymbol{h}(z_k)^\mathsf{T}$, $\boldsymbol{h}(z_k)$, $\nabla \boldsymbol{g}(z_k)^\mathsf{T}$, $\boldsymbol{g}(z_k)$.
4. Determine the search direction $\boldsymbol{p}_{z_k}^*$, $\boldsymbol{\lambda}_{k+1}^*$, and $\boldsymbol{\mu}_{k+1}^*$ by solving the QP.
5. Determine step size $\alpha_k > 0$.
6. Set the new iterate $z_{k+1} = z_k + \alpha_k \boldsymbol{p}_{z_k}^*$, $\boldsymbol{\lambda}_{k+1} = (1 - \alpha_k)\boldsymbol{\lambda}_k + \alpha_k \boldsymbol{\lambda}_{k+1}^*$, $\boldsymbol{\mu}_{k+1} = (1 - \alpha_k)\boldsymbol{\mu}_k + \alpha_k \boldsymbol{\mu}_{k+1}^*$.

7. Check stop criterion, e.g. $\left|\left|\nabla_z L(z_{k+1}, \boldsymbol{\lambda}_{k+1}, \boldsymbol{\mu}_{k+1})\right|\right| \leq \epsilon_L$, $||\boldsymbol{h}(z_{k+1})|| \leq \epsilon_h$, $||\max\{\mathbf{0}, \boldsymbol{g}(z_{k+1})\}|| \leq \epsilon_g$, $\left|\left|\min\{\mathbf{0}, \boldsymbol{\mu}_{k+1}\}\right|\right| \leq \epsilon_\mu$, and $\left|\left|\boldsymbol{\mu}_{i,k+1} \boldsymbol{g}_i(z_{k+1})\right|\right| \leq \epsilon_{\mu g}$.
8. Start new iteration, $k := k + 1$, and continue with step 3.

Example 5.2 (*SQP with inequality constraints*) In the following, the solution of an inequality-constrained optimization problem with the SQP method is investigated. Specifically, one iteration step of the SQP algorithm is detailed. In each iteration step, one QP problem is solved which is illustrated in this example. The optimization problem is given by

$$\begin{aligned} \min_{z \in \mathbb{R}} \quad & J(z) = (1 - z)^2 + (1 - z^2)^2 \\ \text{s.t.} \quad & g(z) = (z - 0.1)^2 - 0.1 \leq 0 \end{aligned} \tag{5.43}$$

In every iteration step of the SQP algorithm, a QP of the following form is solved:

$$\begin{aligned} \min_{z \in \mathbb{R}} \quad & J_{sqp}(z) \\ \text{s.t.} \quad & g_{lin}(z) \leq 0 \end{aligned} \tag{5.44}$$

Three figures are used to illustrate the SQP algorithm at one iteration step, i.e. iteration step k. They show the cost function $J_{sqp}(z)$, the constraint function $g_{lin}(z)$, and the resulting next iterate of z_{k+1}.

In Fig. 5.3, the original cost function J is plotted along with the inequality constraint function $g(z)$. Furthermore, the cost function J_{sqp} is depicted, which is the approximated cost function resulting from the SQP algorithm at iteration step k. With μ_k being the iterate of the Lagrange multiplier of the inequality constraint, it is calculated by

$$J_{sqp}(z) = \frac{1}{2}(\nabla^2_{zz} J(z_k) + \mu_k \nabla^2_{zz} g(z_k))(z - z_k)^2 + \nabla_z J(z_k)(z - z_k) + J(z_k) \tag{5.45}$$

For comparison, the second-order Taylor approximation J_{quad} of the original cost function J at the point z_k is illustrated as well.

$$J_{quad}(z) = \frac{1}{2}\nabla^2_{zz} J(z_k)(z - z_k)^2 + \nabla_z J(z_k)(z - z_k) + J(z_k) \tag{5.46}$$

This shows the difference between the pure quadratic approximation of the cost function and the cost function J_{sqp} resulting from the SQP algorithm. Using

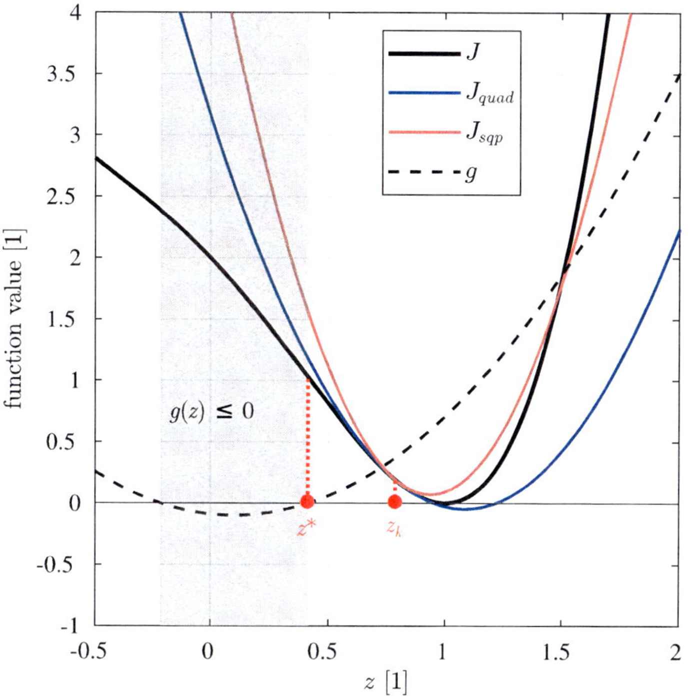

Fig. 5.3 Approximated cost function with SQP algorithm for Example 5.2

$$\nabla_z J(z) = 4z^3 - 2z - 2 \tag{5.47}$$

$$\nabla_{zz}^2 J(z) = 12z^2 - 2 \tag{5.48}$$

$$\nabla_z \boldsymbol{g}(z) = 2z - 0.2 \tag{5.49}$$

$$\nabla_{zz}^2 \boldsymbol{g}(z) = 2 \tag{5.50}$$

the approximated cost functions result to be

$$J_{quad}(z) = 2.73z^2 - 5.9z + 3.16 \tag{5.51}$$

$$J_{sqp}(z) = 5.41z^2 - 10.16z + 4.83 \tag{5.52}$$

Clearly, there is a difference between these two functions, as the function J_{sqp} takes additional curvature information of the constraint into account. The current iterate for the minimizer is $z_k = 0.79$ and the current estimate for the Lagrange multiplier is $\mu_k = 2.69$.

Figure 5.4 shows the function $\boldsymbol{g}_{lin}$ which results from the linearization of the constraint $\boldsymbol{g}$ at the point z_k. Additionally, the feasible region of the original optimization problem is shown along with the feasible region of the approximated QP problem.

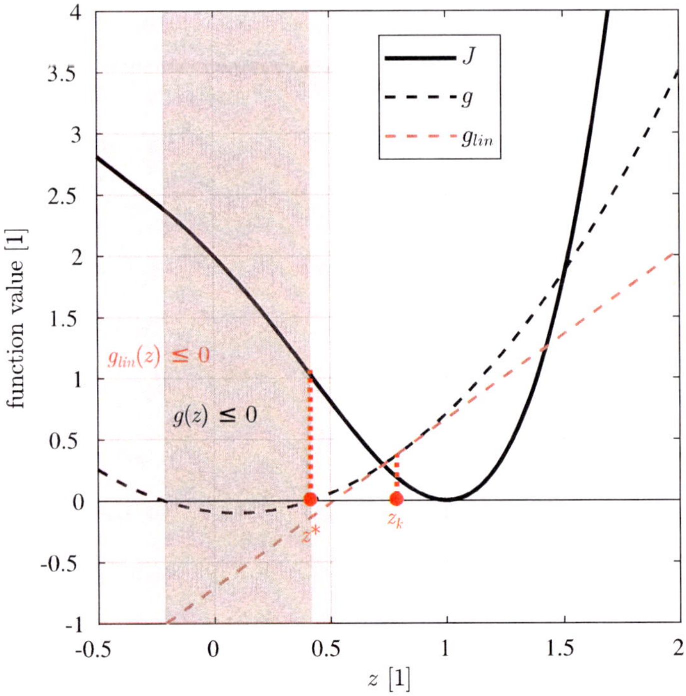

Fig. 5.4 Approximated constraint function with SQP algorithm for Example 5.2

$$\begin{aligned} \boldsymbol{g}_{lin}(z) &= \nabla_z \boldsymbol{g}(z_k)(z - z_k) + \boldsymbol{g}(z_k) \\ &= 1.38z - 0.71 \end{aligned} \tag{5.53}$$

Finally, Fig. 5.5 shows the cost function J_{sqp} along with the constraint function $\boldsymbol{g}_{lin}$ and the resulting feasible region. The solution of (5.44) results in the next iterate z_{k+1} when the full step is used. The figure shows that the iterate z_{k+1} is already very close to the minimizer of the original NLP z^*. For further convergence, within the SQP method, the point z_{k+1} is taken as a new basis for generating an updated quadratic cost function and linear constraint function.

5.2.3 Approximation of Hessian Matrix

Solving an NLP with SQP requires repeated calculations of the Hessian matrix $\nabla^2_{zz} L(z_k, \boldsymbol{\lambda}_k, \boldsymbol{\mu}_k)$ at each SQP step k. The calculation of the Hessian matrix can become computationally expensive. A reduction of the computation time thus can be crucial for real-time applications. To combine a fast computation per iteration step with a fast convergence speed, the Hessian matrix can be approximated using

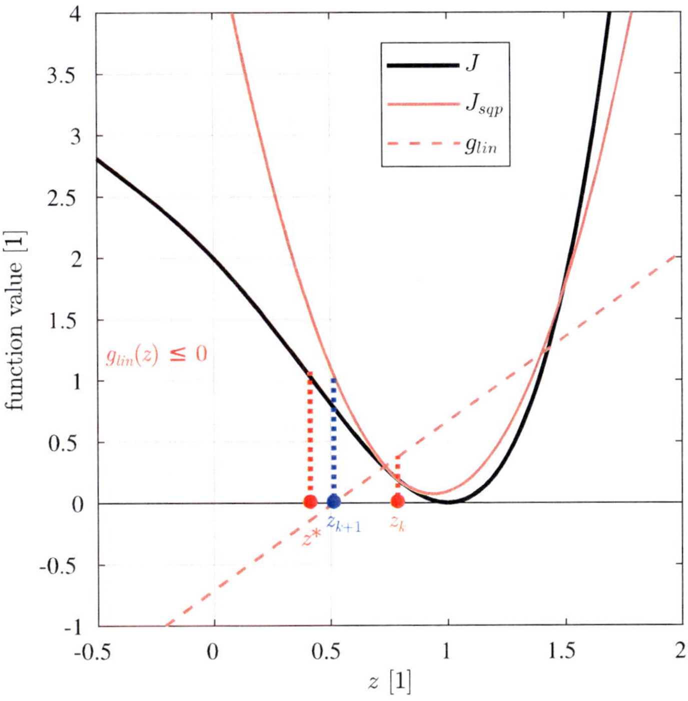

Fig. 5.5 Solution of the QP subproblem within the SQP algorithm for Example 5.2

only first-order derivative information, i.e. $\nabla^2_{zz} L \approx M$. The methods using approximated Hessian matrices are called quasi-Newton methods. For instance, the *Broyden–Fletcher–Goldfarb–Shanno* (BFGS) algorithm is a secant approximation using only Jacobian information [3]. Another commonly used approximation is the *Gauss–Newton* algorithm, which is efficient for nonlinear least-squares problems [14]. Thus, it is well suited for optimization problems occurring in engine NMPC. The cost functions investigated are of the following form:

$$J(z) = \frac{1}{2}||\boldsymbol{f}(z)||^2 = \frac{1}{2}\boldsymbol{f}(z)^{\mathsf{T}}\boldsymbol{f}(z) = \frac{1}{2}\sum_{i=1}^{m} f_i^2(z), \text{ with } \boldsymbol{f} : \mathbb{R}^n \rightarrow \mathbb{R}^m \quad (5.54)$$

The corresponding optimization problem is present as

$$\begin{aligned} \min_{z \in \mathbb{R}^n} \quad & J(z) \\ \text{s.t.} \quad & h_i(z) = \mathbf{0}, \quad i = 1, \ldots, p, \\ & \boldsymbol{g}_i(z) \leq \mathbf{0}, \quad i = 1, \ldots, q \end{aligned} \quad (5.55)$$

For the generalized Gauss–Newton approximation, in each SQP iteration, the following QP problem is solved to determine the search direction p_{z_k}.

$$\begin{aligned}
\min_{p_{z_k}} \quad & \frac{1}{2} p_{z_k}^{\mathsf{T}} \nabla f(z_k) \nabla f(z_k)^{\mathsf{T}} p_{z_k} + p_{z_k}^{\mathsf{T}} \nabla f(z_k) f(z_k) \\
\text{s.t.} \quad & \nabla h(z_k)^{\mathsf{T}} p_{z_k} + h(z_k) = \mathbf{0}, \\
& \nabla g(z_k)^{\mathsf{T}} p_{z_k} + g(z_k) \leq \mathbf{0}
\end{aligned} \tag{5.56}$$

Compared to the SQP iteration with exact Hessian as given in (5.41), only the calculation of the Hessian matrix changes. The gradient calculation of the cost function and the constraints are unaltered. For the calculation of the Hessian matrix, all the second-order derivative terms are neglected. On top of that, there is no information of the Lagrange multipliers λ_k, μ_k entering the Hessian approximation. Hence, they do not need to be calculated. The Hessian is approximated by $M(z_k)$:

$$M(z_k) = \nabla f(z_k) \nabla f(z_k)^{\mathsf{T}} \tag{5.57}$$

Effectively, the Gauss–Newton method uses a linear approximation of the function $f(z)$ within the cost function:

$$\begin{aligned}
J_{GN}(p_{z_k}) &= \frac{1}{2} ||\nabla f(z_k) p_{z_k} + f(z_k)||^2 \\
&= \frac{1}{2} (\nabla f(z_k) p_{z_k} + f(z_k))^{\mathsf{T}} (\nabla f(z_k) p_{z_k} + f(z_k))
\end{aligned} \tag{5.58}$$

The approximation of the Hessian matrix by the Gauss–Newton method has another advantage, besides the reduced computation time. The Hessian approximation is guaranteed to be positive (semi)definite, which is not true for the calculation of the exact Hessian, where a regularization may be required to assure its positive definiteness.

The Gauss–Newton approximation works well if all values of $\nabla^2 f_i(z)$ are small, which means that all functions $f_i(z)$ are close to linear. The approximation is exact if the functions $f_i(z)$ are linear. Additionally, it works well if the residuals $f_i(z)$ are small, which means that the iterate is close to the optimum and $J(z_k^*) \approx 0$ [14].

In contrast to the quadratic convergence of the exact Newton method, the Gauss–Newton method only converges linearly [14]. However, each single iteration step can be computed much faster, which often results in an overall reduction of computation time. Therefore, it can be helpful to trade convergence rate versus computational complexity per iteration.

5.3 Discretization of the OCP via Shooting Methods

There are various strategies to solve the OCP shown in (5.2a)–(5.2e). One possibility is to set up the Hamilton–Jacobi–Bellman (HJB) equation, which results in a partial differential equation (PDE). The HJB poses a sufficient optimality condition for global optimality. This PDE has to be solved in order to find the minimizer. However, the approach can only be used for very small problems in an online context [14]. Besides the HJB, there are the indirect and the direct methods. More details about the various strategies can be found in [14]. For indirect methods, the problem is optimized first and then discretized. The indirect methods are based on Pontryagin's Maximum Principle which poses a sufficient condition for local optimality. A two-point boundary-value problem is solved for finding the optimal solution. For direct methods, the order is reversed, the problem is discretized first, and then optimized. For NMPC, they play a dominant role as they allow to achieve online solutions within the critical timescales of real-time applications. Additionally, the treatment of constraints on the system states is much easier compared to the other approaches. In the following, the direct methods are further investigated.

For the direct methods, the infinite-dimensional OCP is approximated by a finite-dimensional NLP via discretization. Methods that are applied for discretization include *single shooting, multiple shooting*, and *collocation* [14]. In the following, the single shooting and multiple shooting methods are introduced. Necessary background for these discretization schemes are numerical methods for the simulation of ordinary differential equations (ODE).

5.3.1 Numerical Methods for Simulation

For engine control problems, typically the system dynamics are described by nonlinear ODEs. Within NMPC the system states need to be predicted, for which numerical simulation methods are used. They are also called integration methods. The goal of numerical simulation is to compute the trajectory of $\boldsymbol{x}(t)$ which, starting from the initial values, satisfies as best possible the system dynamics given by the ODE. This problem is also referred to as an initial value problem. The system dynamics and the initial value condition can be described by

$$\dot{\boldsymbol{x}}(t) = \boldsymbol{f}(\boldsymbol{x}(t), \boldsymbol{u}(t), t) \tag{5.59}$$

$$\boldsymbol{x}(t_0) = \boldsymbol{x}_0 \tag{5.60}$$

The trajectory $\boldsymbol{u}(t)$ can be assumed to be part of the function $\boldsymbol{f}$. Thus, the following differential equation with simplified notation is examined in the remainder of this section.

$$\dot{\boldsymbol{x}}(t) = \boldsymbol{f}(\boldsymbol{x}(t), t) \tag{5.61}$$

$$\boldsymbol{x}(t_0) = \boldsymbol{x}_0 \tag{5.62}$$

In general, for nonlinear ODEs it is complicated – or even impossible – to find an analytic solution for the trajectory of system states $\boldsymbol{x}(t)$. Instead, numerical integration schemes are used to find approximate solutions of the system states $\boldsymbol{x}$ at a discrete-time grid $k+1, k+2, k+3, \ldots$ which correspond to the time points $t_{k+1} = (k+1)T,\ t_{k+2} = (k+2)T,\ t_{k+3} = (k+3)T, \ldots$ with T being the integration step size.

5.3.1.1 Characteristics of Simulation Methods

There exist many methods for numerical simulation. A good overview of this topic, including proofs for the convergence rate, is given in [6]. In the following, various properties and possibilities to classify numerical simulation methods are detailed.

One-Step Versus Multi-step Methods
One-step methods only use the values $\boldsymbol{x}_k$ at the discrete time instance k to calculate $\boldsymbol{x}_{k+1}$. Multi-step methods also use previous values $\boldsymbol{x}_k, \boldsymbol{x}_{k-1}, \ldots$ at the discrete time instance k for the calculation of $\boldsymbol{x}_{k+1}$. One example for a one-step method is the *Runge–Kutta 4* method. Examples for multi-step schemes are the *Adams–Bashforth* and the *Adams–Moulton* methods.

Implicit Versus Explicit Methods
Explicit methods use for the calculation of $\boldsymbol{x}_{k+1}$ only derivatives at previous values of $\boldsymbol{x}$, e.g. $\boldsymbol{x}_k, \boldsymbol{x}_{k-1}, \boldsymbol{x}_{k-2}$. The function $\boldsymbol{f}_{dis,ex}$ shall represent the discretized system dynamics gained by numerical integration with an explicit method. Using $\boldsymbol{f}_{dis,ex}$, the explicit methods can be characterized by

$$\boldsymbol{x}_{k+1} = \boldsymbol{f}_{dis,ex}(\boldsymbol{x}_i, t_i), \quad i < k+1 \tag{5.63}$$

In implicit methods, $\boldsymbol{x}_{k+1}$ depends also on the derivative at $\boldsymbol{x}_{k+1}$. To determine the values $\boldsymbol{x}_{k+1}$, an iterative solution method as the Newton method has to be used. Implicit methods are especially suited for stiff systems and systems that are described by *differential–algebraic equations* (DAE). The function $\boldsymbol{f}_{dis,im}$ shall represent the discretized system dynamics gained by numerical integration with an implicit method. Implicit methods can be characterized by

$$\boldsymbol{x}_{k+1} = \boldsymbol{f}_{dis,im}(\boldsymbol{x}_i, t_i), \quad i \leq k+1 \tag{5.64}$$

An example is the Euler method which exists as an explicit and also an implicit calculation scheme, as shown below.

Convergence Order of the Simulation Method

An important characteristic to distinguish the various numerical methods is their order p. It furnishes information about the local truncation error. For explicit one-step methods, the simulated value after one integration step at $k+1$ can be described by

$$\boldsymbol{x}_{k+1} = \boldsymbol{f}_{dis}(\boldsymbol{x}_k, t_k) \tag{5.65}$$

The trajectory of the exact solution shall be denoted by $\boldsymbol{x}'(t)$. In this case, the local truncation error is the difference between the exact solution $\boldsymbol{x}'(t_{k+1})$ and the approximate solution $\boldsymbol{x}_{k+1}$. The values $\boldsymbol{x}_{k+1}$ are gained by the numerical integration $\boldsymbol{x}_{k+1} = \boldsymbol{f}_{dis}(\boldsymbol{x}'(t_k), t_k)$ starting from the exact values $\boldsymbol{x}'(t_k)$. The local truncation error follows as

$$e(t_{k+1}) = ||\boldsymbol{x}'(t_{k+1}) - \boldsymbol{x}_{k+1}|| \tag{5.66}$$

With the integration step size $T = t_{k+1} - t_k$, a numerical method is called convergent when the simulated values approach the exact solution for $T \to 0$. The numerical integration scheme is said to have order p if the local truncation error satisfies (5.67).

$$\lim_{T \to 0} e(t_{k+1}) = \mathcal{O}(T^{p+1}) \tag{5.67}$$

Higher-order techniques thus have typically smaller errors for the same integration step size.

5.3.1.2 Runge–Kutta Methods

A well-established family of numerical simulation methods are the *Runge–Kutta* methods, which include explicit and implicit one-step methods for numerical simulation. The general s-stage explicit Runge–Kutta method is defined by

$$\begin{aligned}
\boldsymbol{k}_1 &= \boldsymbol{f}(\boldsymbol{x}_k, t_k) \\
\boldsymbol{k}_2 &= \boldsymbol{f}(\boldsymbol{x}_k + a_{21} T \boldsymbol{k}_1, t_k + c_2 T) \\
\boldsymbol{k}_3 &= \boldsymbol{f}(\boldsymbol{x}_k + a_{31} T \boldsymbol{k}_1 + a_{32} T \boldsymbol{k}_2, t_k + c_3 T) \\
&\vdots \\
\boldsymbol{k}_s &= \boldsymbol{f}(\boldsymbol{x}_k + a_{s1} T \boldsymbol{k}_1 + a_{s2} T \boldsymbol{k}_2 + \cdots + a_{s,s-1} T \boldsymbol{k}_{s-1}, t_k + c_s T) \\
\boldsymbol{x}_{k+1} &= \boldsymbol{x}_k + T \sum_{i=1}^{s} b_i \boldsymbol{k}_i
\end{aligned} \tag{5.68}$$

In order to describe a specific Runge–Kutta method, the number of stages s has to be given along with the constants a_{ij} with $1 \leq j < i \leq s$, b_i with $i = 1, \ldots, s$ and c_i with $i = 2, \ldots, s$.

One very common and simple simulation method is the *explicit Euler* method, which corresponds to a one-stage Runge–Kutta scheme. The explicit Euler method has order $p = 1$ and is defined as

$$\boldsymbol{x}_{k+1} = \boldsymbol{f}_{Euler}(\boldsymbol{x}_k, t_k) = \boldsymbol{x}_k + T\boldsymbol{f}(\boldsymbol{x}_k, t_k) \tag{5.69}$$

Another very common method is the RK4 method, also called classical Runge–Kutta method, which has four stages. The RK4 method has order $p = 4$ and is defined by

$$\begin{aligned}
\boldsymbol{k}_1 &= \boldsymbol{f}(\boldsymbol{x}_k, t_k) \\
\boldsymbol{k}_2 &= \boldsymbol{f}\left(\boldsymbol{x}_k + \frac{T}{2}\boldsymbol{k}_1, t_k + \frac{T}{2}\right) \\
\boldsymbol{k}_3 &= \boldsymbol{f}\left(\boldsymbol{x}_k + \frac{T}{2}\boldsymbol{k}_2, t_k + \frac{T}{2}\right) \\
\boldsymbol{k}_4 &= \boldsymbol{f}\left(\boldsymbol{x}_k + T\boldsymbol{k}_3, t_k + T\right) \\
\boldsymbol{x}_{k+1} &= \boldsymbol{x}_k + \frac{T}{6}(\boldsymbol{k}_1 + 2\boldsymbol{k}_2 + 2\boldsymbol{k}_3 + \boldsymbol{k}_4)
\end{aligned} \tag{5.70}$$

The Runge–Kutta family also contains implicit simulation methods. The simplest one is the *implicit Euler* method, which consists of one stage. It is defined by

$$\boldsymbol{x}_{k+1} = \boldsymbol{x}_k + T\boldsymbol{f}(\boldsymbol{x}_{k+1}, t_k + T) \tag{5.71}$$

As the integration scheme depends on $\boldsymbol{x}_{k+1}$, the solution for $\boldsymbol{x}_{k+1}$ is implicitly defined. The solution thus has to be determined iteratively. The equation can be rearranged to be a root-finding problem. For determining $\boldsymbol{x}_{k+1}$, the Newton method can be applied to this equation:

$$\boldsymbol{0} = \boldsymbol{x}_k + T\boldsymbol{f}(\boldsymbol{x}_{k+1}, t_k + T) - \boldsymbol{x}_{k+1} \tag{5.72}$$

A detailed treatment of implicit methods within NMPC can be found in [12, 13]. Also higher-order implicit Runge–Kutta schemes can be realized. The general s-stage implicit Runge–Kutta methods are defined by

$$\begin{aligned}
\boldsymbol{k}_1 &= f\left(\boldsymbol{x}_k + T\sum_{j=1}^{s} a_{1j}\boldsymbol{k}_j, t_k + \boldsymbol{c}_1 T\right) \\
&\vdots \\
\boldsymbol{k}_s &= f\left(\boldsymbol{x}_k + T\sum_{j=1}^{s} a_{sj}\boldsymbol{k}_j, t_k + \boldsymbol{c}_s T\right) \\
\boldsymbol{x}_{k+1} &= \boldsymbol{x}_k + T\sum_{i=1}^{s} b_i\boldsymbol{k}_i
\end{aligned} \tag{5.73}$$

Example 5.3 (*Numerical simulation methods*) The properties of the simulation methods are compared by the evaluation of an example. A simple, linear mass–spring–damper system is investigated, which is a damped harmonic oscillator. The ODE of the system is denoted as

$$\ddot{x} + \frac{b}{m}\dot{x} + \frac{k}{m}x = 0 \tag{5.74}$$

In this system, x is the position of the mass, $m = 0.1$ is the mass of the pendulum, $b = 0.6$ is the friction coefficient, and $k = 100$ is the spring constant. For this linear system, the analytic solution for the position exists as

$$x'(t) = x_0 e^{-Dt}\left(\cos(\omega t) + \frac{D}{\omega}\sin(\omega t)\right) \tag{5.75}$$

where

$$D = \frac{b}{2m}, \qquad \omega_0 = \sqrt{\frac{k}{m}}, \qquad \omega = \sqrt{\omega_0^2 - D^2} \tag{5.76}$$

The conditions $x(0) = x_0 = 1$ and $\dot{x}(0) = 0$ are used as initial conditions and the step size of the integration is set to $T = 0.005s$. Figure 5.6 shows the solution gained with the Euler method, the RK4 method, and with the analytic solution.

The RK4 method is able to reproduce the analytic solution very accurately. The Euler method, in contrast, already shows a noticeable difference to the analytic solution, although the same step size is used. The increased accuracy of the RK4 method compared to the Euler method comes from the cost of higher computation times.

Within the NMPC context, the choice of the integration scheme is very important. On the one hand, an accurate simulation of the system trajectory has to be ensured. If this is not the case, even for a setup without a model–plant mismatch, the predicted system behavior deviates significantly from the real system behavior. On the other hand, the computational demand is very critical. As depicted, there is a trade-off between the accuracy of the simulation and the computation demands. Both factors are affected by the integration step size and the order of the simulation method. If real-time feasibility is critical, the integration step size for the simulation cannot be chosen arbitrarily small in order to achieve high accuracy. Within real-time NMPC, it is often better to use a higher-order method such as RK4, allowing for larger integration step sizes, than the ones required, e.g. for the Euler method. As the system dynamics of engine control problems often exhibit stiff behavior, implicit methods often must be considered for real-world applications.

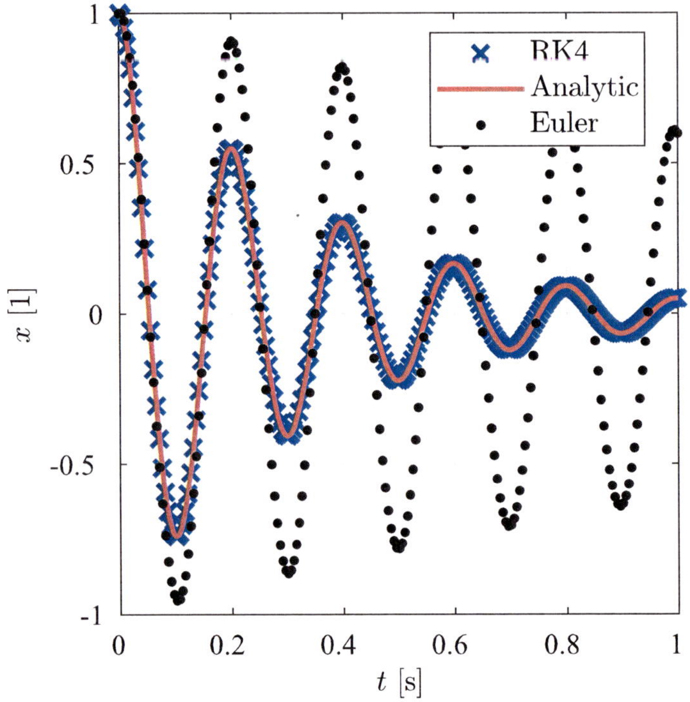

Fig. 5.6 Comparison of various numerical methods for simulation

5.3.2 *Discretization of Actuated Values, Cost Function, and Constraints*

The goal of the discretization is to translate the OCP to a suitable NLP formulation. For obtaining the NLP, the various components of the OCP have to be discretized, namely the actuated values, the cost function, the constraints, and the system model. In the following, all components with the exception of the system model are discussed.

In the context of NMPC, a suitable choice for the discretization of the trajectory of actuated values $\boldsymbol{u}(t)$ is the approximation by piecewise constant functions, i.e. zero order hold.

$$\boldsymbol{u}_k = \boldsymbol{u}(t) = const., \quad t_k \leq t < t_{k+1} \tag{5.77}$$

The entire horizon of actuation is discretized in a fixed grid with N equally long intervals, thus resulting in $\boldsymbol{u} = [\boldsymbol{u}_0, \boldsymbol{u}_1, \ldots, \boldsymbol{u}_{N-1}]$, as shown in Fig. 5.7.

The constraints can be discretized by only evaluating the constraints on some points. Usually, the same grid is taken as for discretization of the actuated values.

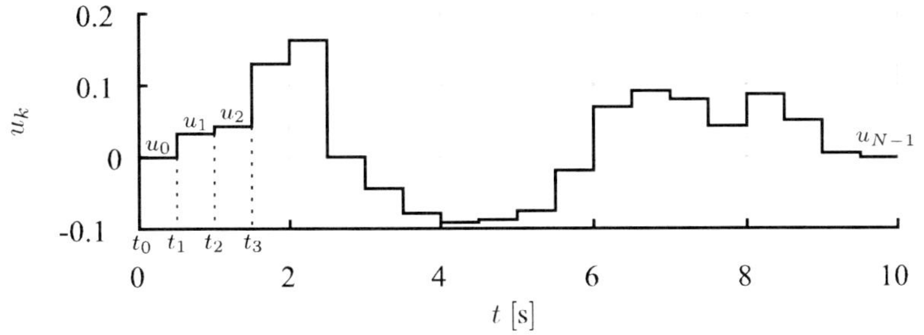

Fig. 5.7 Discretization of the trajectory of actuated values

This obviously shows the approximative behavior of the discretization in that the constraints are only enforced at some discrete points, but no longer in between.

$$\boldsymbol{h}(\boldsymbol{x}(t_f)) = \mathbf{0} \quad \rightarrow \quad \boldsymbol{h}(\boldsymbol{x}_N) = \mathbf{0} \tag{5.78}$$

$$\boldsymbol{g}(\boldsymbol{x}(t), \boldsymbol{u}(t)) \leq \mathbf{0} \quad \rightarrow \quad \boldsymbol{g}(\boldsymbol{x}_k, \boldsymbol{u}_k) \leq \mathbf{0}, \quad \text{k=0,...,N} \tag{5.79}$$

The *rectangular method* can be used for suitable discretization of the cost function, where

$$\min_{\boldsymbol{u}(t),\, \boldsymbol{x}(t)} \quad l_f\left(\boldsymbol{x}(t_f)\right) + \int_{t_0}^{t_f} l_s(\boldsymbol{x}(t), \boldsymbol{u}(t))dt \tag{5.80}$$

leads to

$$\min_{\boldsymbol{u}(\cdot),\, \boldsymbol{x}(\cdot)} \quad l_f(\boldsymbol{x}_N) + \sum_{k=0}^{N-1}(t_{k+1} - t_k) l_s(\boldsymbol{x}_k, \boldsymbol{u}_k) \tag{5.81}$$

However, often it is not meaningful to reproduce the cost function of the OCP in a discretized manner. Instead, a new cost function can be built that directly takes into account the discretized system dynamics.

The influence of the discretized actuated values $\boldsymbol{u}(\cdot)$ on the systems states $\boldsymbol{x}(\cdot)$ is found by numerical simulation methods such as RK4. As the standard engine control problems do not depend on the time t, the parameter t is omitted within the ODE to simplify notation. For a one-step explicit discretization method, the following notation can be used:

$$\dot{\boldsymbol{x}}(t) = \boldsymbol{f}(\boldsymbol{x}(t), \boldsymbol{u}(t)) \quad \longrightarrow \quad \boldsymbol{x}_{k+1} = \boldsymbol{f}_{dis}(\boldsymbol{x}_k, \boldsymbol{u}_k) \tag{5.82}$$

The discretization of the system dynamics with single shooting and multiple shooting is treated in the next sections.

5.3.3 Discretization via Single Shooting

Single shooting is a technique that originally was used for solving a boundary-value problem. Within the single shooting method, the relevant system states $\boldsymbol{x}_{k+i}$ with $i = 1, \ldots, N$ are expressed in terms of the discretized actuated values $\boldsymbol{u} = [\boldsymbol{u}_k, \boldsymbol{u}_{k+1}, \ldots, \boldsymbol{u}_{k+N-1}]$ and the initial condition $\boldsymbol{x}_0$. Using the numerical simulation from (5.82), it follows:

$$\begin{aligned} \boldsymbol{x}_{k+1} &= \boldsymbol{f}_{dis}(\boldsymbol{x}_0, \boldsymbol{u}_k) \\ \boldsymbol{x}_{k+2} &= \boldsymbol{f}_{dis}(\boldsymbol{f}_{dis}(\boldsymbol{x}_0, \boldsymbol{u}_k), \boldsymbol{u}_{k+1}) \\ &\vdots \end{aligned} \tag{5.83}$$

The system states at the time points $\boldsymbol{x}_{k+i}$ can be calculated to be recursively using the function $\boldsymbol{f}_{dis}$. The trajectory over the whole prediction horizon starting from $\boldsymbol{x}_0$ thus can be simulated. Hence, the system states in the cost function can be substituted by functions of the actuated values and the initial condition $\boldsymbol{x}_0$, such that they do not appear explicitly any more in the cost function.

Assume the following OCP is given:

$$\min_{\boldsymbol{x}(t),\, \boldsymbol{u}(t)} \quad J_{ocp}(\boldsymbol{x}(t), \boldsymbol{u}(t)) \tag{5.84a}$$

$$\text{s.t.} \quad \dot{\boldsymbol{x}}(t) = \boldsymbol{f}(\boldsymbol{x}(t), \boldsymbol{u}(t)), \forall t \in [t_0, t_f], \tag{5.84b}$$

$$\boldsymbol{x}(t_0) = \boldsymbol{x}_0, \tag{5.84c}$$

$$\boldsymbol{g}(\boldsymbol{x}(t), \boldsymbol{u}(t)) \leq \boldsymbol{0}, \forall t \in [t_0, t_f] \tag{5.84d}$$

Once the various components are discretized, an NLP results. It can be summarized by the following optimization problem:

$$\begin{aligned} \min_{\boldsymbol{u}(\cdot|k) \in \mathbb{R}^N} \quad & J_{NLP,SS}(\boldsymbol{u}(\cdot|k)) \\ \text{s.t.} \quad & \boldsymbol{g}(\boldsymbol{u}(k+i|k)) \leq \boldsymbol{0}, \quad i = 0, \ldots, N-1 \end{aligned} \tag{5.85}$$

In the case of single shooting, the optimization variables are given by the sequence of actuated values $\boldsymbol{u}(\cdot|k)$. The solution of this optimization problem can be calculated by any NLP solution method, e.g. the SQP method described above. The advantage of the single shooting method is the small number of optimization variables it requires. Only an initial guess for the actuated values $\boldsymbol{u}(\cdot|k)$ is needed, when using an SQP method, for instance. However, if initial guesses for the states $\boldsymbol{x}(\cdot|k)$ are available, they cannot be exploited. Another disadvantage results from the recursive calculation of the state trajectory. Due to the long simulation, the nonlinearity of the function for the system states increases as it is propagated through all the intermediate steps. This is especially critical for unstable systems.

Example 5.4 (*NMPC-based control of the Brusselator with single shooting*) The Brusselator shown in Example 4.5 is revisited to benchmark the NMPC algorithm. The system dynamics of the Brusselator are given by

$$\dot{x}_1 = -8.95x_1 + 3x_1^2x_2 + u \tag{5.86}$$

$$\dot{x}_2 = 4.95x_1 - 3x_1^2x_2 \tag{5.87}$$

The task of the controller is to stabilize the system at the stationary point $\boldsymbol{x}_{SS} = [0.425\ 3.882]^{\mathsf{T}}$. For this purpose, an NMPC algorithm is set up using the single shooting algorithm. The cost function is given by

$$J = \sum_{i=1}^{N} \left|\left|\boldsymbol{x}(k+i|k) - \boldsymbol{x}_{ref}\right|\right|_Q^2 + \sum_{i=0}^{N-1} \left|\left|u(k+i|k) - u_{ref}\right|\right|_R^2 \tag{5.88}$$

The cost function J takes the deviation of the actuated values and the system states to the references into account. The reference values are defined by the stationary point $\boldsymbol{x}_{ref} = [0.425\ 3.882]^{\mathsf{T}}$ and $u_{ref} = 1.7$ which is to be controlled. The horizon is chosen to be $N = 20$, while the sampling time is $T_s = 0.2s$.

The differential equation is discretized by numerical integration. As the sampling step is quite large, intermediate values are necessary for a stable simulation. Four integration steps per control interval thus are used:

$$T_{int} = \frac{T_s}{4} = 0.05s \tag{5.89}$$

Every intermediate integration value is obtained by the RK4 algorithm, which itself requires four function evaluations per step. The actuated values remain constant during each of the four integration steps. Each integration step is calculated by

$$\boldsymbol{x}_{k+1} = \boldsymbol{x}_k + \frac{T}{6}(k_1 + 2k_2 + 2k_3 + k_4) \tag{5.90}$$

The formulation for the calculation of k_i is shown in (5.70). The optimization problem takes additionally the following constraints into account:

$$0 \leq u(\cdot|k) \leq 5 \quad \text{and} \quad -2 \leq \Delta u(\cdot|k) \leq 2 \tag{5.91}$$

$$0 \leq x_1(\cdot|k) \leq 2 \quad \text{and} \quad 0 \leq x_2(\cdot|k) \leq 5 \tag{5.92}$$

After discretization, an NLP results that has the structure of (5.85). The initial states are given by $\boldsymbol{x}_0 = [1\ \ 2]^{\mathsf{T}}$. For the first step of the control algorithm, one NLP results that can be solved using an SQP solution method. The software package CasADi is used for the numerical solution of the NLP [2]. Figure 5.8 shows the converged solution for the resulting NLP with $\boldsymbol{x}_0$ as the initial state. The solution is obtained after 58 iterations.

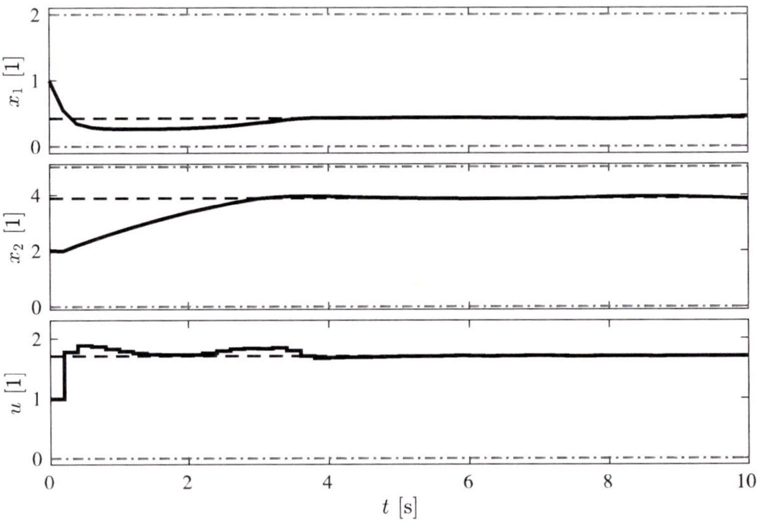

Fig. 5.8 Converged solution obtained with single shooting after 58 iterations

5.3.4 Discretization via Multiple Shooting

The discretization with multiple shooting dates back to the 1980s [5]. For the discretization with multiple shooting, the general procedure is the same as in the single shooting. The various components, the actuated values, the cost function, the constraints, and the system model are discretized in order to obtain a suitable NLP formulation. The difference in single shooting results in the way the information of the system dynamics enters the NLP formulation. With multiple shooting, the simulation and optimization are conducted simultaneously.

The ODE describing the system dynamics is discretized on each interval $[t_k, t_{k+1}]$ starting with initial values $\boldsymbol{x}_k$. The result of the simulation after one integration step is $\boldsymbol{x}_{k+1}$. This simulation is conducted for each interval. Using the numerical simulation from (5.82), it follows:

$$
\begin{aligned}
\boldsymbol{x}_{k+1} &= \boldsymbol{f}_{dis}(\boldsymbol{x}_k, \boldsymbol{u}_k) \\
\boldsymbol{x}_{k+2} &= \boldsymbol{f}_{dis}(\boldsymbol{x}_{k+1}, \boldsymbol{u}_{k+1}) \\
&\vdots
\end{aligned}
\tag{5.93}
$$

Figure 5.9 shows a sketch of the prediction calculated with this approach.

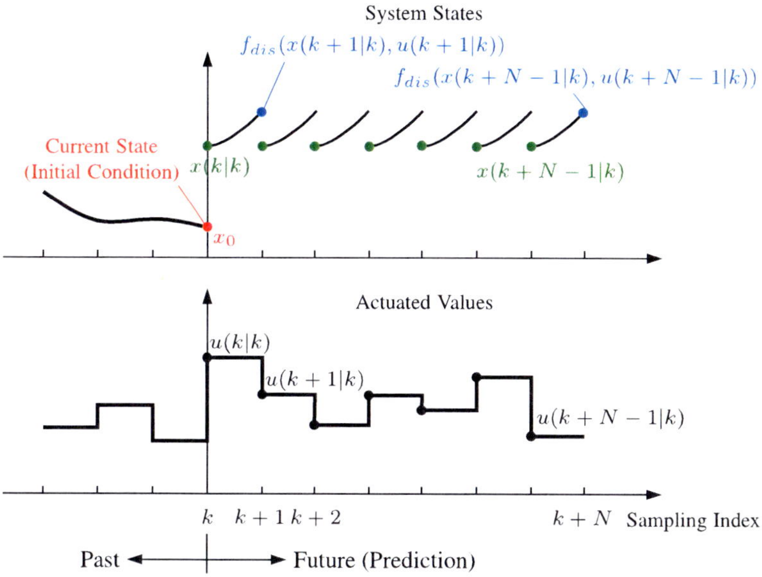

Fig. 5.9 The basic idea of multiple shooting

Assume following OCP is given:

$$\min_{\boldsymbol{x}(t),\,\boldsymbol{u}(t)} \quad J_{ocp}(\boldsymbol{x}(t), \boldsymbol{u}(t)) \tag{5.94a}$$

$$\text{s.t.} \quad \dot{\boldsymbol{x}}(t) = \boldsymbol{f}(\boldsymbol{x}(t), \boldsymbol{u}(t)), \forall t \in [t_0, t_f], \tag{5.94b}$$

$$\boldsymbol{x}(t_0) = \boldsymbol{x}_0, \tag{5.94c}$$

$$\boldsymbol{g}(\boldsymbol{x}(t), \boldsymbol{u}(t)) \leq \boldsymbol{0}, \forall t \in [t_0, t_f] \tag{5.94d}$$

In multiple shooting, the discretized system dynamics are added as equality constraints in the NLP. The transformation with multiple shooting yields the following NLP:

$$\begin{aligned}
\min_{\boldsymbol{x}(\cdot|k),\,\boldsymbol{u}(\cdot|k)} \quad & J_{NLP,MS}(\boldsymbol{x}(\cdot|k), \boldsymbol{u}(\cdot|k)) \\
\text{s.t.} \quad & \boldsymbol{x}(k|k) - \boldsymbol{x}_0 = 0, \\
& \boldsymbol{x}(k+i+1|k) = \boldsymbol{f}_{dis}(\boldsymbol{x}(k+i|k), \boldsymbol{u}(k+i|k), \quad i = 0, \ldots, N-1, \\
& \boldsymbol{g}(\boldsymbol{x}(k+i|k), \boldsymbol{u}(k+i|k)) \leq \boldsymbol{0}, \quad i = 0, \ldots, N
\end{aligned} \tag{5.95}$$

The system dynamics are no longer contained in the cost function; instead, they appear in the equality constraints. With multiple shooting discretization, more optimization variables are present than with single shooting, which increases the size of the optimization problem. The optimization variables are denoted by

$$z := (\boldsymbol{x}(k|k), \boldsymbol{x}(k+1|k), \ldots, \boldsymbol{x}(k+N|k), \boldsymbol{u}(k|k), \boldsymbol{u}(k+1|k), \ldots, \boldsymbol{u}(k+N-1|k)) \tag{5.96}$$

When using multiple shooting for NMPC, the whole formulation can be set up offline. During runtime of the process, in each time step an NLP with the same structure has to be solved. The NLP changes from time step to time step as the current system states are updated, and the reference might change. The resulting NLP can be solved with an iterative solution method, such as the SQP method. For the converged solution, a feasible trajectory of system states results, which implies that the discretized system dynamics are respected. However, for the initial guess, there can be a gap between the initial values and the simulated values, as depicted in Fig. 5.9. The gap can also occur for intermediate steps of the iterative solution.

Analytically, for multiple shooting the same optimization problem as in single shooting is solved, i.e. they have the same global minimizer. However, due to the differences in numerics, the intermediate steps are different and local solution methods such as SQP can converge to different solutions.

The difference in formulation causes additional characteristics, which make multiple shooting often superior to single shooting for numerical solution. The division of the entire simulation into small intervals in multiple shooting leads to a more linear behavior within each simulation step. Along with that, while for single shooting first the entire simulation is done and then the optimization, for multiple shooting the simulation and optimization occur simultaneously. This is advantageous for a fast numerical solution of the optimization problem. The benefits result for nonlinear systems in general and for unstable systems especially. The second advantage results with multiple shooting as knowledge of the system states can be incorporated as initial conditions. This advantage can especially be exploited with NLP solution methods that strongly rely on initial conditions, such as the SQP method. The third advantage is given by the resulting sparsity pattern of the NLP. In multiple shooting, it is block sparse, thus sparsity-exploiting solvers can be used. As a result of all these characteristics, a quick convergence to the solution can be achieved. Additionally, typically after only a few iterations, a very good solution can be obtained with multiple shooting that is close to the converged one [14].

Example 5.5 (*NMPC-based control of the Brusselator with multiple shooting*) To demonstrate the differences between single and multiple shooting, the Brusselator described in Example 5.4 is revisited. The same control task is to be solved, with the only difference that now the multiple shooting scheme is used. Especially for the combination of multiple shooting with SQP, the initialization plays an important role. If no initial guess is available, a suitable choice is the reference point. Within this example, all values along the horizon for $\boldsymbol{x}_k, u_k$ are initialized using the target stationary point $\boldsymbol{x}_{SS} = [0.425\ 3.882]^\mathsf{T}$, $u_{SS} = 1.7$. The initialization step of the algorithm is displayed in Fig. 5.10. The initialization points are shown along with the one-step simulation starting from each point. In this specific case, the continuity

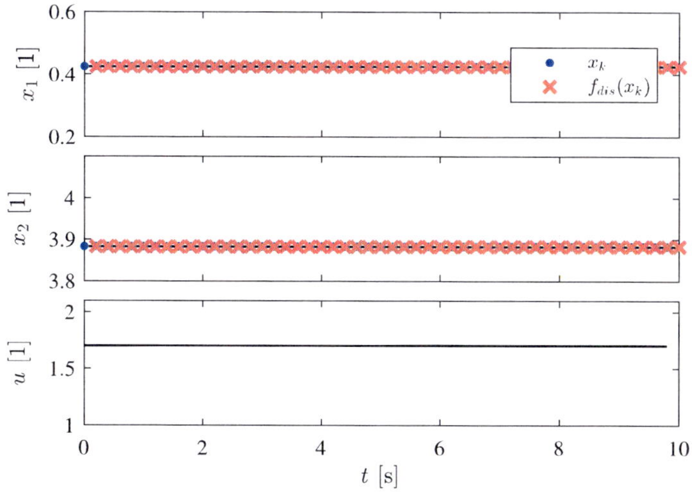

Fig. 5.10 Initialization of the Brusselator with multiple shooting

condition is fulfilled as the initialization values are a stationary point. Thus, there is no gap between the one-step simulation and the subsequent states. However, in general this is not the case; there can be a gap, especially for the initial values.

Figure 5.11 shows the fully converged solution with multiple shooting. The solution is the same as obtained with single shooting. For the converged solution, the continuity conditions are always fulfilled along the entire trajectory. The solution method needs 12 iterations until it reaches this solution, which clearly shows the speedup gained over the single shooting that required 58 iterations.

Another considerable difference between single shooting and multiple shooting results for the very first iteration step. Figure 5.12 thus shows the results after the first iteration obtained by the two methods. The single shooting iterate is still far away from the converged solution. In contrast, the multiple shooting solution is close to the final result. With multiple shooting, the strong nonlinearity is avoided compared to single shooting. This fact can be exploited also by other solution techniques such as IP methods. For solution methods such as SQP, additionally, the possibility of multiple shooting to initialize the system states and the actuated values along the entire trajectory can be exploited.

5.3.5 *Real-Time NMPC*

So far, various methods have separately been discussed that are suited for use within NMPC. In the following, the general procedure for real-time NMPC is described. If the system dynamics are given in a continuous-time representation, the OCP can be

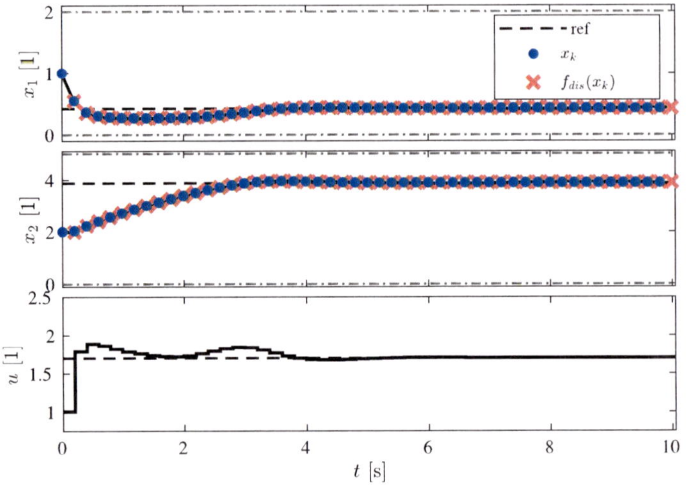

Fig. 5.11 Converged solution obtained with multiple shooting after 12 iterations

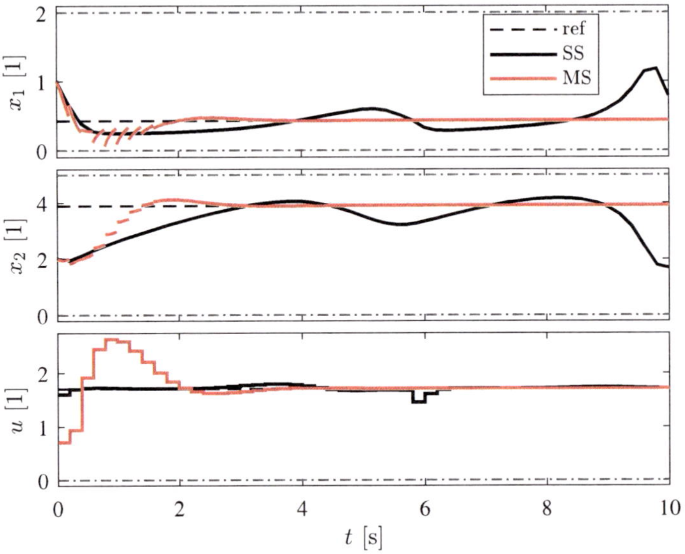

Fig. 5.12 Solution after one iteration of single and multiple shooting

set up. For direct methods, this OCP is discretized to gain an NLP. In most cases, multiple shooting discretization offers significant advantages over the single shooting approach. The discretization step with multiple shooting is conducted offline.

If the system dynamics are present in a discrete-time representation, the NLP can be built directly. Still, there is the choice of using a sequential or a simultaneous approach for setting up the NLP. Within the sequential approach, all system states are replaced in the cost function, as conducted in the single shooting discretization. Alternatively, within the simultaneous approach, all the system states are kept as optimization variables, as conducted in the multiple shooting discretization.

During runtime of the process, in every time step of the control algorithm, a new instance of the NLP has to be solved. From one time step to the next, the structure stays the same, but single parameters like the initial state changes and the reference might change. Especially with the SQP method, the initialization can be exploited in order to obtain a good initial guess. For this purpose, the solution of the last time step can be used. The solution is reused, just shifted by one time step to account for the development of time. The preceding solution for the optimization variables as well as for the Lagrange multipliers can be used for initialization. For the last state and actuated value in the prediction horizon, usually $\boldsymbol{x}_{k+N} := \boldsymbol{x}_{k+N-1}$ and $\boldsymbol{u}_{k+N-1} := \boldsymbol{u}_{k+N-2}$ are used as initial guesses. In the same way, the last Lagrange multipliers can be initialized.

In each time step, the NLP can be solved until complete convergence. From the minimizer, the first actuated values are applied to the process. If the available computation time is limited and if the process is just slightly nonlinear, another approach can be chosen, which is often referred to as the RTI scheme [9]. Instead of solving the optimization problem until convergence in every time step, the RTI only solves the NLP approximately. If suitable algorithms are used, the computation time is reduced significantly, while stability can still be ensured under mild assumptions. The assumptions concern, e.g. the solution quality of the underlying QP problem [8]. Due to the short computation times, the sampling time of the process can be decreased, which leads to a fast disturbance rejection and increased robustness. Often, the RTI scheme is implemented using a direct multiple shooting formulation in combination with the SQP method [14]. In every time step, a single SQP step is performed, and the first actuated values of this approximate solution are applied to the plant. For the subsequent optimization problem, the shifted solution is used as an initial value. As a consequence, instead of converging to the optimal solution in every time step, the convergence is realized over multiple time steps during runtime. To further reduce the computation time, the RTI algorithm often uses a Gauss–Newton approximation for the calculation of the Hessian. Additionally, real-time-feasible NMPC often relies on a constant step size within the line-search algorithm. A common setting is $\alpha = 1$, thus using the full step.

For NMPC, more parameters have to be set than for the LMPC algorithm. They have to be tuned appropriately in order to realize a closed-loop controller that satisfies real-time feasibility and high control quality. Figure 5.13 shows an overview of the most important tuning parameters. Most of the tuning parameters result in a trade-off between control performance and computational complexity. The optimal

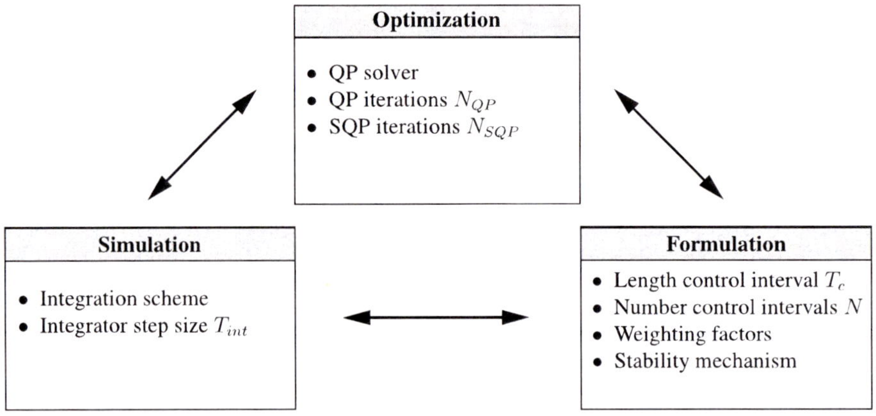

Fig. 5.13 Overview of design choices for the NMPC algorithm – © [2016] IEEE. Reprinted, with permission, from [1]

choices of these parameters have to be analyzed for the specific application to best fit the available computation time and the system nonlinearities. There are the design choices related to the simulation of the process. Important choices are the integration scheme and the integration step size T_{int}. They have to be chosen appropriately to satisfy an accurate simulation of the nonlinear process behavior. This is especially critical for stiff and unstable systems. Other design choices concern the NLP optimization algorithm. First, the optimization algorithm has to be chosen, for example, IP or SQP method. The SQP method relies on solving QP problems, consequently also a QP method has to be chosen. A common combination is the choice of an SQP method with an active set solver. Besides the SQP method and also the QP method, the maximum numbers of iterations N_{QP} and N_{SQP} have to be determined. Finally, there are also parameters concerning the NMPC formulation. This relates to the length of the control interval T_c, the number of control intervals N, the weighting factors, and also the choice of stability mechanism such as terminal cost.

Example 5.6 (*LTI MPC and NMPC-based closed-loop control of the Brusselator*) The Brusselator presented in Example 5.4 is revisited. Now, the performance capabilities of LTI MPC and two different NMPC approaches are compared. All three controllers are applied in a closed-loop control fashion. LTI MPC uses one linear model for the entire simulation. The linear model is obtained by the linearization of the system dynamics at the stationary point. The first NMPC approach uses the RTI scheme. The multiple shooting discretization is applied. One SQP-step is performed in each iteration step. The first actuated value of the iterate is applied to the plant. The latter NMPC approach uses the first actuated value of the fully converged solution in every time step.

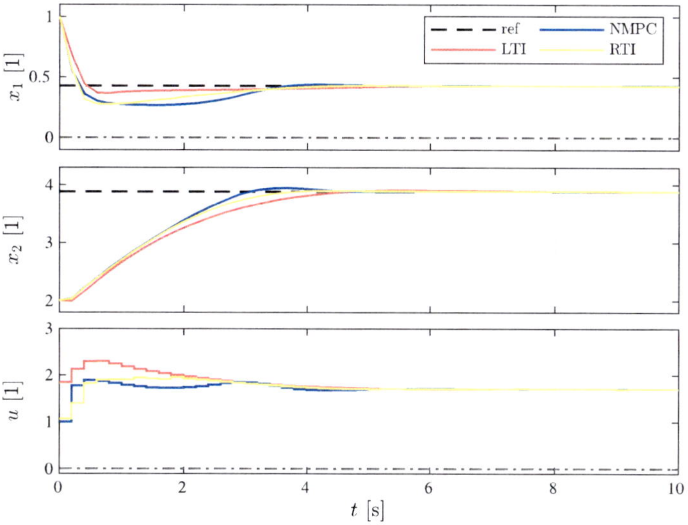

Fig. 5.14 Closed-loop control simulations of LTI MPC, NMPC RTI, and NMPC (fully converged) – state trajectory over time

Figure 5.14 shows the closed-loop control simulations obtained with various MPC algorithms. It shows the trajectories of the system states over time. As expected, the fully converged NMPC performs better than RTI and LTI MPC. The performance improvement concerns especially the settling time of the state x_2. RTI itself performs better than LTI MPC. This behavior is as expected as both NMPC controllers take additional information about the nonlinear system behavior into account. However, the computational demands are in reversed order. The highest computational demands are given for solving the NMPC optimization problem to the converged solution. The RTI scheme only has slightly higher computational demands than LTI MPC. The trade-off between computational demand and performance plays a major role in real-world online engine applications of optimal control.

For the same closed-loop control simulations, Fig. 5.15 shows the trajectories in the phase portrait. The curves of the NMPC-based solutions are longer compared to the ones of the LTI MPC-based solution. However, the NMPC-based solutions go to higher absolute values of the system states. As the system states correspond to a speed, this procedure allows for shorter settling times, showing that they are able to exploit the nonlinear system dynamics.

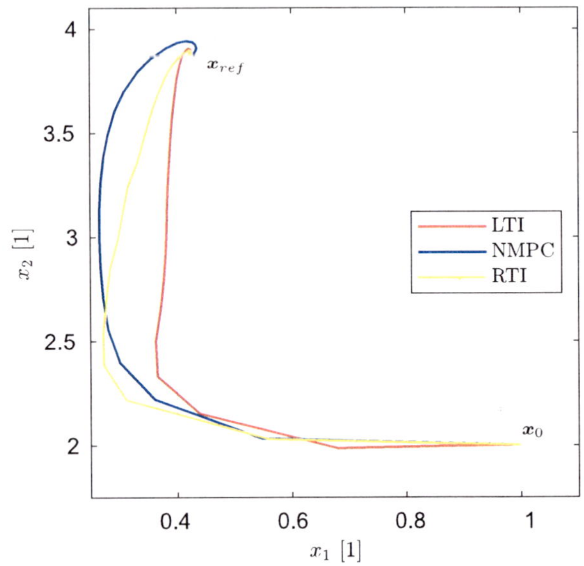

Fig. 5.15 Closed-loop control simulations of LTI MPC, NMPC RTI, and NMPC (fully converged) – phase portrait

References

1. T. Albin, F. Frank, D. Ritter, D. Abel, R. Quirynen, M. Diehl, Nonlinear MPC for combustion engine control: a parameter study for realizing real-time feasibility, in *IEEE Conference on Control Applications* (2016), pp. 311–316
2. J. Andersson, J. Akesson, M. Diehl, CasADi: a symbolic package for automatic differentiation and optimal control, in *Recent Advances in Algorithmic Differentiation* (Springer, 2012), pp. 297–307
3. J.T. Betts, *Practical Methods for Optimal Control and Estimation Using Nonlinear Programming* (SIAM, Philadelphia, 2010)
4. L.T. Biegler, *Nonlinear Programming: Concepts, Algorithms, and Applications to Chemical Processes* (SIAM, Philadelphia, 2010)
5. H.G. Bock, K.J. Plitt, A multiple shooting algorithm for direct solution of optimal control problems. IFAC Proc. Vol. **17**(2), 1603–1608 (1984)
6. S.C. Chapra, R.P. Canale, *Numerical Methods for Engineers* (McGraw-Hill Higher Education, Boston, 2010)
7. M. Diehl, H.J. Ferreau, N. Haverbeke, Efficient numerical methods for nonlinear MPC and moving horizon estimation, in *Nonlinear Model Predictive Control* (Springer, 2009), pp. 391–417
8. M. Diehl, R. Findeisen, F. Allgöwer, H.G. Bock, J.P. Schlöder, Nominal stability of the real-time iteration scheme for nonlinear model predictive control. IEE Proc.-Control Theory Appl. **152**(3), 296–308 (2005)
9. M. Diehl, H.G. Bock, J.P. Schloeder, A real-time iteration scheme for nonlinear optimization in optimal feedback control. SIAM J. Control Optim. **43**(5), 1714–1736 (2005)
10. J. Nocedal, S.J. Wright, *Numerical Optimization* (Springer, Berlin, 2006)
11. M. Papageorgiou, M. Leibold, M. Buss, *Optimierung: Statische, dynamische, stochastische verfahren für die anwendung* (Springer, Berlin, 2015)

12. R. Quirynen, M. Vukov, M. Zanon, M. Diehl, Autogenerating microsecond solvers for nonlinear MPC: a tutorial using ACADO integrators. Optim. Control Appl. Methods **36**, 685–704 (2014)
13. R. Quirynen, M. Vukov, M. Diehl, Auto generation of implicit integrators for embedded nmpc with microsecond sampling times, in *IFAC Nonlinear Model Predictive Control Conference* (2012), pp. 175–180
14. J.B. Rawlings, D.Q. Mayne, M. Diehl, *Model Predictive Control: Theory, Computation and Design* (Nob Hill Publishing, Madison, 2017)

Chapter 6
Formulation of the Optimization Problem

Abstract In this chapter, various aspects concerning the formulation of the model predictive control (MPC) optimization problem are discussed. The formulation of the optimization problem directly determines the closed-loop control characteristics. Hence, it plays a fundamental role in the practical application of MPC. The requirements on the control performance for the specific application have to be translated into a suitable optimization task. An important and common requirement concerns the accuracy of the reference tracking. Usually, a reference shall be tracked by the closed-loop controller while rejecting disturbances. Ideally, the reference is tracked without any control error in the steady state, i.e. offset-free control. The chapter discusses criteria that need to be satisfied such that offset-free control is made viable. Another system characteristic that has to be taken care of by a special formulation of the optimization problem is non-square systems. For non-square systems, the number of inputs is different from the number of outputs. Within engine control, some overactuated systems are present with more actuated than controlled variables. The overactuation can be explicitly used by a suitable consideration within the MPC algorithm. The surplus degree of freedom can be exploited for improvements in the control behavior, e.g. in terms of control performance and robustness. Besides the performance requirements, there are also prerequisites that every MPC-based control algorithm has to fulfill. The two properties that are required to safely employ MPC in practice are the recursive feasibility of the optimization problem and the stability of the closed-loop system. The recursive feasibility ensures that there exists a solution of the optimization problem in every time step. The second property is stability. Even if there exists a solution that is optimal with respect to the optimization problem, the closed-loop system can still be unstable. The stability of a closed-loop system is challenging to prove for practical applications such as in engine control. However, mechanisms can be introduced to improve the stability properties.

T. Albin Rajasingham, *Nonlinear Model Predictive Control of Combustion Engines*, Advances in Industrial Control,
https://doi.org/10.1007/978-3-030-68010-7_6

6.1 Soft Constraints

The MPC algorithm allows constraints to be considered on actuated values, system states, and the system outputs. When constraints are incorporated in the optimization problem, it might become infeasible at a certain time step. This implies that the feasible set of the optimization problem is empty and thus there does not exist any suitable solution of the optimization problem.

Based on three situations, Fig. 6.1 shows an introductory example of infeasibility. Case I schematically depicts an optimization problem with box constraints on the system states. A feasible solution exists to move from the initial states $\boldsymbol{x}_0$ to the final states $\boldsymbol{x}_f$. In the next scenario, named Case II, a disturbance acts on the states at $k = 2$ such that the states $\boldsymbol{x}(k = 3)$ lie outside of the constraints. Still, a solution exists such that all subsequent states can be driven within the box constraints to reach the final states within the prediction. The optimization algorithm can find a solution, although it might come at the expense of high costs in the cost function, e.g. due to high actuated values. In the last scenario, Case III, constraints on u are considered in addition, i.e. $u_{min} \leq u(\cdot|k) \leq u_{max}$. The states $\boldsymbol{x}(k = 3)$ lie so far outside of the constraints such that the intersection of the maximum reachable set for one step and the feasible set defined by the constraints is empty. As a consequence, there is no possibility that the predicted states $\boldsymbol{x}(4|k = 3)$ can lie inside the box-constrained region without exceeding the constraints on the actuated values. The solver thus cannot find a solution anymore.

In an online-control algorithm for an engine, such a situation where no solution can be found has to be strictly avoided. Typical reasons for infeasibility are [2]

- disturbances,
- mismatch between model and plant, and
- inappropriate choice of MPC tuning parameters, e.g. the prediction horizon chosen is too short.

If only actuated values are constrained and these are set reasonably, i.e. the lower limit value smaller than the higher limit value for box constraints, the problem of infeasibility does not occur. The feasible set is directly given by the constrained set of the actuated variables. As a consequence, the constraints on the actuated variables can always be used as hard constraints. This also reflects the technical circumstances. In reality, the actuated variables usually are hard constrained. For instance, the valve cannot be more than 100% open. Whenever the system states or outputs are constrained, the feasibility issues discussed might occur. Additionally, the feasibility of a single MPC step does not automatically guarantee that all of the following steps are feasible, which is a property called *recursive feasibility*.

The most common approach to deal with infeasibility for practical applications is the use of soft instead of hard constraints. The idea of soft constraints is to allow for a violation of the original hard constraints at the expense of a very high cost in the cost function. A suitable penalty function is added to the original cost function. With this augmented cost function, the controller tries to minimize violation of the

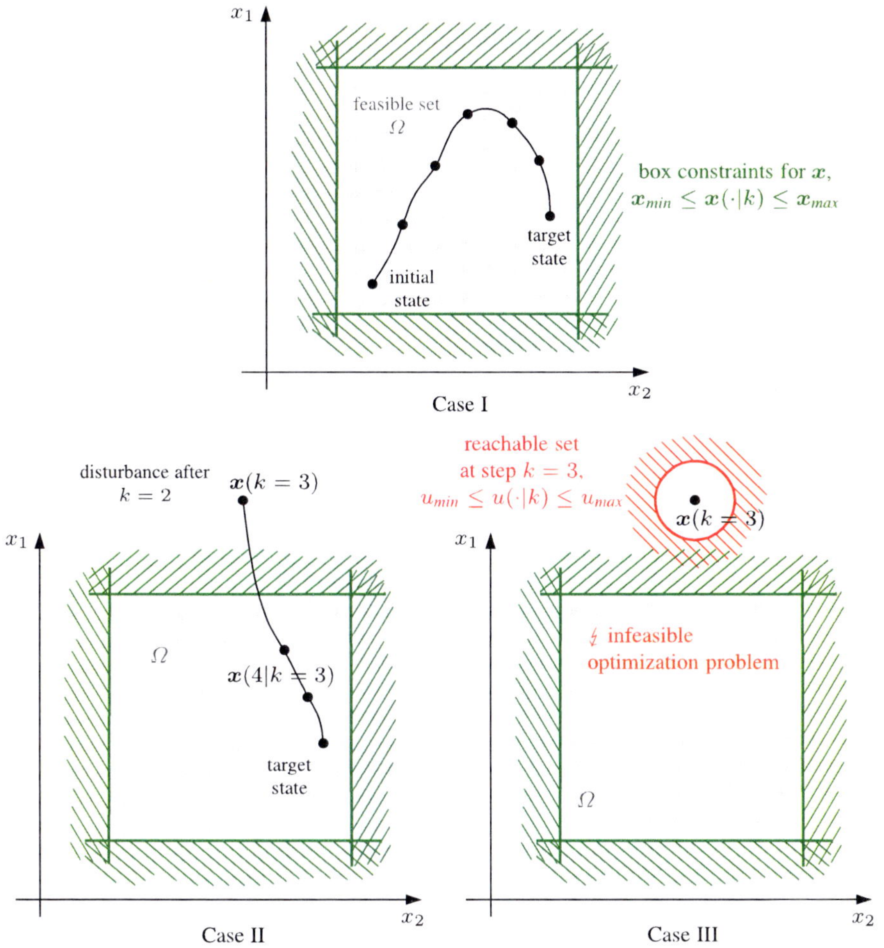

Fig. 6.1 Infeasibility of the optimization problem

originally hard constraints. Ideally, whenever the original optimization problem is feasible, the optimizer of the augmented optimization problem should result in the same solution. If this property is given, the penalty function is called exact [10]. The soft constraints also reflect technical realities, as the constraints on the system states and outputs usually are not hard. The turbocharger speed, as an example of engine control, should be kept below a certain limit value. However, surpassing this limit value for a short time and with low excess is tolerable. For linear MPC (LMPC), the concept of soft constraints is detailed in the following.

Original Optimization Problem

The cost function of the original optimization problem shall be given by (6.1).

$$J_{orig} = \sum_{i=N_1}^{N_2} \|\boldsymbol{x}(k+i|k) - \boldsymbol{x}_{SS}\|_{\boldsymbol{Q}}^2 + \sum_{i=0}^{N_u-1} \|\boldsymbol{u}(k+i|k) - \boldsymbol{u}_{SS}\|_{\boldsymbol{R}}^2 \tag{6.1}$$

The corresponding optimization problem is denoted by (6.2).

$$\begin{aligned} \min_{\boldsymbol{x}(\cdot|k),\, \boldsymbol{u}(\cdot|k)} \quad & J_{orig} \\ \text{s.t.} \quad & \text{system dynamics,} \\ & \text{initial conditions,} \\ & \boldsymbol{u}_{min} \le \boldsymbol{u}(k+i|k) \le \boldsymbol{u}_{max}, \quad i = 0, \dots, N_u - 1, \\ & \boldsymbol{x}_{min} \le \boldsymbol{x}(k+i|k), \quad i = N_1, \dots, N_2, \\ & -\boldsymbol{x}_{max} \le -\boldsymbol{x}(k+i|k), \quad i = N_1, \dots, N_2 \end{aligned} \tag{6.2}$$

The hard constraints on the system states can be relaxed by the introduction of the slack variables $\boldsymbol{\epsilon}$. Within the augmented optimization problem, the slack variables are considered as additional optimization variables which enter the cost function as well as the constraints. In the following, various choices for the penalty function are discussed. Weighting matrices with positive values are used for appropriate penalization.

Quadratic penalty on the slack variables

First, a quadratic penalty on $\boldsymbol{\epsilon}(\cdot|k) \in \mathbb{R}^{n_x(N_2-N_1+1)}$ is investigated. The original cost function is augmented by the penalty function J_{pen}, resulting in $J_{aug} = J_{orig} + J_{pen}$. The resulting optimization problem is represented by (6.3).

$$\begin{aligned} \min_{\boldsymbol{x}(\cdot|k),\, \boldsymbol{u}(\cdot|k),\, \boldsymbol{\epsilon}(\cdot|k)} \quad & \overbrace{J_{orig} + \sum_{i=N_1}^{N_2} \|\boldsymbol{\epsilon}(k+i|k)\|_{\boldsymbol{S}_q}^2}^{=:\, J_{aug}} \\ \text{s.t.} \quad & \text{system dynamics,} \\ & \text{initial conditions,} \\ & \boldsymbol{u}_{min} \le \boldsymbol{u}(k+i|k) \le \boldsymbol{u}_{max}, \quad i = 0, \dots, N_u - 1, \\ & \boldsymbol{x}_{min} - \boldsymbol{\epsilon}(k+i|k) \le \boldsymbol{x}(k+i|k), \quad i = N_1, \dots, N_2, \\ & -\boldsymbol{x}_{max} - \boldsymbol{\epsilon}(k+i|k) \le -\boldsymbol{x}(k+i|k), \quad i = N_1, \dots, N_2, \\ & \boldsymbol{\epsilon}(k+i|k) \ge \boldsymbol{0}, \quad i = N_1, \dots, N_2 \end{aligned} \tag{6.3}$$

One advantage of the quadratic penalty function is that if the original optimization problem is strictly convex, the augmented optimization problem stays strictly convex and the Hessian remains positive definite. However, for a high number of constraints to be softened, many new optimization variables have to be introduced which increase the necessary computation time.

The quadratic penalty function does not introduce any exact penalties. Assume, the minimizer of the original optimization problem is denoted by $\boldsymbol{u}^*_{orig}, \boldsymbol{x}^*_{orig}$. For strictly active linear constraints, the original cost function has a negative gradient at the constrained minimum that is proportional to the Lagrange multipliers $\boldsymbol{\mu}^*_{orig}, \boldsymbol{\lambda}^*_{orig}$. The penalty function has a gradient of $\mathbf{0}$ at the point $\boldsymbol{u}^*_{orig}, \boldsymbol{x}^*_{orig}$. The overall gradient, i.e. the sum of the two gradients, thus can be negative. Hence, the optimizer of the augmented cost function $\boldsymbol{u}^*_{aug}, \boldsymbol{x}^*_{aug}$ can be different from the original minimizer $\boldsymbol{u}^*_{orig}, \boldsymbol{x}^*_{orig}$. The optimizer can be shifted even though there exists a solution without any violation of constraints.

Linear Penalty Function: 1-Norm

Instead of the 2-norm, the 1-norm can be used to penalize $\boldsymbol{\epsilon}(\cdot|k) \in \mathbb{R}^{n_x(N_2-N_1+1)}$. The 1-norm penalizes the sum of the absolute values of the slack variables. For weighting of the penalties, $\boldsymbol{S}_l \in \mathbb{R}^{n_x}$ is used. As the slack variables are defined to be nonnegative, the following optimization problem can be used:

$$
\begin{aligned}
\min_{\boldsymbol{x}(\cdot|k),\, \boldsymbol{u}(\cdot|k),\, \boldsymbol{\epsilon}(\cdot|k)} \quad & J_{orig} + \sum_{i=N_1}^{N_2} \boldsymbol{S}_l^{\mathsf{T}} \boldsymbol{\epsilon}(k+i|k) \\
\text{s.t.} \quad & \text{system dynamics}, \\
& \text{initial conditions}, \\
& \boldsymbol{u}_{min} \leq \boldsymbol{u}(k+i|k) \leq \boldsymbol{u}_{max}, \qquad i = 0, \ldots, N_u - 1, \\
& \boldsymbol{x}_{min} - \boldsymbol{\epsilon}(k+i|k) \leq \boldsymbol{x}(k+i|k), \qquad i = N_1, \ldots, N_2, \\
& -\boldsymbol{x}_{max} - \boldsymbol{\epsilon}(k+i|k) \leq -\boldsymbol{x}(k+i|k), \; i = N_1, \ldots, N_2, \\
& \boldsymbol{\epsilon}(k+i|k) \geq \mathbf{0}, \qquad i = N_1, \ldots, N_2
\end{aligned}
\tag{6.4}
$$

The main advantage of using a linear penalty term by the 1-norm is the fact that exact penalties can be obtained. The additional linear penalty function introduces a gradient of $\boldsymbol{S}_l$. The weightings $\boldsymbol{S}_l$ of the slack variables just need to be chosen sufficiently high, such that the gradient of the penalty function is large enough [10]. When a feasible solution of the original problems exists, the location of the local minimizer thus is not changed by the augmented optimization problem. Details about the calculation of the necessary values for $\boldsymbol{S}_l$ are given in [10].

In the case of adding the linear penalty function, it has to be considered that even if the original problem is strictly convex, the resulting optimization problem no longer is strictly convex. When QP solvers are applied that rely on strictly convex optimization problems, suitable regularization has to be added.

The formulations of (6.3) and (6.4) introduce many optimization variables such as $\boldsymbol{\epsilon}(\cdot|k) \in \mathbb{R}^{n_x(N_2-N_1+1)}$. One possibility to reduce the number of optimization variables is the introduction of the ∞-norm.

Linear Penalty Function: ∞-Norm

An alternative possibility is the penalization of the entire vector of slack variables $\boldsymbol{\epsilon}(\cdot|k)$ by the ∞-norm denoted by $\|\boldsymbol{\epsilon}(\cdot|k)\|_\infty$. The ∞-norm is given by the maximum of the absolute values of the components of the vector $\boldsymbol{\epsilon}(\cdot|k)$. This can be recast to a linear cost term where only one scalar value $\epsilon \in \mathbb{R}$ is considered:

$$
\begin{aligned}
\min_{\boldsymbol{x}(\cdot|k),\, \boldsymbol{u}(\cdot|k),\, \epsilon} \quad & J_{orig} + S_l \epsilon \\
\text{s.t.} \quad & \text{system dynamics}, \\
& \text{initial conditions}, \\
\boldsymbol{u}_{min} \leq \boldsymbol{u}(k+i|k) \leq \boldsymbol{u}_{max}, \quad & i = 0, \ldots, N_u - 1, \\
\boldsymbol{x}_{min} - \epsilon \leq \boldsymbol{x}(k+i|k), \quad & i = N_1, \ldots, N_2, \\
-\boldsymbol{x}_{max} - \epsilon \leq -\boldsymbol{x}(k+i|k), \quad & i = N_1, \ldots, N_2, \\
\epsilon \geq 0, &
\end{aligned}
\tag{6.5}
$$

As in the case of the 1-norm, a sufficiently high weighting of S_l leads to exact penalties. Also, as in the case of the 1-norm, even if the original problem is strictly convex, the resulting augmented optimization problem is no longer strictly convex. Compared to the 1-norm, the formulation of (6.5) only introduces the one additional optimization variable $\epsilon \in \mathbb{R}$. This is computationally much more efficient. However, the exclusive penalization of the maximum value is not sufficient in some applications. Clearly, medium pathways for complexity versus computation time can be used as well, such as penalizing the maximum deviation over the entire prediction horizon separately for each constrained state.

Quadratic Plus Linear Penalty Function

The combination of a quadratic and a linear penalty term combines the advantages of both formulations. The combination sustains a strictly convex optimization problem. Additionally, for sufficiently high values of $\boldsymbol{S}_l$, exact penalties are obtained. In practice, the combination of quadratic and linear penalization functions is used most widely. One example is shown in (6.6) where the maximum deviation, a scalar value, over the entire prediction horizon and all states is penalized.

$$
\begin{aligned}
\min_{\boldsymbol{x}(\cdot|k),\, \boldsymbol{u}(\cdot|k),\, \epsilon} \quad & J_{orig} + S_q \epsilon^2 + S_l \epsilon \\
\text{s.t.} \quad & \text{system dynamics}, \\
& \text{initial conditions}, \\
\boldsymbol{u}_{min} \leq \boldsymbol{u}(k+i|k) \leq \boldsymbol{u}_{max}, \quad & i = 0, \ldots, N_u - 1, \\
\boldsymbol{x}_{min} - \epsilon \leq \boldsymbol{x}(k+i|k), \quad & i = N_1, \ldots, N_2, \\
-\boldsymbol{x}_{max} - \epsilon \leq -\boldsymbol{x}(k+i|k), \quad & i = N_1, \ldots, N_2, \\
\epsilon \geq 0, &
\end{aligned}
\tag{6.6}
$$

For nonlinear MPC (NMPC), the same penalty functions can be used to soften the constraints. In order to not shift the location of the local minimizer, the weighting factors have to be chosen such that the gradient of the penalty function at the original constrained minimizer is sufficiently large. However, the addition of the penalty function can lead to new minima in regions that previously were excluded by the constraints. In general, new local minimizers that exceed the original constraints can be found much easier for NMPC compared to LMPC. Hence, the NLP solver can find unwanted local minima outside of the operating region. This has to be considered when developing NMPC with soft constraints. This affects for instance the process model. It needs to be suitable for usage within NMPC even outside of the feasible region defined by the original hard constraints. This is especially true for data-driven models as their extrapolation capabilities are limited.

6.2 Offset-Free Control

A commonly posed requirement on a closed-loop control system is that it should be free of any offsets. More specifically, usually, it is demanded that for a step in the reference signal and constant disturbances, the tracking error should vanish in steady-state conditions. For MPC, additionally, the reference point has to be reachable despite the existence of constraints. Offset-free control implies the rejection of all arising disturbances. In the case of an engine controller, many unmodeled and unmeasured physical disturbances are acting on the system inputs, states, and outputs. Additionally, there is always a considerable model–plant mismatch. It is shown that the MPC algorithm only leads to offset-free control when the measured outputs match the predictions obtained from the controller-internal model. In practical applications, the different disturbances lead to deviation of the prediction and the measurement. Various schemes will be introduced below that allow to align the prediction, even in the presence of disturbances.

Assume that the controller-internal model of the MPC is given by

$$\begin{aligned} \boldsymbol{x}(k+1) &= \boldsymbol{A}\boldsymbol{x}(k) + \boldsymbol{B}\boldsymbol{u}(k) \\ \boldsymbol{y}(k) &= \boldsymbol{C}\boldsymbol{x}(k) \end{aligned} \tag{6.7}$$

The real system to be controlled is given by the following state-space representation which includes constant disturbances acting on the system output:

$$\begin{aligned} \boldsymbol{x}(k+1) &= \boldsymbol{A}\boldsymbol{x}(k) + \boldsymbol{B}\boldsymbol{u}(k) \\ \boldsymbol{d}'(k+1) &= \boldsymbol{d}'(k) \\ \boldsymbol{y}'(k) &= \boldsymbol{C}\boldsymbol{x}(k) + \boldsymbol{C}_d\boldsymbol{d}'(k) \end{aligned} \tag{6.8}$$

Thus, the MPC controller is aware of the real system model matrices, i.e. $\boldsymbol{A}$, $\boldsymbol{B}$, and $\boldsymbol{C}$, and of the real system states, i.e. $\boldsymbol{x}(k)$, for instance based on full-state measurements. However, the controller-internal model is unaware of the real disturbances $\boldsymbol{d}'$. The prediction of the system states thus is equivalent to the real plant behavior, but the output predictions differ. The predicted outputs from the controller-internal model are denoted by $\boldsymbol{y}$, whereas $\boldsymbol{y}'$ denotes the real, e.g. measured, outputs.

A stabilizing MPC controller allows the closed-loop system to converge to a steady state. The steady-state behavior is defined by the following conditions:

$$\boldsymbol{x} = \boldsymbol{x}_\infty, \quad \Delta\boldsymbol{u}_\infty = \boldsymbol{0}, \quad \boldsymbol{u}_\infty = const. \tag{6.9}$$

With $\boldsymbol{d}'(k) = const.$, the stationary conditions of the real plant behavior can be calculated.

$$\begin{aligned} \boldsymbol{x}_\infty &= \boldsymbol{A}\boldsymbol{x}_\infty + \boldsymbol{B}\boldsymbol{u}_\infty \\ \boldsymbol{y}'_\infty &= \boldsymbol{C}\boldsymbol{x}_\infty + \boldsymbol{C}_d\boldsymbol{d}' \end{aligned} \tag{6.10}$$

The question arises whether offset-free control is realized in steady state. A constant reference value $\boldsymbol{r}(k) = const.$ is to be tracked, such that for offset-free control, $\boldsymbol{r} = \boldsymbol{y}'_\infty$ has to hold. Given a stabilizing MPC controller, also assume that the reference is reachable, i.e. no constraints limit reaching the reference. For an appropriate cost function, e.g. $J = \|\boldsymbol{y}(\cdot|k) - \mathbf{1}\boldsymbol{r}\|_{\boldsymbol{Q}}^2 + \|\Delta\boldsymbol{u}(\cdot|k)\|_{\boldsymbol{R}}^2$, the minimum cost for steady-state conditions is $J^* = 0$. In this case, the predicted error $\left\|\boldsymbol{y}_\infty - \boldsymbol{r}\right\|$ also has to go to zero. In steady state, the prediction of the controller-internal model becomes $\boldsymbol{y}_\infty = \boldsymbol{C}\boldsymbol{x}_\infty$. Thus, the following equation holds for determination of the predicted error:

$$\boldsymbol{e}_\infty = \boldsymbol{y}_\infty - \boldsymbol{r} = \boldsymbol{C}\boldsymbol{x}_\infty - \boldsymbol{r} = \boldsymbol{0} \tag{6.11}$$

The predicted error is given by the difference between the reference and the predicted outputs from the controller-internal model rather than the measured outputs. However, the real plant outputs $\boldsymbol{y}'_\infty$ can differ from the outputs $\boldsymbol{y}_\infty$ predicted by the control-internal model.

$$\boldsymbol{y}_\infty = \boldsymbol{C}\boldsymbol{x}_\infty \quad versus \quad \boldsymbol{y}'_\infty = \boldsymbol{C}\boldsymbol{x}_\infty + \boldsymbol{C}_d\boldsymbol{d}' \tag{6.12}$$

Since the predicted outputs track the reference $\boldsymbol{y}_\infty = \boldsymbol{r}$, these equations can be rearranged to find what the actual system outputs will be in steady state:

$$\boldsymbol{y}'_\infty = \boldsymbol{r} + \boldsymbol{C}_d\boldsymbol{d}' \tag{6.13}$$

As a consequence, an offset-free control system results when the disturbances $\boldsymbol{d}'$ do not influence the output, i.e. when $\boldsymbol{C}_d\boldsymbol{d}' = \boldsymbol{0}$. In all other cases, a steady-state offset error remains.

In general, offset-free control can be realized only when the predicted outputs of the controller-internal model match in steady state the measured outputs. For this example, a specific setting is shown. However, this holds true for the general case; e.g. the disturbances can act anywhere.

Example 6.1 (*MPC with model–plant mismatch*)
The following example demonstrates the resulting closed-loop control error in case of a MPC controller with model–plant mismatch. A plant is considered, whose exact

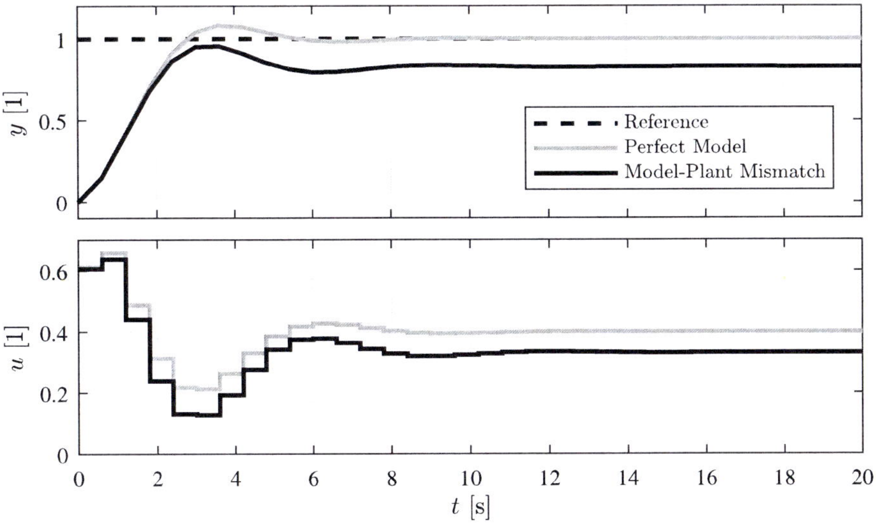

Fig. 6.2 Step response of the closed-loop control system described in Example 6.1 without and with model–plant mismatch

discretization with a sampling time of $T_s = 0.6\,\mathrm{s}$ is given by

$$\text{Plant} \quad G_p(z) = \frac{0.25}{z^2 - 1.65z + 0.75} \tag{6.14}$$

First, the plant is controlled without any model–plant mismatch, i.e. the MPC uses G_p as a controller-internal model. Figure 6.2 shows the response of the closed-loop system for a step-wise change in the reference. As expected, offset-free control can be realized. Now, the controller-internal model is changed such that a model–plant mismatch is present. The controller-internal model is given by a slight offset in one model parameter:

$$\text{MPC-internal Model} \quad G_m(z) = \frac{0.25}{z^2 - 1.65z + 0.70} \tag{6.15}$$

Using G_m as controller-internal model, the MPC controller controls the plant G_p. Figure 6.2 shows the simulative closed-loop control result.

There is a significant steady-state error, even though the offset in the parameter of the model is small. This behavior is due to the considerable difference in the steady-state gain of the plant G_p and the controller-internal model G_m.

$$G_{p,\infty} = G_p(z)\big|_{z=1} = 2.5, \quad G_{m,\infty} = G_m(z)\big|_{z=1} = 5 \tag{6.16}$$

6.2.1 Observer-Based Offset-Free Control

For feedback control loops with classical controllers such as PID-based control algorithms, integral action has to be present in the open-loop system to realize offset-free control, e.g. in the controller or in the plant itself. Within MPC, the approach is different in that it consists of two tasks. First, the disturbances arising are estimated and subsequently, they are accounted for in the prediction of the system behavior. The consideration of the disturbances in the prediction allows to reject them by optimizing for the appropriate actuated values. A suitable observer is needed to estimate the arising disturbances. In the following, the formulas are derived for the linear case.

The observer needs to estimate the unknown disturbances arising. The following state-space system shall be given as a nominal controller-internal model.

$$\boldsymbol{x}(k+1) = \boldsymbol{A}\boldsymbol{x}(k) + \boldsymbol{B}\boldsymbol{u}(k) \tag{6.17}$$

$$\boldsymbol{y}(k) = \boldsymbol{C}\boldsymbol{x}(k) \tag{6.18}$$

This nominal linear model is augmented by the disturbance model as shown in (6.19); see [18].

$$\begin{aligned} \boldsymbol{x}(k+1) &= \boldsymbol{A}\boldsymbol{x}(k) + \boldsymbol{B}\boldsymbol{u}(k) + \boldsymbol{B}_d\boldsymbol{d}(k) \\ \boldsymbol{d}(k+1) &= \boldsymbol{d}(k) \\ \boldsymbol{y}(k) &= \boldsymbol{C}\boldsymbol{x}(k) + \boldsymbol{C}_d\boldsymbol{d}(k) \end{aligned} \tag{6.19}$$

Commonly, the disturbances $\boldsymbol{d}$ are assumed to be constant over the duration of the prediction. If information on the dynamics of the disturbances $\boldsymbol{d}$ is available, they can be incorporated into the disturbance model. The disturbances are represented by $\boldsymbol{d} \in \mathbb{R}^p$. For the following, it is assumed that the measurements of the system correspond to the outputs $\boldsymbol{y}$, and these shall also be the values that are to be tracked offset-free.

This augmented model is applied for the prediction of the system dynamics within the MPC algorithm and it is used within the observer as an internal model. The purpose of the observer is the estimation of the system states and the disturbances. The estimated values are denoted by $\hat{\boldsymbol{x}}, \hat{\boldsymbol{d}}$. A conventional Luenberger observer can be used for instance. It is given by

$$\begin{bmatrix} \hat{\boldsymbol{x}}(k+1) \\ \hat{\boldsymbol{d}}(k+1) \end{bmatrix} = \begin{bmatrix} \boldsymbol{A} & \boldsymbol{B}_d \\ \boldsymbol{0} & \boldsymbol{I} \end{bmatrix} \begin{bmatrix} \hat{\boldsymbol{x}}(k) \\ \hat{\boldsymbol{d}}(k) \end{bmatrix} + \begin{bmatrix} \boldsymbol{B} \\ \boldsymbol{0} \end{bmatrix} \boldsymbol{u}(k) + \begin{bmatrix} \boldsymbol{L}_x \\ \boldsymbol{L}_d \end{bmatrix} \left(\boldsymbol{y}_m(k) - \boldsymbol{C}\hat{\boldsymbol{x}}(k) - \boldsymbol{C}_d\hat{\boldsymbol{d}}(k) \right) \tag{6.20}$$

The most recently measured output is denoted by $\boldsymbol{y}_m(k)$. The goal of the observer is to estimate the states and disturbances in such a way that the estimated output converges to the measured outputs, i.e. $\boldsymbol{C}\hat{\boldsymbol{x}} + \boldsymbol{C}_d\hat{\boldsymbol{d}} \to \boldsymbol{y}_m$. If the observed outputs converge to the measured outputs actually, offset-free control results. A rigorous proof for this statement can be found in [14]. The fundamental approach does not depend on the characteristics of the nominal system. It is as well independent of the cause of the disturbance, e.g. model–plant mismatch or external disturbance parameter and independent of where the disturbances are acting, e.g. on the system states or on the outputs.

The estimated outputs converge to the measured outputs if two criteria are fulfilled. On the one hand, the augmented system model has to be observable. On the other hand, the dynamics of the observer have to be stable, such that estimation errors converge to zero.

The augmented system is observable if the nominal system $(\boldsymbol{A}, \boldsymbol{C})$ is observable and if the following matrix has full column rank, i.e. the rank is equal to the number of systems states n plus the number of disturbances p [14].

$$\operatorname{rank}\begin{bmatrix} \boldsymbol{A}-\boldsymbol{I} & \boldsymbol{B}_d \\ \boldsymbol{C} & \boldsymbol{C}_d \end{bmatrix} = n + p \tag{6.21}$$

The linear algebra implies that this condition can only be satisfied if the number of disturbances p is lower or equal to the number of outputs m, i.e. $p \leq m$. If all output values are to be tracked offset-free, a suitable choice for the amount of disturbance variables is to be equal to the number of outputs $p := m$ [14].

The augmented disturbance model is designed by the matrices $\boldsymbol{B}_d$ and $\boldsymbol{C}_d$. As long as the condition (6.21) is fulfilled, the matrices $\boldsymbol{B}_d$ and $\boldsymbol{C}_d$ can be freely designed. There always exists a pair of matrices $\boldsymbol{B}_d$ and $\boldsymbol{C}_d$ such that the condition (6.21) is fulfilled [18]. In general, the matrices $\boldsymbol{B}_d$ and $\boldsymbol{C}_d$ should reflect the influence of the disturbances on the system states and on the outputs. For instance, if there is a disturbance acting on the input, $\boldsymbol{B}_d := \boldsymbol{B}$ is a suitable choice.

The second criteria concerns the stability of the observer. The closed-loop observer dynamics are given by

$$\begin{aligned} \begin{bmatrix} \hat{\boldsymbol{x}}(k+1) \\ \hat{\boldsymbol{d}}(k+1) \end{bmatrix} &= \begin{bmatrix} \boldsymbol{A} & \boldsymbol{B}_d \\ \boldsymbol{0} & \boldsymbol{I} \end{bmatrix} \begin{bmatrix} \hat{\boldsymbol{x}}(k) \\ \hat{\boldsymbol{d}}(k) \end{bmatrix} + \begin{bmatrix} \boldsymbol{B} \\ \boldsymbol{0} \end{bmatrix} \boldsymbol{u}(k) + \begin{bmatrix} \boldsymbol{L}_x \\ \boldsymbol{L}_d \end{bmatrix} \boldsymbol{y}_m(k) - \begin{bmatrix} \boldsymbol{L}_x \\ \boldsymbol{L}_d \end{bmatrix} \begin{bmatrix} \boldsymbol{C} & \boldsymbol{C}_d \end{bmatrix} \begin{bmatrix} \hat{\boldsymbol{x}}(k) \\ \hat{\boldsymbol{d}}(k) \end{bmatrix} \\ &= \begin{bmatrix} \boldsymbol{A}-\boldsymbol{L}_x\boldsymbol{C} & \boldsymbol{B}_d-\boldsymbol{L}_x\boldsymbol{C}_d \\ -\boldsymbol{L}_d\boldsymbol{C} & \boldsymbol{I}-\boldsymbol{L}_d\boldsymbol{C}_d \end{bmatrix} \begin{bmatrix} \hat{\boldsymbol{x}}(k) \\ \hat{\boldsymbol{d}}(k) \end{bmatrix} + \begin{bmatrix} \boldsymbol{B} \\ \boldsymbol{0} \end{bmatrix} \boldsymbol{u}(k) + \begin{bmatrix} \boldsymbol{L}_x \\ \boldsymbol{L}_d \end{bmatrix} \boldsymbol{y}_m(k) \end{aligned} \tag{6.22}$$

The pole location of the observer can be determined by evaluating the eigenvalues of the closed-loop state transition matrix of the observer. The transition matrix is given by

$$\begin{bmatrix} \boldsymbol{A}-\boldsymbol{L}_x\boldsymbol{C} & \boldsymbol{B}_d-\boldsymbol{L}_x\boldsymbol{C}_d \\ -\boldsymbol{L}_d\boldsymbol{C} & \boldsymbol{I}-\boldsymbol{L}_d\boldsymbol{C}_d \end{bmatrix} \tag{6.23}$$

For an asymptotically stable observer, the poles need to lie within the unit disc. The matrices $\boldsymbol{L}_x$ and $\boldsymbol{L}_d$ of the observer have to be designed such that the observer becomes asymptotically stable. This implies that the estimation error vanishes, i.e. $\boldsymbol{C}\hat{\boldsymbol{x}} + \boldsymbol{C}_d\hat{\boldsymbol{d}} \rightarrow \boldsymbol{y}_m$. There are various ways to design the observer matrices $\boldsymbol{L}_x$, $\boldsymbol{L}_d$. In the classical Luenberger observer, $\boldsymbol{L}_x$, $\boldsymbol{L}_d$ are designed to achieve specific pole locations. Another popular choice is to use the theory of optimal linear estimation, resulting in a Kalman filter. More information on this approach in the context of MPC can be found in [19].

To sum up, the resulting MPC procedure for obtaining offset-free control is as follows: In every time step, a new measurement of the outputs $\boldsymbol{y}_m(k)$ is taken. Based on these measurements, the estimates of the disturbances and of the system states are calculated. These disturbance estimates subsequently are used within MPC for the internal prediction. In steady-state operation, this ensures that the predicted values of $\boldsymbol{y}$ used in the MPC are equal to the measured values $\boldsymbol{y}_m$. In the final step, the optimal actuated values $\Delta\boldsymbol{u}^*(\cdot|k)$ are calculated under consideration of the disturbance estimate.

The general algorithm for offset-free control is similar for LMPC and NMPC. In fact, the same procedure is used in both cases. For NMPC, a nonlinear observer is needed, for instance, an extended Kalman filter. The disturbance model itself can be chosen as in the linear case. The generalization for the nonlinear case is not treated within this section but can be found in [17].

6.2.2 Offset-Free Control Using a Deadbeat Observer

In some cases already a simple measure is sufficient to realize offset-free control in MPC. Due to the ease of implementation, it is widely applied in practical applications. First, the implementation is detailed, followed by an explanation that shows how this measure is actually a special case of the estimator framework introduced in Sect. 6.2.1 with a deadbeat observer. The estimator framework derivation delivers some further insight into this procedure.

The following state-space system is taken as a nominal controller-internal model.

$$\boldsymbol{x}(k+1) = \boldsymbol{A}\boldsymbol{x}(k) + \boldsymbol{B}\boldsymbol{u}(k) \tag{6.24}$$

$$\boldsymbol{y}(k) = \boldsymbol{C}\boldsymbol{x}(k) \tag{6.25}$$

The disturbance estimate $\hat{\boldsymbol{d}}(k+1)$ is calculated by

$$\hat{\boldsymbol{d}}(k+1) = \boldsymbol{y}_m(k) - \boldsymbol{C}\boldsymbol{x}(k) \tag{6.26}$$

The current measurement $\boldsymbol{y}_m(k)$ is used along with the states $\boldsymbol{x}(k)$ to calculate $\hat{\boldsymbol{d}}(k+1)$. The states $\boldsymbol{x}(k)$ can be derived by measurements of the states or by an open-loop estimation. In the prediction of the system behavior, the disturbance is assumed to be constant.

$$\hat{\boldsymbol{d}}(\cdot|k) = \hat{\boldsymbol{d}}(k+1) \tag{6.27}$$

For the prediction within MPC, the calculated disturbances are used as follows:

$$\boldsymbol{y}(\cdot|k) = \boldsymbol{\Gamma}\boldsymbol{x}(\cdot|k) + \hat{\boldsymbol{d}}(\cdot|k) \tag{6.28}$$

In many cases, this simple procedure already realizes offset-free control. The main reason for the simplicity comes from the fact that no observer needs to be designed explicitly. Instead, just the difference between the predicted outputs and the measurements has to be calculated.

To analyze this procedure and show its limitations, the algorithm is converted into an equivalent structure in the estimator framework as given in Sect. 6.2.1. The strategy described implicitly uses the following augmented model:

$$\begin{aligned} \boldsymbol{x}(k+1) &= \boldsymbol{A}\boldsymbol{x}(k) + \boldsymbol{B}\boldsymbol{u}(k) \\ \boldsymbol{d}(k+1) &= \boldsymbol{d}(k) \\ \boldsymbol{y}(k) &= \boldsymbol{C}\boldsymbol{x}(k) + \boldsymbol{d}(k) \end{aligned} \tag{6.29}$$

Compared to (6.19), the disturbance model here is characterized by the choices $\boldsymbol{B}_d = \boldsymbol{0}$ and $\boldsymbol{C}_d = \boldsymbol{I}$. Next, an observer gain matrix $\boldsymbol{L}$ is chosen that produces a closed-loop behavior that is equivalent to the strategy described above. To resemble (6.26), the observer gain matrix $\boldsymbol{L}$ has to be chosen as follows:

$$\boldsymbol{L} := \begin{bmatrix} \boldsymbol{L}_x \\ \boldsymbol{L}_d \end{bmatrix} = \begin{bmatrix} \boldsymbol{0} \\ \boldsymbol{I} \end{bmatrix} \tag{6.30}$$

Knowing these choices, the observer dynamics can be evaluated. The dynamics of the closed-loop observer become

$$\begin{bmatrix} \boldsymbol{x}(k+1) \\ \hat{\boldsymbol{d}}(k+1) \end{bmatrix} = \begin{bmatrix} \boldsymbol{A} & \boldsymbol{0} \\ \boldsymbol{0} & \boldsymbol{I} \end{bmatrix} \begin{bmatrix} \boldsymbol{x}(k) \\ \hat{\boldsymbol{d}}(k) \end{bmatrix} + \begin{bmatrix} \boldsymbol{B} \\ \boldsymbol{0} \end{bmatrix} \boldsymbol{u}(k) + \underbrace{\begin{bmatrix} \boldsymbol{0} \\ \boldsymbol{I} \end{bmatrix}}_{=:\, \boldsymbol{L}} \left(\boldsymbol{y}_m(k) - \boldsymbol{C}\boldsymbol{x}(k) - \hat{\boldsymbol{d}}(k) \right) \tag{6.31}$$

The strategy described above thus can be transformed into the estimator framework using the specific choices for the observer gain matrix $\boldsymbol{L}_x = \boldsymbol{0}$, $\boldsymbol{L}_d = \boldsymbol{I}$ and for the augmented model $\boldsymbol{B}_d = \boldsymbol{0}$ and $\boldsymbol{C}_d = \boldsymbol{I}$. To further analyze this strategy, especially to derive its limitations, the criteria for suitable disturbance estimation mentioned above can be evaluated. With these choices, the closed-loop state transition matrix of the observer is given by

$$\begin{bmatrix} \boldsymbol{A} & \boldsymbol{0} \\ -\boldsymbol{C} & \boldsymbol{0} \end{bmatrix} \tag{6.32}$$

The calculation of the eigenvalues of the state transition matrix shows that the poles of the observer are the poles of the nominal system and a number of poles at the origin. Clearly, using this method of estimation, an unstable nominal system will render an unstable observer. As asymptotic stability would be lost, the procedure cannot be applied for systems that are open-loop unstable or exhibit an integrative behavior. The additional poles at the origin make it a deadbeat observer. This is exactly where another limitation arises: as is shown in Example 6.2, all arising disturbances are directly fed through to the actuated values. This causes a problem for measurements with considerable noise. However, in the case of a constant disturbance, the maximum possible speed for a state estimation is given.

Example 6.2 (*Offset-free control with a deadbeat disturbance observer*) The system of Example 6.1 is revisited. A mismatch between the plant G_p and the MPC-internal model G_m is considered.

$$\text{Plant} \quad G_p = \frac{0.25}{z^2 - 1.65z + 0.75} \tag{6.33}$$

$$\text{MPC-internal Model} \quad G_m = \frac{0.25}{z^2 - 1.65z + 0.70} \tag{6.34}$$

Without additional measures, there is a steady-state offset for reference tracking, as shown in Example 6.1. Now, a deadbeat observer such as the one shown in Sect. 6.2.2 is employed to achieve offset-free control. Figure 6.3 shows the step response of a closed-loop system that includes a deadbeat observer. Two cases are depicted; measurements without and with noise. The model mismatch can be compensated using the disturbance estimation. As a result, offset-free reference tracking can be achieved. With the deadbeat observer, a very desirable control performance is achieved when no measurement noise is present. When measurement noise exists, the actuated values exhibit high fluctuations due to the deadbeat fashion of the observer estimation. For two reasons, this behavior is undesirable. It can damage the actuators, and the noise is propagated through the actuation onto the states and therefore onto the output.

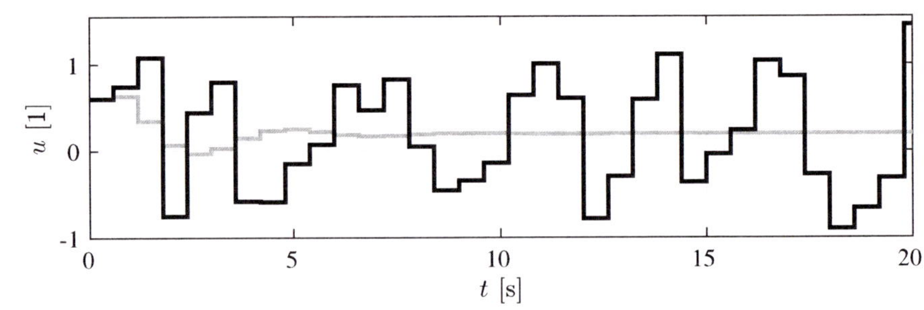

Fig. 6.3 Step response of the closed-loop system described in Example 6.2 using a deadbeat observer

To overcome the problem of measurement noise, the observer has to be designed in order to explicitly account for the measurement noise, i.e. by use of a Kalman filter. It reduces the convergence speed in favor of other desirable properties, such as a minimum variance estimation for a normally distributed noise.

6.3 Reference Tracking

In this section, the topic of reference tracking is detailed by the introduction of two methods. First, the delta formulation is explained which leads to reference tracking in a quite natural fashion. Second, a two-layered approach is presented. It is computationally slightly more demanding but offers advantages in certain applications, e.g. for non-square systems. Non-square systems are systems where the number of inputs and the number of outputs are different. The ideas are presented for the linear case. However, they can be extended to the nonlinear case. The same optimization problems can be used, just subject to the nonlinear instead of the linear model. Throughout the section, the reference values are assumed to be constant over time.

6.3.1 Delta Formulation

Appropriate reference tracking can be achieved in many cases by the delta formulation. The optimization problem takes into account the change of the actuated values $\Delta \boldsymbol{u}$. For instance, the following optimization problem can be used, where the disturbances are assumed to be constant in the prediction, i.e. $\boldsymbol{d}(\cdot|k) = \boldsymbol{d}(k+1)$.

$$\begin{aligned} \min_{\boldsymbol{x}(\cdot|k),\, \Delta\boldsymbol{u}(\cdot|k)} \quad & \sum_{i=N_1}^{N_2} \|\boldsymbol{C}\boldsymbol{x}(k+i|k) + \boldsymbol{C}_d\boldsymbol{d}(k+1) - \boldsymbol{r}\|_{\boldsymbol{Q}}^2 + \sum_{i=0}^{N_u-1} \|\Delta\boldsymbol{u}(k+i|k)\|_{\boldsymbol{R}}^2 \\ \text{s.t.} \quad & \text{system dynamics,} \\ & \text{initial conditions,} \\ & \text{constraints} \end{aligned} \tag{6.35}$$

For a stable closed-loop control system, the system can converge to a stationary point $(\boldsymbol{u}, \boldsymbol{x}, \boldsymbol{y}) \rightarrow (\boldsymbol{u}_\infty, \boldsymbol{x}_\infty, \boldsymbol{y}_\infty)$ with $\Delta\boldsymbol{u}_\infty = \boldsymbol{0}$. The stationary point is characterized by being the minimizer of the optimization problem. For offset-free stationary conditions, a cost of $J^* = 0$ is present. A higher value $J^* > 0$ results at stationary conditions if e.g. the actuated values are constrained such that the reference is not reachable. If the stationary point $\boldsymbol{r}$ is reachable and the closed-loop system is stable, the point is attained by a MPC controller with the given cost function. As a result, the delta formulation can quite naturally achieve suitable reference tracking. The disturbance variables $\boldsymbol{d}(k+1)$ can be calculated as detailed in Sect. 6.2.

Alternative formulations do not that easily solve the task of reference tracking. As counterexample to the delta formulation, a cost function is investigated that penalizes the absolute actuated values instead of its change. The cost function shall be given by

$$
\begin{aligned}
\min_{\boldsymbol{x}(\cdot|k),\,\boldsymbol{u}(\cdot|k)} \quad & \sum_{i=N_1}^{N_2} \|\boldsymbol{C}\boldsymbol{x}(k+i|k) + \boldsymbol{C}_d\boldsymbol{d}(k+1) - \boldsymbol{r}\|_{\boldsymbol{Q}}^2 + \sum_{i=0}^{N_u-1} \|\boldsymbol{u}(k+i|k)\|_{\boldsymbol{R}}^2 \\
\text{s.t.} \quad & \text{system dynamics,} \\
& \text{initial conditions,} \\
& \text{constraints}
\end{aligned}
\tag{6.36}
$$

The controller is able to regulate a linear system to the offset-free steady state with $\boldsymbol{u}_\infty = \boldsymbol{0}$ and $\boldsymbol{r} = \boldsymbol{C}\boldsymbol{x}_\infty + \boldsymbol{C}_d\boldsymbol{d}_\infty$, as long as this point is reachable. At this stationary point, the cost function reaches a minimum of $J^* = 0$. However, for reference values that can only be achieved with $\boldsymbol{u}_\infty \neq \boldsymbol{0}$, there is a steady-state offset, even if no constraints are considered. In steady-state conditions, a trade-off exists between the cost of tracking the reference values $\|\boldsymbol{C}\boldsymbol{x}_\infty + \boldsymbol{C}_d\boldsymbol{d}_\infty - \boldsymbol{r}\|_{\boldsymbol{Q}}^2$ and the penalization of the absolute value of the actuated values $\|\boldsymbol{u}_\infty\|_{\boldsymbol{R}}^2$. The two terms cannot simultaneously be reduced below a certain threshold. Hence, the steady-state solution depends on the weighting matrices $\boldsymbol{Q}$, $\boldsymbol{R}$. In contrast to the delta formulation shown in (6.35), the formulation with absolute values $\boldsymbol{u}$ in (6.36) does not allow for an offset-free tracking of arbitrary references, even if the references are reachable.

However, in specific applications, the delta formulation in (6.35) does not allow for enough flexibility. Consider a general linear system given in state-space representation, where the reference variables are denoted by $\boldsymbol{r}$. Clearly, the steady state $(\boldsymbol{u}_\infty, \boldsymbol{x}_\infty)$ of the system must satisfy the following condition for offset-free reference tracking:

$$
\boldsymbol{x}_\infty = \boldsymbol{A}\boldsymbol{x}_\infty + \boldsymbol{B}\boldsymbol{u}_\infty + \boldsymbol{B}_d\boldsymbol{d}_\infty \tag{6.37}
$$

$$
\boldsymbol{C}\boldsymbol{x}_\infty + \boldsymbol{C}_d\boldsymbol{d}_\infty = \boldsymbol{r} \tag{6.38}
$$

This can be written as follows:

$$
\begin{bmatrix} \boldsymbol{I}-\boldsymbol{A} & -\boldsymbol{B} \\ \boldsymbol{C} & \boldsymbol{0} \end{bmatrix} \begin{bmatrix} \boldsymbol{x}_\infty \\ \boldsymbol{u}_\infty \end{bmatrix} = \begin{bmatrix} \boldsymbol{B}_d\boldsymbol{d}_\infty \\ \boldsymbol{r} - \boldsymbol{C}_d\boldsymbol{d}_\infty \end{bmatrix} \tag{6.39}
$$

The structure of the solution of this equation system depends on the number of controlled outputs m compared to the number of actuated values l.

Systems with as many actuated values as controlled values $m = l$ are called *square systems*. With a full-row rank matrix, there exists one unique solution. The delta formulation is regulating the system to this specific point, as long as the closed-loop system is stable and the point is reachable.

Systems with more controlled than actuated values $m > l$ are called *underactuated* and belong to the category of non-square systems. For a solution to exist for all $\boldsymbol{r}$, the matrix in (6.39) needs to have linearly independent rows and thus $m \leq l$. As a consequence, in the case of underactuated systems, no solution can be guaranteed for arbitrary values of $\boldsymbol{r}$, even if no constraints are active. In general, it is not guaranteed that the reference point can be reached. The system is regulated to a point that minimizes the given cost function. For the delta formulation, the resulting solution in steady state depends on the initial states, i.e. no unique steady states result.

Systems with more actuated than controlled values $l > m$ are called *overactuated* and belong to the category of non-square systems. With a full-row rank matrix, there can exist multiple instead of one unique solution. Thus, there exist degrees of freedom for achieving offset-free control. The delta formulation is regulating the system to the reference point, i.e. $\boldsymbol{C}\boldsymbol{x} + \boldsymbol{C}_d\boldsymbol{d} \rightarrow \boldsymbol{r}$, as long as the closed-loop system is stable and this point is reachable. However, for overactuated systems, the resulting states in steady state $\boldsymbol{x}_\infty$ and the steady-state actuated values $\boldsymbol{u}_\infty$ depend on the initial states, i.e. no unique steady states result.

The delta formulation, such as in (6.35), leads to steady-state values that minimize the cost function. However, for the non-square systems, for two different initial conditions but same reference values, two different solutions in terms of $\boldsymbol{x}_\infty, \boldsymbol{u}_\infty$ might be found. This is shown in Example 6.3. For non-square systems often a particular solution is preferred. For instance, while one actuator is expensive to use, the other one is cheap to use. Thus, it is desirable to apply the cheap one as much as possible. Another example is the classical mid-ranging control algorithm where the actuated values are tried to be kept at medium values in steady-state conditions. The regulation of the actuated values to medium values allows a high control authority in both directions, e.g. in order to quickly reject disturbances. These specific requirements cannot be handled with the delta formulation. Instead, a two-layered approach has to be chosen, which offers more flexibility at the expense of higher computational demands, as explained below.

6.3.2 Two-Layered Control Structure

In some applications, a desired steady-state value for the actuated values or for the system states does exist. The steady-state target values $\boldsymbol{x}_{SS}, \boldsymbol{u}_{SS}$ need either to be known or need to be computed explicitly. Typically, there exist requirements on them, e.g. the absolute value should be as small as possible to decrease energy consumption. However, they are usually not known a priori, especially as the values $\boldsymbol{x}_{SS}, \boldsymbol{u}_{SS}$ change in dependence of the reference values and the recent disturbance states $\boldsymbol{d}$. Hence, these values need to be recalculated in every time step.

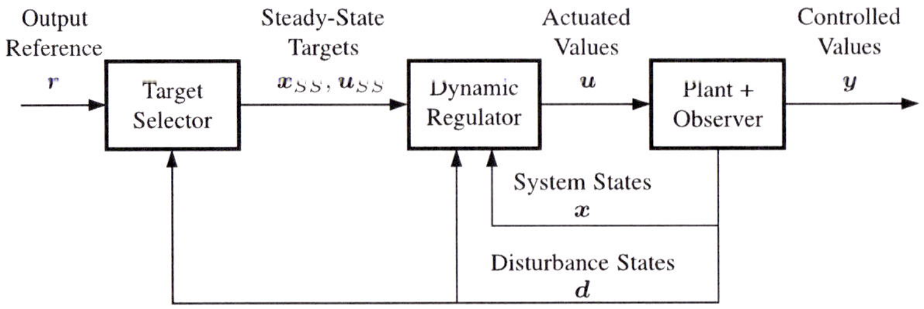

Fig. 6.4 Two-layered control structure with target selector

In the following, a two-layered control structure is introduced that addresses these requirements [19]. The structure of this two-layered control approach is shown in Fig. 6.4. The actuated values are calculated by solving two optimization problems sequentially in each time step. First, the ideal steady-state target values are calculated by the so-called target selector. The target selector only takes the steady-state dynamics into account. Subsequently, within the dynamic regulator, the actuated values are calculated which are applied to the plant. The solution of the target selector serves as input for the dynamic regulator where the entire system dynamics including the transient behavior are considered. As two optimization problems need to be solved, this formulation incurs additional overhead in terms of computational cost compared to the delta formulation.

Target Selector

The objective of the target selector is to find the optimal steady-state values for the actuated values $\boldsymbol{u}_{SS}$ and for the system states $\boldsymbol{x}_{SS}$. The values need to be steady-state values. The steady-state values are supposed to lead to the reference values, fulfilling (6.39). Additionally, the constraints on the system states, defined by $\boldsymbol{x}_{SS} \in \mathcal{X}$, and the constraints on the actuated values, defined by $\boldsymbol{u}_{SS} \in \mathcal{U}$, have to be considered. The definition of the "optimal" target values depends on the specific task. In Example 6.3, the actuated values are associated with energy consumption, such that a solution with a minimal absolute value of actuated values is requested. For this specific case, a possible optimization problem for a target selector is given by

$$\begin{aligned} \min_{\boldsymbol{u}_{SS}, \boldsymbol{x}_{SS}} \quad & \|\boldsymbol{u}_{SS}\|^2_{\boldsymbol{S}_T} \\ \text{s.t.} \quad & \begin{bmatrix} \boldsymbol{I} - \boldsymbol{A} & -\boldsymbol{B} \\ \boldsymbol{C} & \boldsymbol{0} \end{bmatrix} \begin{bmatrix} \boldsymbol{x}_{SS} \\ \boldsymbol{u}_{SS} \end{bmatrix} = \begin{bmatrix} \boldsymbol{B}_d \boldsymbol{d}_\infty \\ \boldsymbol{r} - \boldsymbol{C}_d \boldsymbol{d}_\infty \end{bmatrix}, \\ & \boldsymbol{x}_{SS} \in \mathcal{X}, \\ & \boldsymbol{u}_{SS} \in \mathcal{U} \end{aligned} \tag{6.9}$$

The optimization problem takes into account the steady-state value of the disturbance $\boldsymbol{d}_\infty$. As the disturbances are assumed to be constant in the prediction, the value $\boldsymbol{d}_\infty := \boldsymbol{d}(k+1)$ is typically used. A weighting matrix $\boldsymbol{S}_T$ is introduced to allow prioritization of the various actuated variables. The definition of the cost function enables calculation of a unique steady state, even for an overactuated system. In this formulation, constraints on actuated values and system states are considered while simultaneously the outputs are enforced to be equal to the reference values. In this case, feasibility issues can arise, i.e. the references $\boldsymbol{r}$ may not be reachable in steady state with the constraints imposed. As a result, an infeasible optimization problem is present. This can be handled via relaxation of the equality constraints by adding slack variables as detailed in Sect. 6.1.

Dynamic Regulator

The objective of the dynamic regulator is to regulate the system values to the steady-state values $\boldsymbol{x}_{SS}$, $\boldsymbol{u}_{SS}$ that are calculated by the target selector. The regulation is conducted such that the transient behavior toward the steady-state values is considered by including the system dynamics. A possible optimization problem for the dynamic regulator is given by (6.41). All in all, this procedure allows to track the reference without any offset if the reference is reachable. Additionally, even for an overactuated system, the system is always steered to the same steady-state values, independent of its initial condition. Assuming the disturbances to be constant, i.e. $\boldsymbol{d}(\cdot|k) = \boldsymbol{d}(k+1)$, the following optimization problem results:

$$\begin{aligned} \min_{\boldsymbol{x}(\cdot|k),\,\boldsymbol{u}(\cdot|k)} \quad & \sum_{i=N_1}^{N_2} \|\boldsymbol{C}(\boldsymbol{x}(k+i|k) - \boldsymbol{x}_{SS}) + \boldsymbol{C}_d\boldsymbol{d}(k+1)\|_{\boldsymbol{Q}}^2 + \sum_{i=0}^{N_u-1} \|\boldsymbol{u}(k+i|k) - \boldsymbol{u}_{SS}\|_{\boldsymbol{R}}^2 \\ \text{s.t.} \quad & \text{system dynamics,} \\ & \text{initial conditions,} \\ & \text{constraints} \end{aligned} \tag{6.41}$$

Example 6.3 (*Reference tracking with the delta formulation*) In this example, the properties of the delta formulation are exemplified. For this purpose, an overactuated system is closed-loop-controlled with an MPC algorithm. The MPC uses the real system model and no disturbances are considered, i.e. $\boldsymbol{d} = \boldsymbol{0}$. As a plant, a system with two actuated values and only one controlled value is given. The exact discretization with a sampling time of $T_s = 0.05$ s leads to the following discrete-time system in state-space representation:

$$\begin{aligned} \boldsymbol{A}_D &= \begin{bmatrix} 1 & 0.05 \\ 0 & 1 \end{bmatrix} & \boldsymbol{B}_D &= \begin{bmatrix} 0.0013 & 0.0013 \\ 0.05 & 0.05 \end{bmatrix} \\ \boldsymbol{C}_D &= \begin{bmatrix} 1 & 0 \end{bmatrix} & \boldsymbol{D}_D &= \begin{bmatrix} 0 & 0 \end{bmatrix} \end{aligned} \tag{6.42}$$

The MPC controller is applied for tracking a constant reference value r. As reference value $r = 1$ is chosen. A prediction and a control horizon of $N = 20$ are applied. The cost function of the MPC is given by the delta formulation. Additionally, constraints on the actuated values are considered. The optimization problem is given by

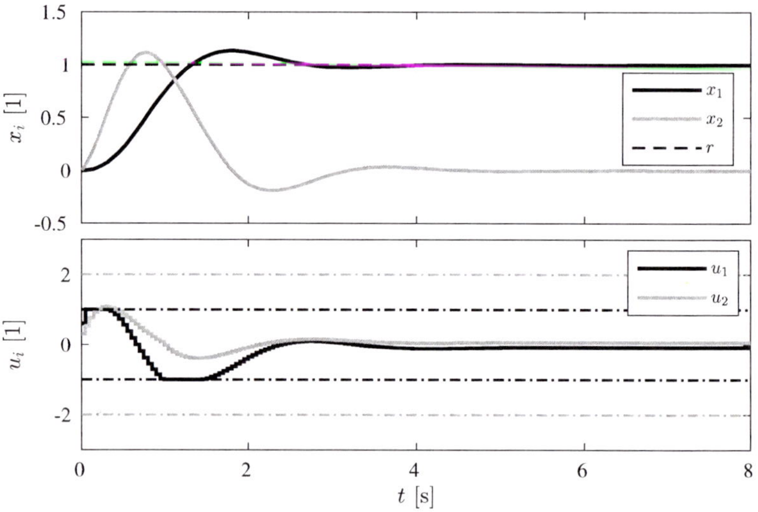

Fig. 6.5 Simulative closed-loop control result from Example 6.3 with $\boldsymbol{x}_0 = [0 \ \ 0]^\mathsf{T}$

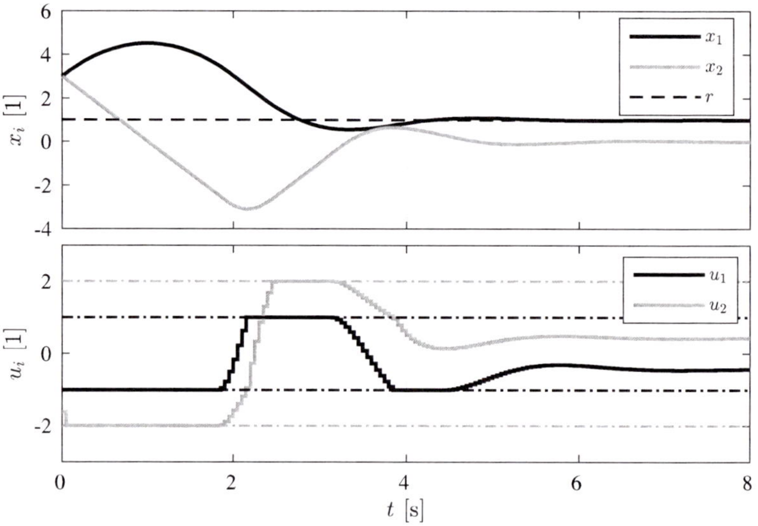

Fig. 6.6 Simulative closed-loop control result from Example 6.3 with $\boldsymbol{x}_0 = [3 \ \ 3]^\mathsf{T}$

$$
\begin{aligned}
\min_{\boldsymbol{x}(\cdot|k),\, \boldsymbol{u}(\cdot|k),\, \Delta\boldsymbol{u}(\cdot|k)} \quad & \sum_{i=1}^{N} \|\boldsymbol{C}\boldsymbol{x}(k+i|k) - r\|_{\boldsymbol{Q}}^2 + \sum_{i=0}^{N-1} \|\Delta\boldsymbol{u}(k+i|k)\|_{\boldsymbol{R}}^2 \\
\text{s.t.} \quad & \boldsymbol{x}(k+i+1|k) = \boldsymbol{A}\boldsymbol{x}(k+i|k) + \boldsymbol{B}\boldsymbol{u}(k+i|k),\ i = 0, ..., N-1, \\
& \boldsymbol{x}(k|k) = \boldsymbol{x}_0, \\
& \boldsymbol{u}(k+i|k) = \boldsymbol{u}(k-1+i|k) + \Delta\boldsymbol{u}(k+i|k),\ i = 1, ..., N-1, \\
& \boldsymbol{u}(k|k) = \Delta\boldsymbol{u}(k|k), \\
& \begin{bmatrix} -1 \\ -2 \end{bmatrix} \leq \boldsymbol{u}(k+i|k) \leq \begin{bmatrix} 1 \\ 2 \end{bmatrix},\ i = 0, ..., N-1
\end{aligned}
\tag{6.43}
$$

The closed-loop control simulation is conducted with the initial condition $\boldsymbol{x}_0 = [0\ 0]^\mathsf{T}$. Figure 6.5 shows the simulative closed-loop control result. Another simulation with the initial state $\boldsymbol{x}_0 = [3\ \ 3]^\mathsf{T}$ is shown in Fig. 6.6, where all other settings are identical to the preceding case. The two examples illustrate the properties of the delta formulation; the reference value is reached in both cases in steady state. However, the steady-state actuated values depend on the initial condition. If the actuation is associated with energy consumption, it is preferable that the actuated values are rather set to the resting position $\boldsymbol{u}_\infty = \boldsymbol{0}$ in steady state. The resting position is a feasible solution for tracking the reference, i.e. the system exhibits an integrating behavior.

Example 6.4 (*Reference tracking with the two-layered control approach*)
The system of Example 6.3 is revisited. Now, the MPC approach is based on the two-layered control structure. The cost function of the dynamic regulator is extended by an additional term which takes the deviation from the steady-state actuated values into account.

$$
J = \sum_{i=1}^{N} \|\boldsymbol{C}\boldsymbol{x}(k+i|k) - \boldsymbol{C}\boldsymbol{x}_{SS}\|_{\boldsymbol{Q}}^2 + \sum_{i=0}^{N-1} \|\boldsymbol{u}(k+i|k) - \boldsymbol{u}_{SS}\|_{\boldsymbol{S}}^2 + \sum_{i=0}^{N-1} \|\Delta\boldsymbol{u}(k+i|k)\|_{\boldsymbol{R}}^2 \tag{6.44}
$$

For the dynamic regulator, all other settings such as weighting parameters, are used as those given in Example 6.3. The optimization problem of the dynamic regulator is given by

$$
\begin{aligned}
\min_{\boldsymbol{x}(\cdot|k),\, \boldsymbol{u}(\cdot|k),\, \Delta\boldsymbol{u}(\cdot|k)} \quad & J \\
\text{s.t.} \quad & \boldsymbol{x}(k+i+1|k) = \boldsymbol{A}\boldsymbol{x}(k+i|k) + \boldsymbol{B}\boldsymbol{u}(k+i|k),\ i = 0, ..., N-1, \\
& \boldsymbol{x}(k|k) = \boldsymbol{x}_0, \\
& \boldsymbol{u}(k+i|k) = \boldsymbol{u}(k-1+i|k) + \Delta\boldsymbol{u}(k+i|k),\ i = 1, ..., N-1, \\
& \boldsymbol{u}(k|k) = \Delta\boldsymbol{u}(k|k), \\
& \begin{bmatrix} -1 \\ -2 \end{bmatrix} \leq \boldsymbol{u}(k+i|k) \leq \begin{bmatrix} 1 \\ 2 \end{bmatrix},\ i = 0, ..., N-1
\end{aligned}
\tag{6.45}
$$

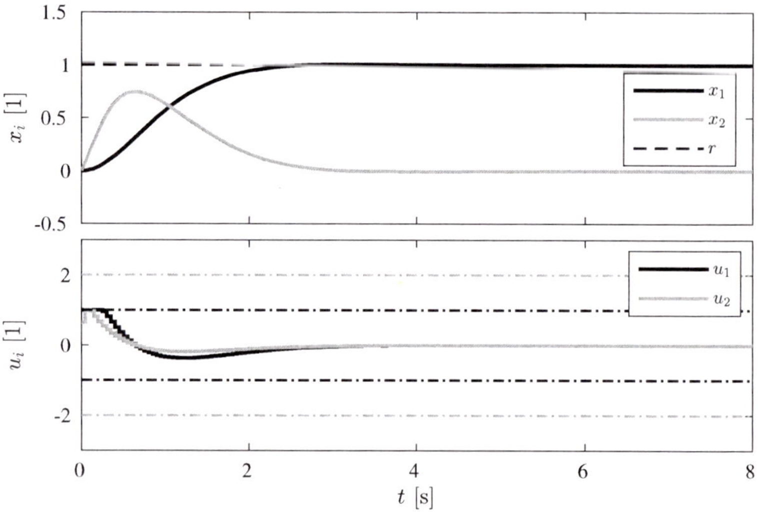

Fig. 6.7 Simulative closed-loop control result from Example 6.4 with $\boldsymbol{x}_0 = [0, 0]^\mathsf{T}$

The steady-state actuated values $\boldsymbol{u}_{SS}$ are calculated by a target selector where the following optimization problem is solved. The optimization problem considers that the actuated values are associated with energy cost. Thus, the absolute values of both actuated values are minimized.

$$\begin{aligned} \min_{\boldsymbol{u}_{SS},\, \boldsymbol{x}_{SS}} \quad & \|\boldsymbol{u}_{SS}\|^2_{\boldsymbol{S}_T} \\ \text{s.t.} \quad & \begin{bmatrix} \boldsymbol{I} - \boldsymbol{A} & -\boldsymbol{B} \\ \boldsymbol{C} & \boldsymbol{0} \end{bmatrix} \begin{bmatrix} \boldsymbol{x}_{SS} \\ \boldsymbol{u}_{SS} \end{bmatrix} = \begin{bmatrix} \boldsymbol{0} \\ \boldsymbol{r} \end{bmatrix}, \\ & \begin{bmatrix} -1 \\ -2 \end{bmatrix} \leq \boldsymbol{u}_{SS} \leq \begin{bmatrix} 1 \\ 2 \end{bmatrix} \end{aligned} \tag{6.46}$$

The optimal target values for a reference of $r = 1$ are given by $\boldsymbol{u}_{SS} = [0\ \ 0]^\mathsf{T}$, $\boldsymbol{x}_{SS} = [1\ \ 0]^\mathsf{T}$. Figures 6.7 and 6.8 show the closed-loop simulation starting with the initial condition $\boldsymbol{x}_0 = [0\ \ 0]^\mathsf{T}$ and $\boldsymbol{x}_0 = [3\ \ 3]^\mathsf{T}$. The steady-state actuated values now are independent of the initial conditions. As desired, they move to the resting position $\boldsymbol{u}_\infty = \boldsymbol{u}_{SS} = [0\ \ 0]^\mathsf{T}$ in steady state.

6.4 Stability Mechanisms

In order to operate a system with MPC, in addition to the recursive feasibility of the optimization problem, the closed-loop system has to be stable. While the actuated values are optimal with respect to the optimization problem formulated, the closed-

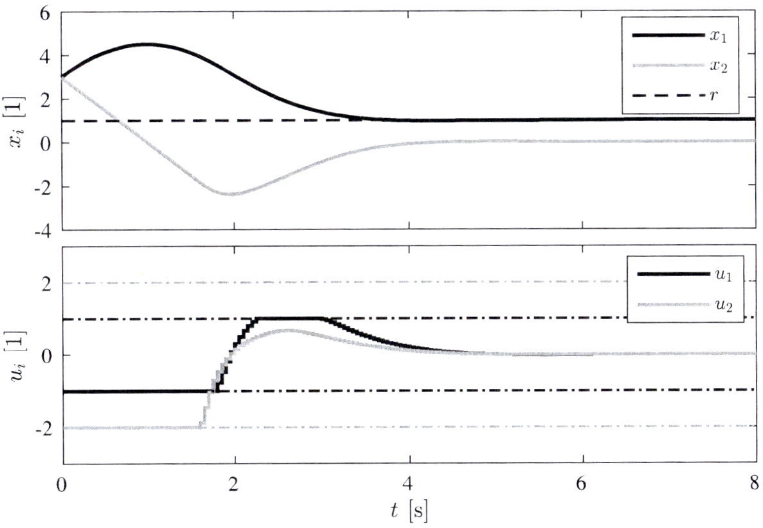

Fig. 6.8 Simulative closed-loop control result from Example 6.4 with $\boldsymbol{x}_0 = [3, 3]^\mathsf{T}$

loop system can still be unstable. This holds true even for nominal stability, which is the focus of this section. For nominal stability, nominal conditions are assumed, i.e. the exact system behavior and the system states are known and no disturbances are present. The nominal stability of the closed-loop system depends on many factors of the MPC algorithm, such as the weighting factors, the length of the prediction horizon, and the constraints considered.

An obvious complication about proving the stability of the closed-loop system is the fact that the actuated values are computed during the runtime of the process. Therefore, the closed-loop system dynamics are not known beforehand which complicates its system analysis. Since a constrained MPC-controlled system is necessarily nonlinear, nonlinear stability theory has to be applied. Almost all stability proofs for MPC schemes are based on the Lyapunov stability, thus in the remainder of this section, the notion of Lyapunov stability is used.

There exist several formulations of the MPC optimization problem that are theoretically well founded with respect to their stability properties. They offer rigorous stability proofs such that, under certain assumptions, nominal closed-loop stability is guaranteed. However, these MPC formulations are typically not suited for practical applications. They can become computationally very complex and can be very limiting concerning the initial states that result in a feasible solution of the optimization problem [20]. Still, the investigation of nominal stability allows for mechanisms to be developed that are advantageous for achieving closed-loop stability in practical applications, although the guarantees are often lost. In the following, some possible formulations for achieving closed-loop stability are sketched. Instead of a rigorous treatment including proofs and all technicalities, the ideas of the formulations are depicted and the implications for practical usage are discussed. There are good monographs treating the topic of stability for MPC in detail, especially [6, 19].

Lyapunov Stability

In the following, the Lyapunov stability is introduced. It is concerned with the stability of the equilibrium point $\boldsymbol{x}_{eq}$ of an autonomous, nonlinear dynamical system [11]. The discrete-time system shall be given by $\boldsymbol{x}(k+1) = \boldsymbol{f}(\boldsymbol{x}(k))$ with $\boldsymbol{f} : \mathcal{D} \rightarrow \mathbb{R}^n$, and f shall be locally Lipschitz in $\mathcal{D} \subset \mathbb{R}^n$. In this section, the origin is assumed to be an equilibrium point, i.e. $\boldsymbol{x}_{eq} = \boldsymbol{0}$. This assumption holds without loss of generality since the coordinate transformation $\tilde{\boldsymbol{x}} = \boldsymbol{x} - \boldsymbol{x}_{eq}$ can be used to shift any equilibrium point to the origin, i.e. $\tilde{\boldsymbol{x}}_{eq} = \boldsymbol{0}$. An equilibrium point $\boldsymbol{x}_{eq}$ of a system is Lyapunov-stable if any solution that starts close to the equilibrium stays close to it for all time. Otherwise, the point $\boldsymbol{x}_{eq}$ is unstable in the sense of Lyapunov. More formally, the equilibrium point $\boldsymbol{x}_{eq} = \boldsymbol{0}$ is *stable* if for all $\epsilon > 0$, there is a $\delta(\epsilon) > 0$ such that

$$||\boldsymbol{x}(0)|| < \delta(\epsilon) \rightarrow ||\boldsymbol{x}(k)|| < \epsilon, \ \forall k \in \{0, 1, \dots\} \tag{6.47}$$

The point $\boldsymbol{x}_{eq}$ is *asymptotically stable* if it is stable and additionally, there is a region around $\boldsymbol{x}_{eq}$ from where the solution converges to the equilibrium point with increasing time. Thus, a region $r > 0$ needs to exist, such that

$$||\boldsymbol{x}(0)|| < r \rightarrow \lim_{k \rightarrow \infty} \boldsymbol{x}(k) = \boldsymbol{0} \tag{6.48}$$

If the equilibrium point is stable and additionally for any $\boldsymbol{x}(0) \in \mathbb{R}^n$, the system converges to the origin, it is called *globally asymptotic stable*. The following criteria need to be fulfilled:

$$\forall \boldsymbol{x}(0) \rightarrow \lim_{k \rightarrow \infty} \boldsymbol{x}(k) = \boldsymbol{0} \tag{6.49}$$

To prove stability of an equilibrium point, the so-called Lyapunov function can be used. The existence of a Lyapunov function provides a sufficient but not necessary condition for stability. A Lyapunov candidate function is a continuous function $V : \mathcal{D} \rightarrow \mathbb{R}$ for which the following holds:

$$V(\boldsymbol{0}) = 0 \text{ and } V(\boldsymbol{x}) > 0, \forall \boldsymbol{x} \in \mathcal{D} \backslash \{\boldsymbol{0}\} \tag{6.50}$$

The equilibrium point $\boldsymbol{x}_{eq} = \boldsymbol{0}$ is stable if a Lyapunov candidate function V exists such that

$$V(\boldsymbol{f}(\boldsymbol{x})) - V(\boldsymbol{x}) \leq 0, \forall \boldsymbol{x} \in \mathcal{D} \backslash \{\boldsymbol{0}\} \tag{6.51}$$

Moreover, the equilibrium point $\boldsymbol{x}_{eq} = \boldsymbol{0}$ is asymptotically stable if a Lyapunov candidate function V exists such that

$$V(\boldsymbol{f}(\boldsymbol{x})) - V(\boldsymbol{x}) < 0, \forall \boldsymbol{x} \in \mathcal{D} \backslash \{\boldsymbol{0}\} \tag{6.52}$$

In the following, several MPC formulations are discussed which lead to guaranteed closed-loop stability. These MPC schemes construct a cost function that is at the same time a Lyapunov function for the closed-loop system.

6.4.1 *Stability of the Infinite-Horizon MPC*

First, a MPC problem with infinite horizon is investigated. The analysis of this problem helps to understand the various formulations that can be employed to show stability of the conventional finite-horizon MPC scheme. The resulting cost for infinite prediction $N = \infty$ is denoted as $J_\infty(\boldsymbol{x}(k))$. Let the stage cost $l(\boldsymbol{x}, \boldsymbol{u})$ be a non-negative function which is decrescent and where $l = 0$ holds only for the steady-state conditions, which shall be given by $\boldsymbol{x} = \boldsymbol{0}, \boldsymbol{u} = \boldsymbol{0}$. A suitable and common stage cost for reference tracking the origin is given by the weighted quadratic 2-norm, i.e. $l(\boldsymbol{x}, \boldsymbol{u}) = \|\boldsymbol{x}(k+i|k)\|_{\boldsymbol{Q}}^2 + \|\boldsymbol{u}(k+i|k)\|_{\boldsymbol{R}}^2$ with positive definite weighting matrices $\boldsymbol{Q}, \boldsymbol{R}$. The infinite-horizon problem shall be given by

$$
\begin{aligned}
\min_{\boldsymbol{x}(\cdot|k),\, \boldsymbol{u}(\cdot|k)} \quad & J_\infty(\boldsymbol{x}(k)) = \sum_{i=0}^{\infty} \left(\|\boldsymbol{x}(k+i|k)\|_{\boldsymbol{Q}}^2 + \|\boldsymbol{u}(k+i|k)\|_{\boldsymbol{R}}^2 \right) \\
\text{s.t.} \quad & \boldsymbol{x}(k+i+1|k) = \boldsymbol{f}(\boldsymbol{x}(k+i|k), \boldsymbol{u}(k+i|k)),\ i = 0, ..., \infty, \\
& \boldsymbol{x}(k|k) = \boldsymbol{x}(k), \\
& \boldsymbol{x}(k+i|k) \in \mathcal{X} \qquad i = 1, ..., \infty, \\
& \boldsymbol{u}(k+i|k) \in \mathcal{U} \qquad i = 0, ..., \infty
\end{aligned}
\tag{6.53}
$$

For nominal conditions, i.e. perfect knowledge of the system and without disturbances, the open-loop trajectory is the same as the closed-loop trajectory for infinite predictions. The reason behind this is Bellman's principle of optimality. It states that any subtrajectory of an optimal trajectory must itself be optimal. This property facilitates to show that the cost function is a Lyapunov function [9, 15]. If some conditions are fulfilled, such as stabilizability, the infinite-horizon problem ensures an asymptotically stable closed-loop system. Additionally, a feasible solution of the first optimization problem (6.53) needs to exist, which directly leads to recursive feasibility of the optimization problem. The feasible set of the first optimization problem can be empty, i.e. there does not exist any solution of the infinite-horizon MPC. However, if there does exist a solution, it is indeed found by solving the infinite-horizon MPC. In fact, in this case, the optimization problem needs to be solved only once, since all subsequent iterations lead to the same solution, shifted by the actuated values already applied. This procedure assumes that a solution can be found; however, the optimization problem cannot be solved in practice due to the infinite number of optimization variables. Still, the setup helps to develop stability mechanisms for the finite-horizon MPC.

6.4.2 *Stability of Finite-Horizon MPC*

There exists a key difference between the finite-horizon and the infinite-horizon MPC. In general, due to the finite-horizon prediction in MPC, the open-loop trajectory no longer is the same as the closed-loop trajectory. This is even true in nominal conditions. The common way to still guarantee nominal stability for MPC is the introduction of a stability mechanism in the optimization problem, i.e. by a modification of the cost function or the constraints. This can be interpreted as an approximation of the infinite-horizon MPC. The rigorous inclusion of these mechanisms offers a sufficient condition for nominal stability. As these are sufficient conditions, stable closed-loop control systems do exist which do not include these stability mechanisms. This is a major conceptual difference from classical control loops where necessary conditions for stability exist. The approaches for stabilizing MPC can be divided into two subdomains:

1. By the use of stability constraints within the MPC formulation, e.g. by the inclusion of a terminal cost and/or a terminal constraint.
2. Approaches without stability constraints, which are also called "unconstrained MPC". This refers to the fact that neither a terminal constraint nor a terminal cost is used, still conventional constraints can be considered.

These subdomains can be combined into hybrid strategies which result in a large number of distinct MPC formulations that each have their advantages and drawbacks. In the following, these two subdomains are introduced.

Stability-Constrained MPC with Terminal Equality Constraint

In the following, finite-horizon MPC formulations with stability constraints are discussed. In this case, the optimization problem is modified by the inclusion of terminal costs and/or terminal states. The term stability constraints refers to the fact that these are added not due to physical reasons of the application, but in order to achieve stability.

First, a simple mechanism is discussed where nominal stability is ensured by the inclusion of a terminal equality constraint. The goal is to track a certain reference. Without loss of generality, the regulation to the origin is exemplified where the origin is assumed to be a steady state with $\boldsymbol{x} = \mathbf{0}, \boldsymbol{u} = \mathbf{0}$. As discussed above, for points that are not at the origin, a shift can always be included for $\boldsymbol{x}$ and $\boldsymbol{u}$. The idea is to use a finite prediction horizon and to introduce an equality constraint at the end of the prediction horizon, i.e. $\boldsymbol{x}(k + N|k) = \mathbf{0}$. The resulting optimization problem thus reads as follows:

$$
\begin{aligned}
\min_{\boldsymbol{x}(\cdot|k),\,\boldsymbol{u}(\cdot|k)} \quad & \sum_{i=0}^{N-1} (\|\boldsymbol{x}(k+i|k)\|_{\boldsymbol{Q}}^2 + \|\boldsymbol{u}(k+i|k)\|_{\boldsymbol{R}}^2) \\
\text{s.t.} \quad & \boldsymbol{x}(k+i+1|k) = \boldsymbol{f}(\boldsymbol{x}(k+i|k), \boldsymbol{u}(k+i|k)),\ i = 0, ..., N-1, \\
& \boldsymbol{x}(k|k) = \boldsymbol{x}(k), \\
& \boldsymbol{x}(k+i|k) \in \mathcal{X} \qquad i = 1, ..., N-1, \\
& \boldsymbol{u}(k+i|k) \in \mathcal{U} \qquad i = 0, ..., N-1, \\
& \boldsymbol{x}(k+N|k) = \boldsymbol{0}
\end{aligned}
\tag{6.54}
$$

Since the origin is assumed to be an equilibrium point of the system, the additional constraint $\boldsymbol{x}(k+N|k) = \boldsymbol{0}$ enforces that $\boldsymbol{x}(k+N+i|k) = \boldsymbol{0}\, \forall i \geq N$, i.e. the system stays at the origin without any control action. Thus, no additional cost is incurred after reaching this terminal constraint. This property already renders the cost function resulting from (6.54) a Lyapunov function that ensures asymptotic stability [9]. This holds true for every initial state where a feasible solution of the optimization problem does exist. Recursive feasibility is directly given: since there exists a solution for the initial iteration ending at the origin, there always exists a feasible solution for the subsequent iterations. Namely, the solution found in the preceding iteration shifted by the first element. However, this is not necessarily the solution of the subsequent iteration, since a trajectory with lower cost may be found.

The major drawback of this method is the fact that the inclusion of the equality constraint can be unnecessarily restrictive. With the infinite-horizon formulation, every theoretically possible initial state will lead to a feasible solution of the optimization problem. In contrast, with the equality-constrained MPC formulation, the region of attraction may become very small, as the final state has to be reached in N time steps. The region of attraction is defined as all initial states that result in a feasible trajectory. Its size depends strongly on the prediction horizon N. If the prediction horizon is rather short, the feasible set can even be empty. Additionally, the stability mechanism assumes that a numerical solution of the optimization algorithm can be found in each time step. However, the inclusion of the equality constraints makes numerical problem solving harder. In conclusion, the terminal equality constraint offers a simple, sufficient condition for asymptotic stability, but it significantly reduces the initial states that result in a feasible solution.

Stability-Constrained MPC with Terminal Set and Terminal Cost

To increase the region of attraction, a terminal set and/or a terminal cost can be introduced. The modified MPC contains the terminal set around the origin denoted by $\mathcal{X}_f$ and the terminal cost l_f. A typical choice for terminal cost is given by $l_f = \|\boldsymbol{x}(k+N|k)\|_{\boldsymbol{P}}^2$. The modified optimization problem results to be

$$
\begin{aligned}
\min_{\boldsymbol{x}(\cdot|k),\, \boldsymbol{u}(\cdot|k)} \quad & \|\boldsymbol{x}(k+N|k)\|_{\boldsymbol{P}}^2 + \sum_{i=0}^{N-1} \left(\|\boldsymbol{x}(k+i|k)\|_{\boldsymbol{Q}}^2 + \|\boldsymbol{u}(k+i|k)\|_{\boldsymbol{R}}^2\right) \\
\text{s.t.} \quad & \boldsymbol{x}(k+i+1|k) = \boldsymbol{f}(\boldsymbol{x}(k+i|k), \boldsymbol{u}(k+i|k)),\ i = 0, ..., N-1, \\
& \boldsymbol{x}(k|k) = \boldsymbol{x}(k), \\
& \boldsymbol{x}(k+i|k) \in \mathcal{X} \qquad i = 1, ..., N-1, \\
& \boldsymbol{u}(k+i|k) \in \mathcal{U} \qquad i = 0, ..., N-1, \\
& \boldsymbol{x}(k+N|k) \in \mathcal{X}_f
\end{aligned}
\tag{6.55}
$$

Various measures exist to design the terminal set $\mathcal{X}_f$, the terminal cost l_f, or the combination of both to guarantee stability [15]. The measures typically give an upper bound on the infinite-horizon cost and thus can guarantee a decrease in the cost function in each time step which makes the cost function a Lyapunov function [1].

One idea for the terminal set is to drive the states into the set $\mathcal{X}_f$ [16]. Once inside the terminal set, the controller is switched to a precomputed control law $\boldsymbol{u} = \kappa_f(\boldsymbol{x})$ which is stabilizing within $\mathcal{X}_f$. This approach can only work if $\mathcal{X}_f$ is reachable in at most N steps from the initial condition. This condition is less restrictive than reaching the terminal equality constraints in N steps. Many approaches use a combination of a terminal state with a terminal cost. Due to the inclusion of the terminal cost, the controller does not need to be switched within the terminal set; see for instance [3]. If the terminal costs are chosen appropriately, they act as a stabilizing controller within the terminal set $\mathcal{X}_f$.

There also exist approaches that rely on the terminal cost, without using a terminal set [8, 12]. The effect of using the terminal cost l_f is that it drives the final state closer to the origin. The terminal cost can be seen as the tail of the infinite-horizon MPC cost, as it approximates the cost of the time steps from $N+1$ to ∞. If the tail of the infinite-horizon cost is captured appropriately in l_f, then the stability guarantees of the infinite-horizon MPC are recovered [8]. The cost function again constitutes a Lyapunov function.

For linear systems, the terminal cost $\boldsymbol{P}$ and the terminal states $\mathcal{X}_f$ can be chosen to guarantee stability and recursive feasibility based on some results of the unconstrained infinite-horizon optimal controller, i.e. the linear quadratic regulator (LQR) [2]. The terminal cost can be designed by calculating the unique stable solution of the discrete algebraic Riccati equation (DARE). The DARE arises in unconstrained infinite-horizon optimal control with $\boldsymbol{A}$, $\boldsymbol{B}$ being the system matrices and $\boldsymbol{Q}, \boldsymbol{R}$ the weighting matrices. The terminal cost $\boldsymbol{P}$ can be calculated by solving the following equation:

$$
\boldsymbol{P} = \boldsymbol{A}^{\mathsf{T}}\boldsymbol{P}\boldsymbol{A} + \boldsymbol{Q} - \boldsymbol{A}^{\mathsf{T}}\boldsymbol{P}\boldsymbol{B}\left(\boldsymbol{B}^{\mathsf{T}}\boldsymbol{P}\boldsymbol{B} + \boldsymbol{R}\right)^{-1}\boldsymbol{B}^{\mathsf{T}}\boldsymbol{P}\boldsymbol{A}
\tag{6.56}
$$

If a positive definite solution $\boldsymbol{P}$ exists, then this matrix can be seen as the weight matrix for the infinite-horizon cost-to-go $\boldsymbol{x}_N{}^{\mathsf{T}}\boldsymbol{P}\boldsymbol{x}_N$ for the linearly optimally controlled system. The solution of the DARE can be chosen as terminal cost matrix for the finite-horizon MPC problem, where $\boldsymbol{P}$ reflects the optimal cost from N to ∞. The terminal set $\mathcal{X}_f$ can be chosen to be the maximum invariant set of the closed-loop system. As a controller, an LQR controller $\boldsymbol{u}(k) = \boldsymbol{K}_{LQR}\boldsymbol{x}(k)$ is chosen, which results in the following system dynamics considered:

$$\boldsymbol{x}(k+1) = \boldsymbol{A}\boldsymbol{x}(k) + \boldsymbol{B}\boldsymbol{K}_{LQR}\boldsymbol{x}(k) \tag{6.57}$$

The maximum invariant set $\mathcal{X}_f$ is the set that contains all points $\boldsymbol{x}$ for which holds $\boldsymbol{x}(k) \in \mathcal{X}_f$ while satisfying for all $k \geq 0$:

$$\boldsymbol{x}(k+1) = \boldsymbol{A}\boldsymbol{x}(k) + \boldsymbol{B}\boldsymbol{u}(k) \tag{6.58}$$

$$\boldsymbol{u}(k) = \boldsymbol{K}_{LQR}\boldsymbol{x}(k) \tag{6.59}$$

$$\boldsymbol{x} \in \mathcal{X} \tag{6.60}$$

$$\boldsymbol{u} \in \mathcal{U} \tag{6.61}$$

Intuitively, this means that once a trajectory of the system enters the maximum invariant set, it will never leave it again. The state feedback matrix $\boldsymbol{K}_{LQR}$ can be derived according to the unconstrained LQR controller:

$$\boldsymbol{K}_{LQR} = -(\boldsymbol{B}^{\top}\boldsymbol{P}\boldsymbol{B} + \boldsymbol{R})^{-1}(\boldsymbol{B}^{\top}\boldsymbol{P}\boldsymbol{A}) \tag{6.62}$$

with $\boldsymbol{P}$ being the positive definite solution of the DARE. The design of an linear MPC controller with the terminal cost weight matrix $\boldsymbol{P}$ and the terminal set $\mathcal{X}_f$ leads to guaranteed stability and recursive feasibility; the closed-loop system is stable for every initial state that has a feasible solution of the optimization problem.

In conclusion, compared to the use of a terminal equality constraint, the region of attraction is increased by using the terminal cost and/or the terminal state. The resulting region of attraction by the inclusion of a terminal set might still be small. Depending on the investigated system, the computation of the terminal set and the terminal cost might be very complicated. Additionally, the appropriate terminal set and/or terminal cost need to be calculated in dependence of parameters such as the reference value. The inclusion of a terminal cost might put a lot of weight on the last states, so that performance tuning can become difficult.

MPC Without Stability Constraints

Another approach is the "unconstrained" MPC, where no stability constraints are included. An infinite prediction horizon $N \to \infty$ guarantees recursive feasibility and leads to $\boldsymbol{x}(k) \to \boldsymbol{0}$ as $k \to \infty$, if the optimization problem is feasible for the initial state. The idea is to emulate this behavior by using a long but finite prediction horizon. Controllability assumptions are used to prove that a sufficiently long prediction horizon is sufficient to ensure stability [4, 5]. If the prediction horizon is chosen to be bigger than a certain limit value $N > N_{lim}$, the MPC cost function acts as a Lyapunov function for the system and guarantees stability. A sufficiently long prediction horizon implicitly fulfills the same purpose as a terminal set; namely to enforce $\boldsymbol{x}(k+N|k)$ to be close to the equilibrium point.

This scheme is attractive for practical usage because of its simplicity. The cost function does not need to be modified, i.e. no terminal set or terminal cost needs to be calculated. Additionally, it is less restrictive concerning the region of attraction compared to other stability mechanisms. However, for a specific setting, it is hard to find the exact minimum length of the prediction horizon that leads to stability. This number also depends on parameters such as the reference values. Usually, this value is determined numerically or by using simulations. It also has to be considered that a sufficiently long prediction horizon may lead to unacceptably large computation time.

Conclusions for Practical Applications
There exist many schemes that guarantee nominal stability for the closed-loop system. However, they are usually not applied in practice due to their disadvantages, such as the small region of attraction. Still, several conclusions can be drawn by investigation of the stability mechanisms: The use of a prediction horizon that is as long as possible, while satisfying the computational constraints, can be helpful to realize stability. If due to the computational restrictions, the prediction cannot be chosen very long, a terminal cost can be added. The authors of [20] show that the addition of weighting terms in the unconstrained MPC controller yields improved stability conditions. Additionally, the inclusion of terminal costs aside from a long prediction horizon does not reduce the initial states that result in feasible solutions. In reality, there is always a model–plant mismatch so that nominal conditions are never given. Thus, besides the consideration of these stability mechanisms, it is inevitable to test the convergence of the numerical algorithms, the recursive feasibility, and also the stability thoroughly in simulations as well as in experiments.

Example 6.5 (*Stability of MPC*) This example investigates the effects of various stability mechanisms. For this purpose, an open-loop unstable system is controlled with an MPC algorithm. The discrete-time linear system is given by $\boldsymbol{x}(k+1) = \boldsymbol{A}\boldsymbol{x}(k) + \boldsymbol{B}u(k)$ and has a sampling time of $T_s = 0.5$ s. The system is characterized by

$$\boldsymbol{A} = \begin{bmatrix} 1.1 & 0.5 \\ 0 & 1 \end{bmatrix}, \quad \boldsymbol{B} = \begin{bmatrix} 0.1 \\ 0.5 \end{bmatrix} \tag{6.63}$$

The MPC algorithm considers constraints on the actuated values and on the system states. The optimization problem is given by

$$
\begin{aligned}
\min_{\boldsymbol{x}(\cdot|k),\, \boldsymbol{u}(\cdot|k)} \quad & \|\boldsymbol{x}(k+N|k)\|_{\boldsymbol{P}}^2 + \sum_{i=0}^{N-1} \left(\|\boldsymbol{x}(k+i|k)\|_{\boldsymbol{Q}}^2 + \|\boldsymbol{u}(k+i|k)\|_{\boldsymbol{R}}^2\right) \\
\text{s.t.} \quad & \boldsymbol{x}(k+i+1|k) = \boldsymbol{A}\boldsymbol{x}(k+i) + \boldsymbol{B}u(k+i),\ i = 0, ..., N-1, \\
& \boldsymbol{x}(k|k) = \boldsymbol{x}_0, \\
& -1 \le u(k+i|k) \le 1, \qquad i = 0, \ldots, N-1, \\
& -3 \le x_1(k+i|k) \le 3, \qquad i = 1, \ldots, N-1, \\
& -5 \le x_2(k+i|k) \le 5, \qquad i = 1, \ldots, N-1, \\
& \boldsymbol{x}(k+N|k) \in \mathcal{X}_f
\end{aligned}
\tag{6.64}
$$

For the closed-loop control system, nominal conditions are assumed; the MPC has perfect model knowledge, there are no disturbances present, and full-state measurement is available. The weighting factors are set to $\boldsymbol{Q} = \boldsymbol{I},\ R = 100$, and the initial state is given by $\boldsymbol{x}_0 = [2.5,\ -2.5]^{\mathrm{T}}$. For implementation of the closed-loop controller, the software packages YALMIP [13] and SeDuMi [21] have been used.

The first approach does not use any stability mechanism, i.e. $\mathcal{X}_f = \mathcal{X}$ and $\boldsymbol{P} = \boldsymbol{0}$. The control horizon is chosen to be $N = 3$. The closed-loop control result is depicted in Fig. 6.9. The closed-loop system is unstable although nominal conditions are present. The figure shows for $k = 0$, the current state in black and the predicted values for the system states in red. For $k = 1$, the states have evolved. As no disturbances are active, the new current system state corresponds to the one predicted in the previous time step. The new predicted state trajectory is shown accordingly. As the prediction horizon is finite, the new open-loop prediction can differ from the ones of the previous time step. In this example, there is quite a big difference between open-loop prediction and the resulting closed-loop trajectory. At the time point $k = 3$, there is no possibility to satisfy all constraints on the system states and the actuated values. Consequently, the optimization problem is infeasible which is clearly unwanted.

The second MPC formulation uses a stability mechanism, i.e. stability constraints are introduced. It relies on the same, small prediction horizon of $N = 3$, and additionally, terminal cost and a terminal set are added. The final cost $\boldsymbol{P}$ is calculated by solving the DARE according to (6.56). It results to be

$$
P = \begin{bmatrix} 25.6 & 49.3 \\ 49.3 & 141.5 \end{bmatrix} \tag{6.65}
$$

The terminal set $\mathcal{X}_f$ is chosen to be the maximum invariant set of the closed-loop control system with an LQR. The LQR $\boldsymbol{K}_{LQR}$ is designed with the system matrices $\boldsymbol{A}, \boldsymbol{B}$ and the weighting factors $\boldsymbol{Q}, R$ according to (6.62). The maximum invariant set is calculated using the software package Multi-Parametric Toolbox 3.0 [7]. Figure 6.10 shows the closed-loop control result using this MPC formulation. The MPC control algorithm allows for an asymptotic stable behavior in this setup. This can be achieved although a small prediction horizon is used for the unstable system,

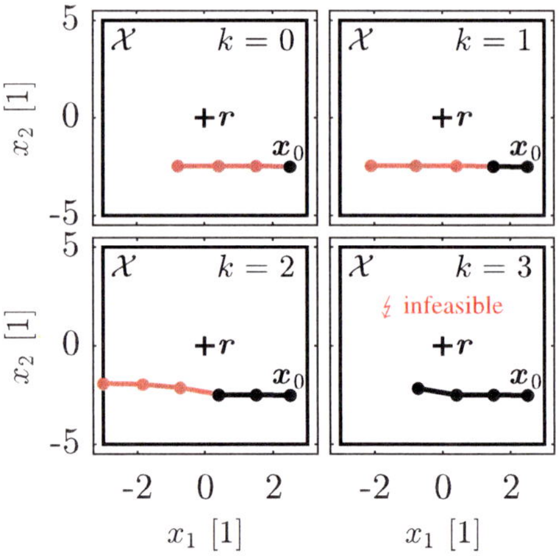

Fig. 6.9 MPC without any stability mechanism

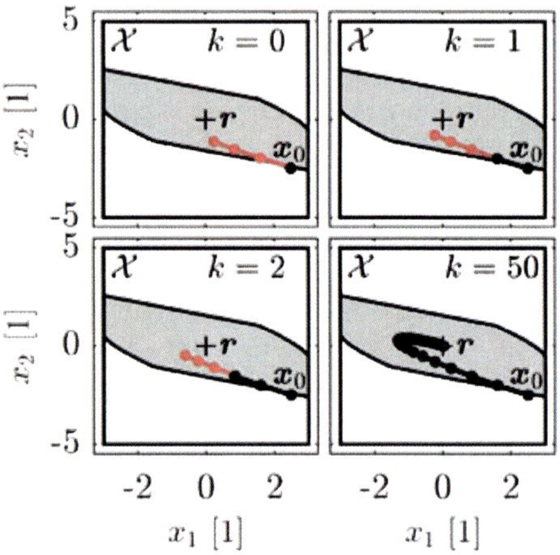

Fig. 6.10 MPC with terminal cost and terminal set

thus resulting in low computation times. The major drawback of this approach is the reduced region of attraction. Moreover, the stability constraints and the region of attraction depend on the reference values. All in all, the addition of the terminal set and costs has benefits if closed-loop stability is critical, i.e. when controlling an unstable plant.

The third formulation mimics the MPC approach mentioned without stability constraints. It uses a long control horizon of $N = 20$, which is chosen heuristically. The stability constraints are neglected, i.e. $\mathcal{X}_f = \mathcal{X}$ and $\boldsymbol{P} = \boldsymbol{0}$. All other weighting factors are kept constant. Figure 6.11 shows the resulting closed-loop behavior. The problem of infeasibility is avoided and instead, an asymptotically stable behavior can be achieved. The MPC algorithm is able to steer the system states into the origin and keep it there. Due to the long prediction horizon, the closed-loop trajectory is also quite similar to the open-loop trajectories, although it is still not exactly the same.

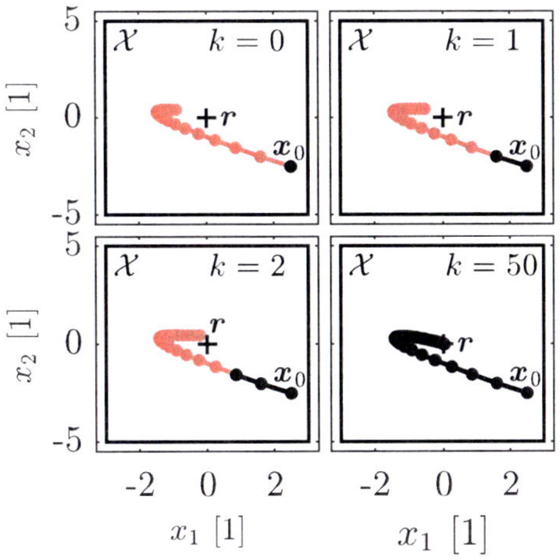

Fig. 6.11 MPC with a long prediction horizon

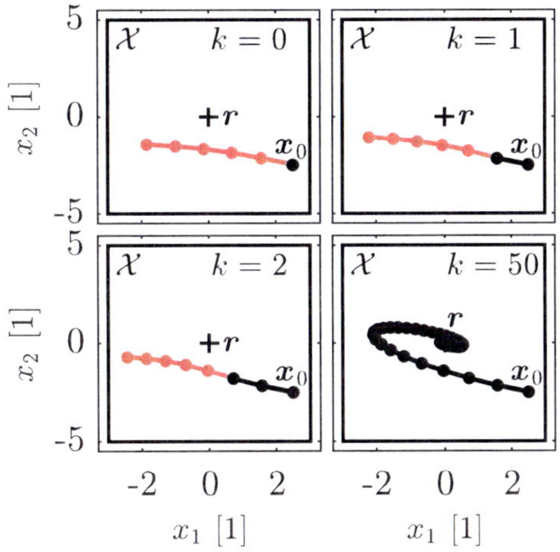

Fig. 6.12 MPC with slightly enlarged prediction horizon and terminal costs

The major drawback of this stability mechanism is the presence of many optimization variables within the optimization problem that has to be solved in each time step.

Lastly, a combination of the long prediction horizon and the terminal costs is investigated. The terminal costs are added to prevent the need for an excessively long prediction horizon. On the other side, the longer prediction horizon allows to choose lower terminal costs, which is beneficial for performance tuning. In this example, a heuristic choice of $N = 5$ and $\boldsymbol{P} = 20\boldsymbol{I}$ gives good results. The closed-loop control results are shown in Fig. 6.12. An asymptotic stable behavior can be observed. Just increasing the prediction horizon to $N = 5$ without any terminal costs does not allow for asymptotically stable behavior, neither does the inclusion of the terminal costs $\boldsymbol{P} = 20\boldsymbol{I}$ without any increase in prediction horizon. Only the combination of both leads to an asymptotic stable behavior.

References

1. F. Allgöwer, R. Findeisen, Z.K. Nagy, Nonlinear model predictive control: from theory to application. J.-Chin. Inst. Chem. Eng. **35**(3), 299–315 (2004)
2. F. Borrelli, A. Bemporad, M. Morari, *Predictive Control for Linear and Hybrid Systems* (Cambridge University Press, Cambridge, 2017)
3. H. Chen, F. Allgöwer, A quasi-infinite horizon nonlinear model predictive control scheme with guaranteed stability. Automatica **34**(10), 1205–1217 (1998)
4. L. Grüne, Analysis and design of unconstrained nonlinear MPC schemes for finite and infinite dimensional systems. SIAM J. Control Optim. (2009)
5. L. Grüne, NMPC without terminal constraints. IFAC Proc. Vol. **45**(17), 1–13 (2012)
6. L. Grüne, J. Pannek, *Nonlinear Model Predictive Control* (Springer, Berlin, 2017)
7. M. Herceg, M. Kvasnica, C. N. Jones, and M. Morari, Multi-parametric toolbox 3.0, in *2013 European Control Conference (ECC)* (2013), pp. 502–510
8. A. Jadbabaie, J. Hauser, On the stability of receding horizon control with a general terminal cost. IEEE Trans. Autom. Control **50**(5), 674–678 (2005)
9. S. Keerthi, E.G. Gilbert, Optimal infinite-horizon feedback laws for a general class of constrained discrete-time systems: Stability and moving-horizon approximations. J. Optim. Theory Appl. **57**(2), 265–293 (1988)
10. E. Kerrigan, J.M. Maciejowski, Soft constraints and exact penalty functions in model predictive control, in UKACC International Conference (2000)
11. H. Khalil, J. Grizzle, *Nonlinear Systems* (Prentice Hall, 1996)
12. D. Limón, T. Alamo, F. Salas, E.F. Camacho, On the stability of constrained mpc without terminal constraint. IEEE Trans. Autom. control **51**(5), 832–836 (2006)
13. J. Lofberg, Yalmip: a toolbox for modeling and optimization in matlab, in *2004 IEEE International Conference on Robotics and Automation*. IEEE, pp. 284–289
14. U. Maeder, F. Borrelli, M. Morari, Linear offset-free model predictive control. Automatica **45**(10), 2214–2222 (2009)
15. D.Q. Mayne, J.B. Rawlings, C.V. Rao, P.O. Scokaert, Constrained model predictive control: stability and optimality. Automatica **36**(6), 789–814 (2000)
16. H. Michalska, D.Q. Mayne, Robust receding horizon control of constrained nonlinear systems. IEEE Trans. Autom. Control **38**(11), 1623–1633 (1993)
17. M. Morari, U. Maeder, Nonlinear offset-free model predictive control. Automatica **48**(9), 2059–2067 (2012)
18. G. Pannocchia, J.B. Rawlings, Disturbance models for offset-free model-predictive control. AIChE J. **49**(2), 426–437 (2003)
19. J.B. Rawlings, D.Q. Mayne, M. Diehl, *Model Predictive Control: Theory, Computation and Design* (Nob Hill Publishing, 2017)
20. M. Reble, Model predictive control for nonlinear continuous-time systems with and without time-delays, *Ph.D. Thesis, University of Stuttgart* (2013)
21. J.F. Sturm, Using sedumi 1.02, a matlab toolbox for optimization over symmetric cones. Optim. Methods Softw. **11**(1–4), 625–653 (1999)

Part II
Introduction to Combustion Engine Control

Chapter 7
SI and CI Engine Control Architectures

Abstract In this chapter, the engine control structure for spark ignition (SI) and compression ignition (CI) engines is examined. First, a general overview of engine control is given. A suitable architecture is presented which can be used to handle the demanding requirements on process control. By the use of hierarchization and modularization, the complex interaction of the various components can be tackled. For both the SI and the CI engine, a typical hardware setup is presented. Based on the two examples, the main control loops are introduced for both combustion concepts. The goals of the control tasks are outlined and exemplary sensitivities of the controlled values on the actuated values are shown. Specifically, the tasks within the air path, the ignition path, the combustion path, and the aftertreatment path are investigated.

7.1 Overview of Engine Control Tasks

The engine controller has to manage all components of the engine to ensure the desired performance, as discussed in Sect. 1.2. In order to handle the complex and strong interactions between the components, the structure of the controller plays an important role. A common approach to tackle the complexity of the arising tasks is the use of hierarchization and modularization. The engine controller is often structured by dividing the tasks into four layers, as shown in Fig. 7.1.

The first layer is the torque manager which calculates the requested torque, i.e. the torque the engine should provide. An important input to this layer is the gas pedal position, as by this interface the drivers express their torque demand for driving. However, additional factors affect the requested torque, such as the torque needed to drive the auxiliary devices. Also, the driving situation is evaluated, where modes such as idle speed or fuel cutoff are considered, along with the stability of the vehicle

T. Albin Rajasingham, *Nonlinear Model Predictive Control of Combustion Engines*, Advances in Industrial Control,
https://doi.org/10.1007/978-3-030-68010-7_7

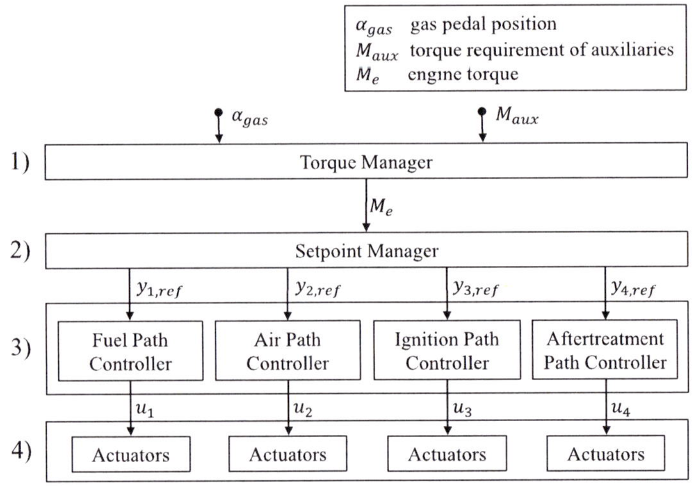

Fig. 7.1 Engine control architecture consisting of four layers

dynamics. The torque manager usually is a static map that is implemented via look-up tables [17].

The second layer is the setpoint manager where desired references for physical quantities are calculated. The requested torque calculated by the torque manager serves as the input for the setpoint manager. The calculated setpoints are subsequently tracked by the third layer, where the main engine control loops are present. From a control point of view, the setpoint manager is a supervisory control layer for the various open-loop and closed-loop control tasks. The setpoints have to be calculated in such a way that the requested torque is delivered, while additionally considering requirements such as emissions and component protection. For this purpose, external parameters such as the ambient temperature and pressure have to be accounted for. The setpoint manager can be designed as a static feedforward controller. However, such a design can lead to a suboptimal performance during the transients, especially when dynamics with different timescales are present. One example is the interaction between the air path and the combustion path where the dynamics of the air path are much slower than that of the combustion path. If the setpoints for both paths are optimized for the steady-state values only, emission peaks can occur during the transients. In CI engines, the controller has to avoid soot peaks in the transients. This is done by ensuring that the combination of air and fuel paths leads to an air–fuel charge in the combustion chamber that is lean enough during the entire transients [2, 28]. The higher the engine performance is, the more complex is the interaction of the various components that have to be accounted for, not only in a steady-state operation, but also during the transient dynamics. Consequently, a model predictive control (MPC) solution offers several advantages for this layer [39]. One distinctive advantage MPC can offer on this layer is its ability to include preview information. For example, the knowledge of the future torque profile due to a prediction of the elevation and speed profile can be exploited for the engine control system [22, 38, 45]. The MPC approach can systematically account for this by anticipatory action.

The third layer contains the main engine control loops. These track the setpoints calculated in the second layer. Typically, the overall task is divided into four modules: air path, fuel path, ignition path, and aftertreatment path. Usually, separate controllers are designed for each path. The division into separate controllers has the advantage that each control task is easier to manage. However, the drawback is that optimality is lost in terms of performance. The general control task of the four paths can be summarized as follows.

- **Air Path Controller**
  It ensures that the intake gas with appropriate properties concerning overall mass, temperature, and chemical composition, especially with respect to the ratio of O_2 (as an oxidizer) and burnt gas, is delivered.
- **Fuel Path Controller**
  It ensures that fuel is delivered into the combustion chamber with appropriate properties concerning amount, pressure, and timing.
- **Ignition Path Controller**
  It initializes the combustion by ignition, e.g. via a spark plug.
- **Aftertreatment Path Controller**
  It ensures the regulation of the emissions via control of the exhaust gas aftertreatment systems.

The control tasks arising in the third layer represent the core of the engine controller. Often, these control tasks are quite complex. They can be characterized by multiple-input multiple-output (MIMO) dynamics with nonlinearities and the need to consider constraints. For many of these control tasks, the MPC approach offers the best solutions.

The fourth layer is the actuator layer, where low-level control actions take place. They include the conversion from physical reference values to electrical quantities and the closed-loop control of the actuator itself for the rejection of disturbances on the actuator level. One example is the control of a valve, where the reference is a certain opening position and a closed-loop controller sets the appropriate electric voltage. In the low-level control layer, often either feedforward or simple feedback controllers such as PID controllers are sufficient. While these plants usually are characterized by rather simple single-input single-output (SISO) system dynamics, there are exceptions, where the actuator shows distinctive complex nonlinear behavior. An example are solenoid valves used for the air path [34] or electromagnetic actuators [8].

The control architecture depicted is responsible for the fundamental control loops. However, a multitude of other control loops is present in an engine system. These include the control of auxiliary components, such as coolant temperature control [25], control of the fan [42] as well as control of the fuel pump [6]. For these control loops usually, simple control techniques are sufficient.

Obviously, the specific control tasks required depend on the individual engine. A major difference exists in the combustion concept, e.g. SI versus CI. But there are also other factors, such as the hardware applied, e.g. naturally aspirated versus turbocharged engine, and the operation purpose of the engine, e.g. stationary versus dynamic operation. These characteristics have an influence on which control tasks are required, how complex they are, and also on the requirements for the closed-loop-controlled system. The following overview of combustion concepts is presented as they substantially affect the control loops necessary as well as the hardware involved.

A multitude of combustion concepts is used in ICEs. This book follows a classification proposed in [23]. According to [23], there are three fundamental combustion concepts, namely the SI concept, the CI concept, and the low-temperature combustion (LTC) concept; see Fig. 7.2. The SI concept is characterized by the combustion of a premixed charge which is triggered by a spark plug. Another important characteristic is the air-to-fuel equivalence ratio λ. The value for λ can be calculated by setting the air-to-fuel ratio AFR in relation to the stoichiometric air-to-fuel ratio AFR_s.

$$\lambda = \frac{AFR}{AFR_s} \tag{7.1}$$

The AFR itself is defined as the ratio between the mass of air and the mass of fuel used within the combustion process:

$$AFR = \frac{m_{air}}{m_{fuel}} \tag{7.2}$$

The value AFR_s describes the ratio of air to fuel which is needed for stoichiometric combustion, i.e. without excess of air. For gasoline combustion $AFR_s = 14.7$ holds, thus for the combustion of 1 kg of gasoline fuel 14.7 kg of air is needed.

In the SI concept, stoichiometric combustion is given, such that $\lambda = 1$. In the CI concept, a diffusive (non-premixed) combustion of a lean charge is used, which is triggered by a late injection of the fuel directly into the cylinder. The LTC concept is characterized by the combustion of a premixed charge via autoignition, where the charge can be lean or stoichiometric. All other combustion concepts, besides the three fundamental ones, can be classified along the triangle shown. The non-fundamental combustion concepts comprise properties of two different fundamental concepts. One example is the concept of a lean burning SI engine, which would be classified between the SI and the CI engine concepts [23]. As in an SI engine, it uses a spark ignition to trigger the combustion, but it is operated with a lean charge like that of a CI engine. In the following, the basic control loops of the three fundamental combustion concepts are introduced.

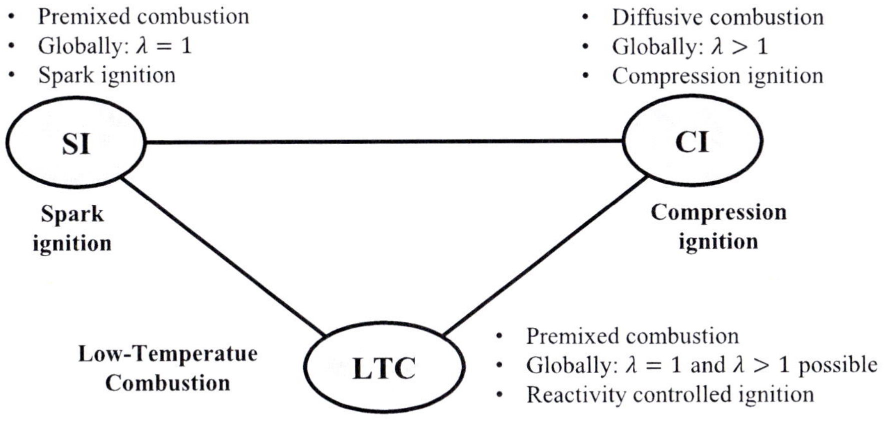

Fig. 7.2 Three fundamental engine combustion concepts – reproduced with permission from [23] © 2016 Wiley-VCH Verlag GmbH & Co

7.2 SI Engine Control Systems

The SI concept is characterized by the spark ignition of a homogeneous charge. In order to achieve a homogeneous mixture, the fuel is either injected into the intake port or directly into the combustion chamber, typically during the intake stroke or early in the compression stroke. As a consequence, there is enough time to mix the air and the fuel, which results in a premixed charge. The combustion is triggered by a spark plug. This initiates a small flame kernel which is propagated through the combustion chamber as a flame front [20]. This combustion mode relies on a fuel that evaporates at sufficiently low temperatures. Currently, most SI engines run on gasoline fuel. However, there are also SI engines which run with other fuels, such as natural gas.

The homogeneous mixture has favorable properties concerning pollutant emissions. For instance, only a little soot is produced. The amount is low enough that often no aftertreatment for soot is necessary. However, especially for direct-injecting gasoline engines, more and more gasoline particulate filters are used to filter soot emissions. The remaining pollutant emissions are usually aftertreated by a three-way catalyst (TWC) which combats nitrogen oxides NO_x, carbon monoxide CO, and unburned hydrocarbon HC emissions simultaneously. The pollutant emissions are converted into nitrogen dioxide N_2, carbon dioxide CO_2, and water H_2O [26]. Compared to a CI engine, the exhaust aftertreatment of an SI engine is cheaper and simpler. The TWC only works effectively when the combustion takes place within a narrow range of $\lambda = 1 \pm 0.005$ [18]. To achieve a high conversion efficiency of the TWC and thus low tailpipe pollutant emissions, the air-to-fuel equivalence ratio λ is close-loop-controlled to be one. A *quantitative load control* concept is thus used, i.e. the load is changed by adjusting the amount of air which is inducted into the cylinder. The fuel is adapted to achieve a stoichiometric operation.

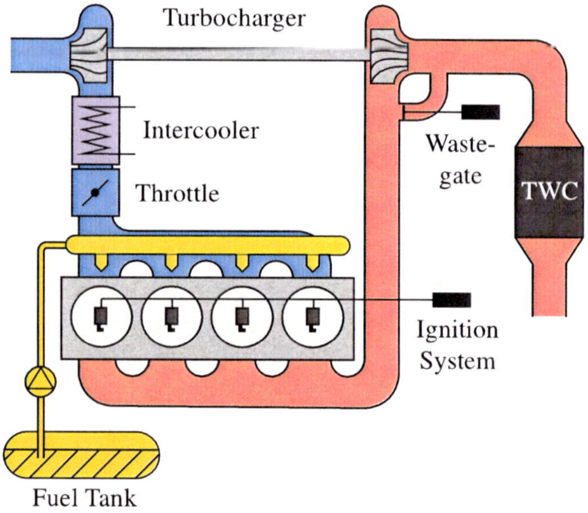

Fig. 7.3 Exemplary structure of a turbocharged SI engine

The main control loops of an SI engine are illustrated by an exemplary setup, which is depicted in Fig. 7.3. In this setup, the fuel is injected into the intake port and the charge is ignited by a spark plug. The engine is equipped with a throttle and a single turbocharger. For the exhaust gas aftertreatment, a TWC is used. In the following, the fundamental control tasks arising in the air path, the fuel path, and the ignition path are discussed. For this SI engine, no aftertreatment path control system is necessary. As it is a passive device, the TWC needs no dedicated control loop.

7.2.1 Air Path Controller

The main goal of an air path controller is to manage the conditions in the intake port. In the given setup, only fresh air is present in the intake path, i.e. no external exhaust gas recirculation is considered. The pressure in the intake manifold can be used as a controlled variable. It correlates to the air mass inducted into the combustion chamber, which is the relevant variable that has to be set. Due to the quantitative load control in SI engines, the fuel mass injected is adapted to the air mass inducted, which itself is determined by the intake manifold pressure, in order to fulfill $\lambda = 1$. Thus, the air path controller directly determines the torque produced. The use of the intake manifold pressure as a controlled value has the advantage that it can be measured by a pressure sensor. In some cases, instead of a pressure sensor, an air mass flow sensor is used along with a model to estimate the intake manifold pressure.

At least a throttle is needed in the air path of an SI engine to control the load. In the setup investigated, in addition to the throttle, a turbocharger equipped with a wastegate is installed. The parameters involved for actuation are the variables u_{th} which represents the opening position of the throttle and u_{wg} which represents the

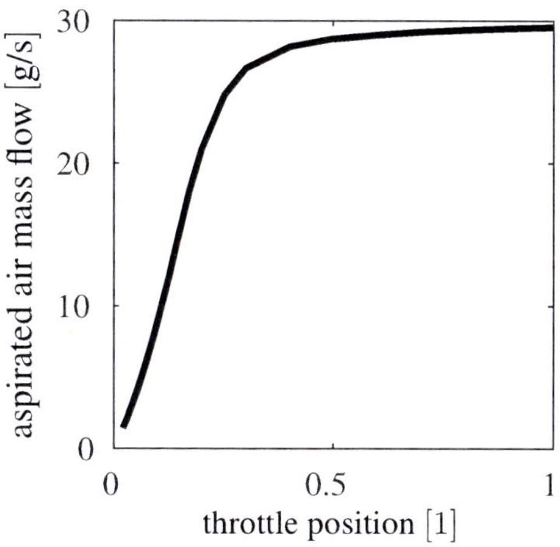

Fig. 7.4 Influence of throttle position on steady-state inducted air mass flow

wastegate opening position. Both are used to adjust the amount of air that is inducted into the cylinder. As the buildup of torque is directly determined by the air path controller, its proper design is a crucial aspect of driveability. Usually, the requirements on driveability translate into a target time constant of an equivalent first-order system for the closed-loop control of the intake manifold pressure; see [40].

By changing the position of the throttle plate, the flow into the intake manifold can be restricted, and thus the amount of air entering the cylinder is adjusted. Figure 7.4 shows measurement data of the air mass flow aspirated in dependence of the throttle position at steady-state conditions. The throttle can only act in one direction. Compared to the upstream conditions of the throttle, the pressure can only be decreased. Usually, electronic actuators are used for the throttle, which allows for closed-loop control of the throttle position. The control of the load with the throttle inherently decreases the efficiency of the entire engine process, as the closing of the throttle causes pumping losses [20]. Using the throttle, the intake manifold pressure can be changed with dynamics in the order of milliseconds, which is a fast change that leads to an almost immediate torque response.

One concept to reduce fuel consumption is the application of "downsizing" by the use of turbochargers. The goal of downsizing is the improvement of the engine operating points concerning efficiency while maintaining the engine power. Using a turbocharger, the engine size can be reduced without impairing the engine power. The reduction of the engine size leads to a reduction of friction losses. Additionally, in SI engines the throttle can be opened wider, thus reducing the pumping losses. The turbocharger consists of a turbine and a compressor that are connected by a common shaft. The turbine is powered by energy extracted from the hot exhaust gas. The intake air of the combustion engine thus can be compressed to higher densities than those achievable with a naturally aspirated engine. Compared to the upstream conditions of the compressor, the pressure can only be increased.

To adjust the turbocharging to the specific operating point, variability is introduced in these concepts. In SI engines, typically wastegates on the turbine side are used. The wastegate adjusts the amount of mass flow that bypasses the turbine. The closing of the wastegates leads to a higher mass flow through the turbine and thus higher turbocharger speeds, higher intake manifold pressures, and higher exhaust gas back pressures. The wastegate actuators are usually electro-pneumatic or electric.

As a direct coupling between the intake and exhaust paths exists with turbocharging, the process control has to address these couplings. In modern engines, the focus is set on fuel efficiency; thus, the use of the throttle is avoided as much as possible to reduce pumping losses. As a result, for turbocharged operation, the throttle is not used for control but is fully opened. By the use of turbocharging, the intake manifold pressure can be changed with dynamics in the order of seconds. This time lag is perceptible to the driver and is known as "turbo lag". The lag results from the inertia of the turbocharger and the dynamics of the volumes to build up pressure.

Another important aspect in the control of turbochargers is the consideration of their operation limits. The controller has to ensure that an operation at excessive turbocharger speeds is avoided as well as an operation in the surge and the choking mode [27]. These limits have to be obeyed in order to ensure a safe operation of a turbocharger. Additionally, the exhaust gas back pressure has to be considered as it influences the efficiency of the engine. For the load adjustment, usually two different controllers are used. Whenever the load cannot be set with an open throttle, i.e. the reference for the intake manifold pressure is lower than the base charging pressure, the wastegate is fully opened and the throttle is used for control. For a turbocharged operation, the throttle is fully opened and the wastegate is used for control.

7.2.2 Fuel Path Controller

In the setup investigated, the fuel is injected into the intake port. A common choice for a gasoline injector are solenoid-actuated valve plungers with multiple fuel outlet holes that improve spray atomization. The time point of injection needs to allow for enough time to mix the air and the fuel, such that a homogeneous charge results. Within a reasonable range, the injection timing has only a negligible influence on the performance of the combustion process. The duration of injection u_{doi} is used as an actuated value.

The duration of injection u_{doi} correlates in a nonlinear fashion with the fuel mass injected. It is driven by the requirements of the TWC. The engine-out emissions are lowest at slightly lean conditions, i.e. $\lambda \approx 1.2$ [18]. However, the highest TWC efficiency is achieved at $\lambda \approx 1$. As a result, the lowest tailpipe emissions are achieved for $\lambda \approx 1$ as well. To keep the charge within the demanded narrow range of $\lambda = 1 \pm 0.005$, an air-to-fuel ratio controller is used. It has a crucial task in an SI engine in that it has to set the amount of fuel according to the air mass inducted into the cylinder. It thus requires an accurate estimation of the air mass inducted. This estimate can be used for feedforward control, which is essential during a transient operation.

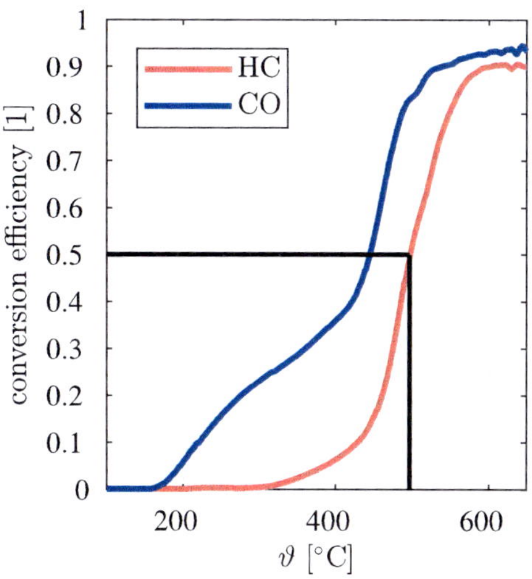

Fig. 7.5 Conversion efficiency of the three-way catalyst in dependence of its temperature

Additionally, lambda sensors are used to measure the air-to-fuel ratio in the exhaust gas and therefore also the λ value. Often a combination of sensors is used. A wide-band lambda sensor installed before the TWC furnishes continuous lambda values, while a switch-type lambda sensor located after the TWC furnishes discrete values, i.e. $\lambda < 1$ or $\lambda > 1$ [36]. The information from both sensors are processed in the feedback controller. For the rejection of disturbances and thus for higher accuracy, a feedback controller is necessary in addition to the feedforward controller. The value of u_{doi} is set on a cycle-to-cycle basis.

For an appropriate control result, the specific characteristics of the TWC have to be taken into account. The fact that the TWC has some oxygen storage capacity can serve the closed-loop controller by damping small deviations from $\lambda = 1$ for a short time. Another characteristic of the TWC is the dependence of its conversion efficiency on its temperature. At low TWC temperatures, i.e. at cold-start conditions, only a small fraction of the pollutant emissions NO_x, CO, and HC are combated due to the reaction kinetics [4]. Figure 7.5 shows the conversion efficiency of a TWC for the emission levels of CO and HC in dependence of its temperature. One characteristic point is the so-called light-off temperature, which is defined as the temperature where 50% of the emissions of the HC are converted [18]. In this example, the light-off temperature is 500°C. Therefore, until the light-off temperature is reached, the controller should focus on increasing the TWC temperature by hot exhaust gases rather than on achieving the highest combustion efficiency.

7.2.3 Ignition Path Controller

The combustion of the charge is triggered by a spark plug. For actuation, the spark ignition timing u_{si} results, which has a strong effect on the process of combustion. The spark ignition timing has an influence on the trajectory of the heat release and

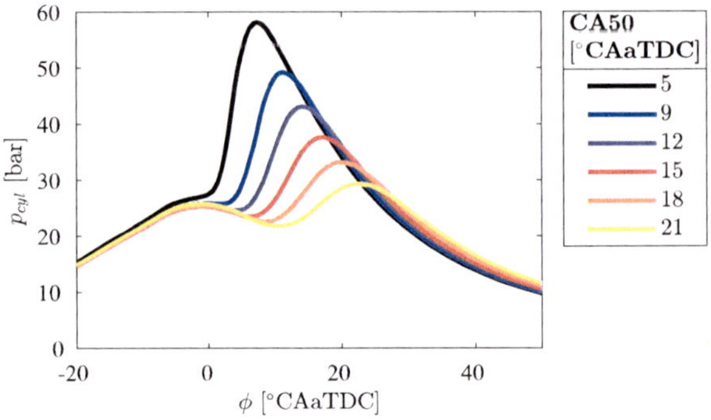

Fig. 7.6 Influence of the combustion average CA50 on cylinder pressure trace

with that on the combustion average (CA50). The combustion average corresponds to the time point where 50% of the fuel is burnt; its calculation will be detailed in Sect. 14.3. The combustion average strongly influences combustion efficiency. In a normal operation, the spark timing is set to achieve high combustion efficiency. Often it is set such that a combustion average of CA50 $\approx$ 8 °CA results, as this represents high efficiency. Influencing factors of the optimal ignition timing are the engine speed, the load, and ambient conditions such as ambient temperature, pressure, and air humidity [13]. Figure 7.6 shows the cylinder pressure trace in dependence of the crank angle position for various timings of the combustion average. All of the traces are operated at the same load by closed-loop control of the fuel amount. Besides the efficiency, the allowable maximum cylinder pressure and the maximum exhaust temperature have to be respected, as these pose a mechanical limit.

In some cases, the ignition timing is shifted on purpose to points with lower efficiency. This retardation of the spark timing results in lower efficiency and in higher exhaust gas temperatures. It can be used, for instance, to heat up the TWC. For cold-start control, fast heating is required, such that light-off temperatures are reached as quickly as possible [7, 19]. The retardation of the ideal spark timing is also used as an actuated value in knock control. For high-load operations, the so-called knock can occur in an SI engine [47]. Knock is an unwanted autoignition phenomenon with drawbacks in terms of noise and mechanical stress. First, the knocking combustion has to be detected [3]. Based on this detection, the spark timing is adjusted by a feedback controller in order to shift operation toward non-knocking combustion again [32]. For knock control, the spark timing is adjusted on a cycle-to-cycle basis.

7.2.4 Control Structure

Figure 7.7 shows the structure of the resulting SI engine controller. All the main control tasks in normal operation are shown. It consists of the air-to-fuel ratio controller via the duration of injection. Additionally, within the ignition path, the combustion average CA50 is controlled via ignition timing. Within the air path, the intake man-

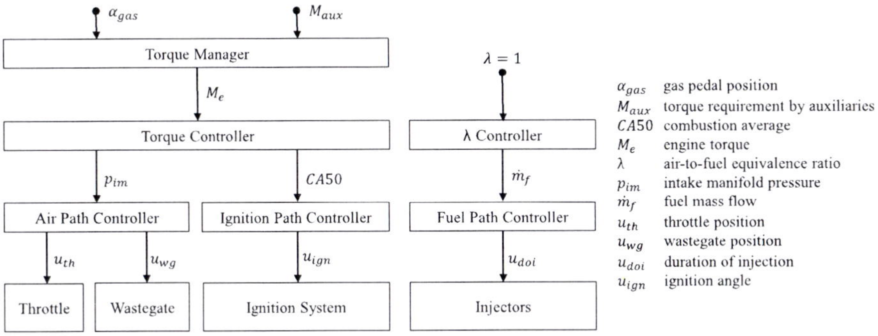

Fig. 7.7 SI engine control architecture (normal operation)

ifold pressure is controlled via the throttle and the wastegate of the turbocharger. Although separate controllers are used, the important cross-couplings of the various paths have to be considered. This is especially important when the system dynamics occur on different timescales. For instance, the fuel amount always has to match the inducted air mass which is managed by the air path controller. The TWC needs no dedicated control loop, as it is a passive device.

Many more control loops are necessary to run an SI engine. This concerns the low-level controllers such as the controllers for the wastegate opening position as well as the auxiliary functions such as lubrication control, as mentioned in Sect. 7.1. Additionally, there exists for instance the idle speed controller. Its goal is to provide certain minimum engine torque to avoid engine stall. This is ensured by appropriate closed-loop control of the engine speed [18]. For idle speed control also MPC-based solutions are investigated [9, 10]. Some SI engines additionally contain devices for waste heat recovery which have to be controlled as well [35].

7.3 CI Engine Control

The CI engine is characterized by a diffusion-controlled combustion of a lean mixture. The fuel is usually injected directly into the cylinder. In a CI engine, the fuel is injected much later than in a direct-injected SI engine, i.e. the injection timing is at the end of the compression stroke, close to the top-dead center. Due to the compression, at high pressures at this point, hot gas is present in the combustion chamber. A typical condition is given by a gas temperature of 900°C and pressure of 70 bar. As a result, autoignition conditions exist, such that combustion starts already during the injection process. As the time available for mixing is very limited, the charge is stratified such that a high portion of the fuel is combusted in a diffusion-controlled manner, where the mixing of the fuel and the air as well as the combustion occurs simultaneously. Most CI engines run on diesel fuel. An alternative is the use of bio-fuels [1].

For the CI engine, a so-called *qualitative load control* is used. The load is not set by adjusting the amount of air, as in SI engines, but instead by adjusting the amount of fuel. Excessive air is present, as the engine is always operating in a lean mode. The main advantage of qualitative load control is the possibility of an unthrottled operation. This is very fuel-efficient as pumping losses are avoided. However, this combustion approach has disadvantages concerning pollutant emissions, e.g. a high amount of engine-out soot emissions can result. Especially for conditions with $\lambda < 1.3$, excessive soot emissions result. These operating conditions thus are generally avoided [18]. As CI engines run in a lean operating mode, a TWC can no longer be used to combat NO_x, CO, and HC simultaneously. As a result, the aftertreatment of the exhaust gases is more complex than in SI engines. Usually, a single device is not enough for the exhaust gas aftertreatment, i.e. a combination of several aftertreatment devices is used to combat the produced emissions, which causes higher costs and higher complexity of the aftertreatment system.

Figure 7.8 shows an overview of a standard setup of a CI engine. It consists of a direct-injection system with exhaust gas recirculation and turbocharging. For the exhaust gas aftertreatment, a diesel particulate filter (DPF) is used in combination with a selective catalytic reduction (SCR) system. In the following, the fundamental control loops typical of the fuel path, the air path, and the aftertreatment path are discussed. The ignition is controlled by the fuel path controller. Hence, no dedicated control loop for the ignition path is necessary.

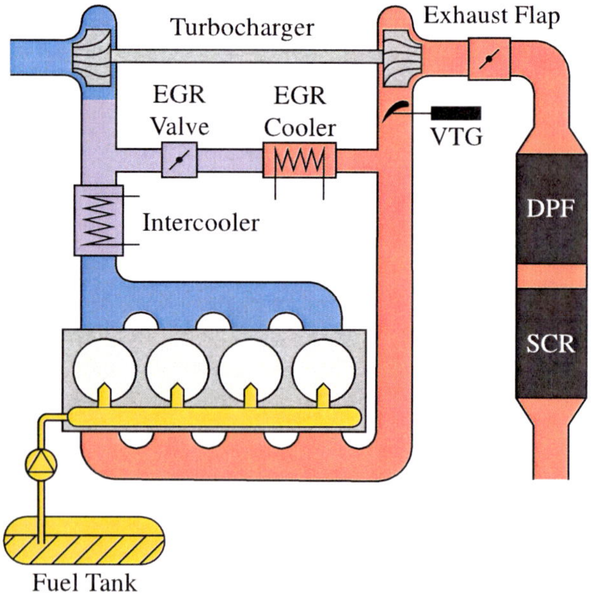

Fig. 7.8 Exemplary structure of a turbocharged CI engine

7.3.1 Fuel Path Controller

The central actuator of the fuel path in a CI engine is the fuel injector. In the setup shown, the fuel is injected directly into the cylinder with a single injection. At the late injection timings, the combustion chamber pressure is so high already that autoignition occurs. Various technologies are available for fuel injection, such as piezo-actuated valves or solenoid-actuated injectors [12]. Additionally, a fuel pump is used to set the rail pressure of the injector [29]. The main parameters which are used as actuated values are the fuel pump control valve position u_{fp} and parameters from the injector, i.e. the start of injection u_{soi} and the duration of injection u_{doi}. The actuated values allow to strongly influence the combustion process. Influences can be made in terms of the shape of the heat release profile, e.g. concerning combustion average, the maximum pressure, and the maximum pressure rise gradient.

The injection pressure is one of the values that is closed-loop-controlled by low-level controllers. For this purpose, a rail pressure sensor is available. The high-pressure fuel pump is used to set the rail pressure as necessary. Sometimes, additionally, a pressure release valve is used. The control loop dynamics for controlling the rail pressure are in the order of milliseconds. Modern injection systems are able to attain rail pressures of more than 2000 bar. An increased injection pressure results in better fuel evaporation, a lower ignition delay, and faster burn rates. Figure 7.9 shows the influence of the injection pressure on the emissions levels of NO_x and particulate matter PM. The PM can be decreased by higher rail pressures, while the NO_x emissions are increased. At the same time, the increased rail pressure also leads to decreased overall efficiency due to the energy needed for the fuel pump. The rail pressure reference is usually set in a feedforward manner, depending on parameters such as engine speed and load [21].

Besides the rail pressure, the start of injections u_{soi} and the duration of injections u_{doi} are used as actuated values inside the control algorithm. Earlier injection timings lead to earlier combustion, a higher peak temperature, and higher peak pressures. This effect leads to increased NO_x emissions and lower levels of PM emissions. Figure 7.9 shows the sensitivity of NO_x and PM emissions for single injection on start of energizing of the injector u_{soe}, based on measurements taken on an engine test bench. The parameter start of energizing of the injector u_{soe} is related to the start of injection u_{soi} by a simple dead-time which accounts for the physical delay. The parameter duration of injection u_{doi} has to be set such that the requested torque is delivered.

Both the parameters u_{soi} and u_{doi} can either be set in a feedforward manner or can be adjusted via feedback control. In the case of feedback control, a sensor is needed, for instance, an in-cylinder pressure sensor can be applied. State of the art in combustion control is the control of a few cycle-integral surrogate parameters. Most commonly, the indicated mean effective pressure (IMEP) and the combustion average CA50 are used. These values can be extracted from the cylinder pressure signal. The injection parameters u_{soi} and u_{doi} are set on a cycle-to-cycle basis for control. In more

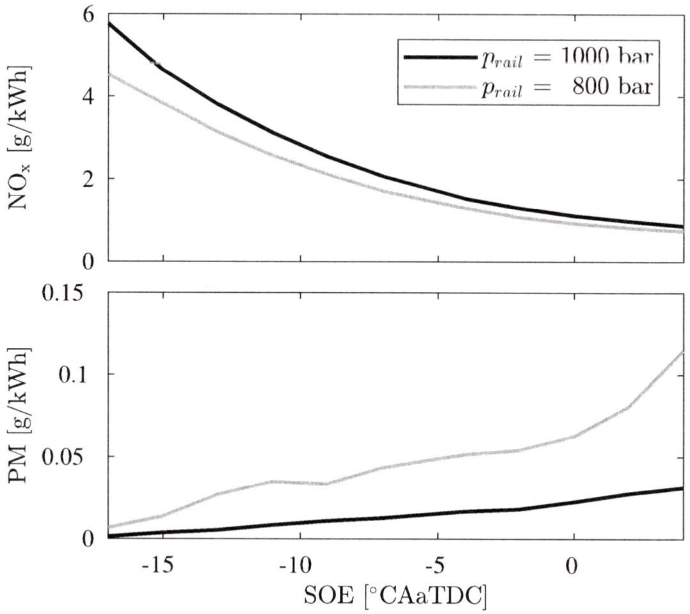

Fig. 7.9 Influence of the injection parameters start of energizing and rail pressure on pollutant emissions

advanced combustion control concepts, the quasi-continuous combustion profile is controlled. Both control concepts are introduced in more detail in Sect. 14.1.

7.3.2 Air Path Controller

In the given CI setup, exhaust gas recirculation is used in combination with turbocharging. The setup allows the guide vane position u_{vtg} and the EGR valve position u_{egr} to be adjusted. These two actuated values have an influence on the intake manifold pressure and the chemical composition of the intake in terms of the burnt gas ratio and, respectively, the oxygen concentration.

The general functionality of the turbocharging was described in Sect. 7.2.1. A common difference between SI and CI turbocharging is the function principle of actuation. In CI engines, variable turbine geometry (VTG) is the standard for the introduction of variability. The VTG allows the guide vane position to be changed, which adjusts the flow conditions. Narrow vane openings lead to a higher intake manifold pressure, while wide vane opening reduces them. With a VTG actuation, no mass flow is bypassed, the flow characteristics are adjusted instead. The VTG allows for a more accurate and easier control compared to a wastegate [14]. For durability reasons, in SI engines VTG is usually not used as SI engines have higher exhaust gas temperatures than CI engines. The VTG actuator can be electric or electro-pneumatic and typically incorporates a position sensor that allows for closed-loop control of the guide vane position.

In the given setup, the exhaust gas can be recirculated from the exhaust manifold to the intake manifold where it is mixed with fresh air. Due to the lean combustion conditions, the exhaust gas consists of burnt gas and (unburnt) air. The burnt gas, which consists mainly of N_2, H_2O, and CO_2, is an inert gas. It lowers the oxygen concentration of the gas inducted by the cylinder and increases the heat capacity of the gas. As a result, the combustion temperatures, especially the peak temperature, are lowered. High temperatures and excess oxygen, i.e. high air-to-fuel ratios lead to the formation of NO_x emissions [37]. Increasing burnt gas ratios x_{bg} thus lead to lower levels of NO_x emissions. However, at the same time, the emission levels of PM tend to increase. Figure 7.10 shows the measured sensitivity of the PM and the NO_x emissions to the burnt gas ratio x_{bg} and the intake manifold pressure p_{im}. The CI-inherent PM versus NO_x trade-off is clearly visible. If PM emissions are lowered, NO_x is increased and vice versa. The actuated value u_{egr} allows the valve position in the EGR path to be adjusted. As the gas flow to the intake manifold is driven by the pressure difference, the valve allows it to be controlled. Often, poppet valves with electric or electro-pneumatic actuation are used as EGR valves. Increased EGR rates also slow down the combustion, which reduces the noise levels.

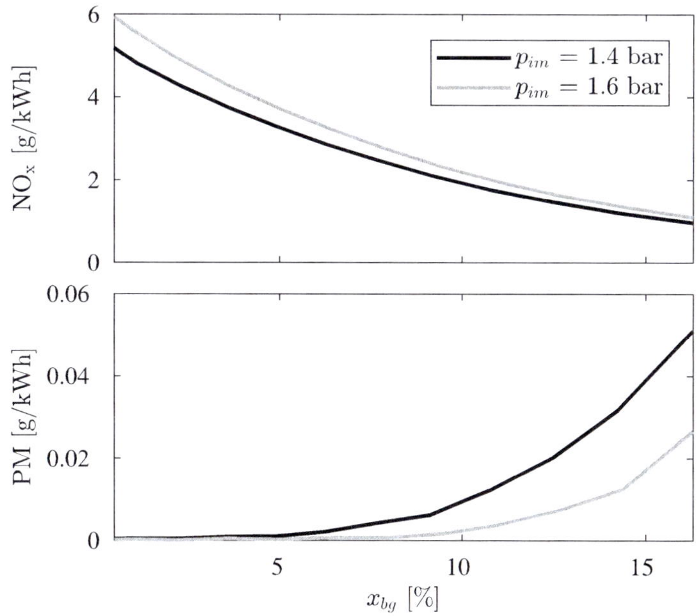

Fig. 7.10 Influence of the air path parameters burnt gas ratio and intake manifold pressure on pollutant emissions

Along with the burnt gas ratio x_{bg}, the intake manifold pressure p_{im} can be used as controlled values. These two values are sufficient to specify the two combustion relevant parameters, namely the amount of oxygen O_2 inducted into the cylinder and the amount of burnt gas inducted into the cylinder. An increased intake manifold pressure leads to more mass in the cylinder and higher cylinder pressures, and it influences the shape of the heat release profile and thus parameters such as the combustion average. The turbocharger speed, i.e. the mechanical limit of the turbocharger, and the back pressure of the exhaust gas have to be considered as well. Either an intake manifold pressure sensor or an air mass flow sensor can be used. Additionally, a pressure difference sensor across the EGR valve is used. Together with the information on the EGR valve position, its data is used to estimate the EGR mass flow and thus also the burnt gas ratio. The relevant dynamics for turbocharging are in the order of seconds. They are dominated by the turbocharger inertia and the pressure dynamics of the volumes. For EGR, the relevant timescale for the dynamics is slightly faster. Due to the strong cross-couplings, the MIMO system dynamics have to be accounted for in order to obtain an appropriate control system.

7.3.3 Aftertreatment Path Controller

For achieving low tailpipe pollutant emissions, a combination of various exhaust gas aftertreatment systems is usually used in CI engines. In the given setup, a DPF and an SCR system are combined.

The DPF is used to mechanically filter the PM emissions. For its proper operation, the fact has to be considered that the PM is stored in the filter. As soon as the maximum allowable PM loading is exceeded, the PM emissions need to be oxidized. The regeneration can either be passive, with an operation at high loads, or an active regeneration can be applied [41]. For the latter, the exhaust gases have to be at a temperature level of at least 550 °C [5]. A pressure difference sensor is used across the DPF to estimate its loading. The critical timescale is given by the dynamics of the DPF loading. The repetition of the active regeneration has to be conducted in the timescale of hours to days.

Various in-cylinder as well as out-of-cylinder measures exist to increase the exhaust gas temperature. The temperature raised is used to heat up the DPF and the SCR system. In the given setup, an exhaust flap is used for actuation. Increasing the exhaust temperature using the exhaust flap leads to higher exhaust gas back pressure and subsequently to higher gas exchange losses and reduced overall efficiency. By using the exhaust flap, the exhaust gas temperature can be adjusted in the order of seconds. The change in the exhaust gas temperature leads to an increase in the DPF temperature, which itself changes in the order of minutes.

To combat the NO_x emissions, a urea solution is injected upstream of the SCR catalyst with a dosing valve. The urea decomposes to ammonia NH_3 which is absorbed on the SCR surface. There, the ammonia reacts together with NO_x to form N_2 and H_2O. The NO_x conversion efficiency η_{SCR} depends in a nonlinear fashion on the tem-

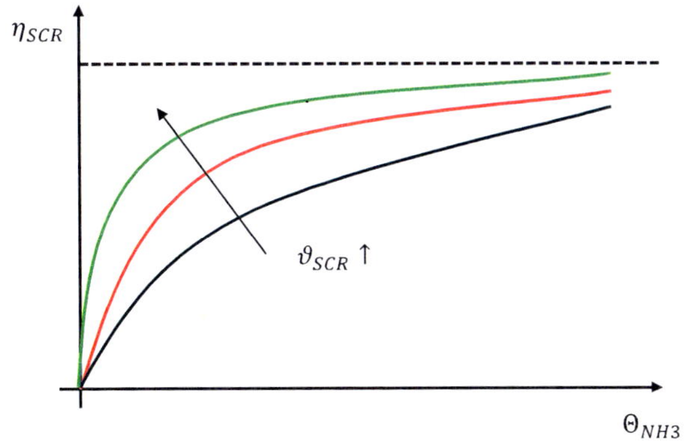

Fig. 7.11 Schematic SCR conversion efficiency in dependence of SCR temperature and ammonia surface coverage ratio

perature of the SCR ϑ_{SCR} and on the surface coverage ratio Θ_{NH3} [31]. Figure 7.11 schematically shows the conversion efficiency in dependence of the two parameters.

The management of an appropriate SCR temperature and ammonia surface coverage poses a complex closed-loop control task [30, 43, 44]. Due to the complex MIMO system behavior also MPC-based approaches are investigated [33]. High conversion efficiency is required to achieve a sufficiently low NO_x emission level. Simultaneously, a temperature-dependent maximum ammonia storage capacity has to be considered as a constraint in the controller. If this maximum capacity is exceeded, harmful NH_3 slip results. The various sensors used to estimate the ammonia surface coverage ratio include NO_x sensors mounted before and after the SCR, an NH_3 sensor, and temperature sensors [46]. The dynamics of changing the exhaust gas temperature are in the order of seconds. As a result, the SCR temperature changes in the order of minutes. The urea injection, in contrast, is a fast actuator, which allows the surface coverage ratio to be adjusted within seconds. Usually, a combined controller is used that considers both actuated values.

7.3.4 Control Structure

Figure 7.12 summarizes the structure of the CI engine controller with the main control tasks for normal operation. The fuel path controller influences the fuel mass injected, the combustion average, and the rail pressure. The air path controller sets the conditions in the intake manifold in terms of the pressure and the burnt gas ratio. The aftertreatment path controller adjusts the temperature of the exhaust gas and the SCR conversion efficiency. The ignition is controlled by the fuel path controller, thus no separate control loop is present. Many more control loops are necessary to run a CI engine. They include the low-level controllers such as those that control the valve positions and those required for the auxiliary functions such as lubrication control. Besides these, there are additional control tasks such as idle speed control and the cold-start control. For cold-start control, on the one hand, the control system actively

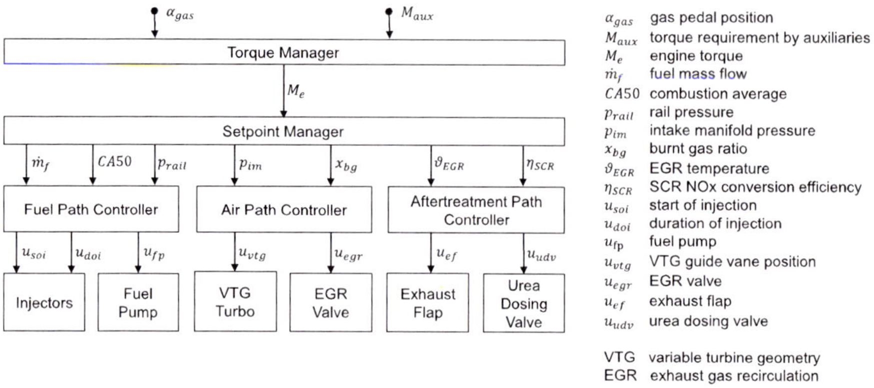

Fig. 7.12 CI engine control architecture (normal operation)

heats up the SCR catalyst such that the light-off temperature is reached quickly. On the other hand, the injection parameters are adjusted to ensure a combustion process with low pollutant emissions even at cold ambient conditions [11]. In some CI engine applications, especially in heavy-duty trucks, additional devices for waste heat recovery are used. In these applications, closed-loop control plays an essential role. Due to the complexity of the task also MPC approaches are investigated [15, 16, 24].

References

1. C. Arcoumanis, C. Bae, R. Crookes, E. Kinoshita, The potential of di-methyl ether (DME) as an alternative fuel for compression-ignition engines: a review. Fuel **87**(7), 1014–1030 (2008)
2. J. Asprion, O. Chinellato, L. Guzzella, Optimal control of diesel engines: numerical methods, applications, and experimental validation. Mathematical Problems in Engineering (2014)
3. P. Bares, D. Selmanaj, C. Guardiola, C. Onder, A new knock event definition for knock detection and control optimization. Applied Thermal Engineering **131**, 80–88 (2018)
4. D. Chatterjee, O. Deutschmann, J. Warnatz, Detailed surface reaction mechanism in a three-way catalyst. Faraday Discussions **119**, 371–384 (2002)
5. P. Chen, J. Wang, Control-oriented model for integrated diesel engine and aftertreatment systems thermal management. Control Engineering Practice **22**, 81–93 (2014)
6. M. De Cesare, M. Parotto, F. Covassin, S. Sgatti, Electric low pressure fuel pump control for fuel saving, *SAE Technical Paper*, no. 2013-01-0339 (2013)
7. Y. Deng, H. Liu, X. Zhao, E. Jiaqiang, J. Chen, Effects of cold start control strategy on cold start performance of the diesel engine based on a comprehensive preheat diesel engine model. Appl. Energy **210**, 279–287 (2018)
8. S. Di Cairano, A. Bemporad, I.V. Kolmanovsky, D. Hrovat, Model predictive control of magnetically actuated mass spring dampers for automotive applications. Int. J. Control **80**(11), 1701–1716 (2007)
9. S. Di Cairano, D. Yanakiev, A. Bemporad, I.V. Kolmanovsky, D. Hrovat, An mpc design flow for automotive control and applications to idle speed regulation, in *IEEE Conference on Decision and Control* (2008), pp. 5686–5691

10. S. Di Cairano, D. Yanakiev, A. Bemporad, I.V. Kolmanovsky, D. Hrovat, Model predictive idle speed control: design, analysis, and experimental evaluation. IEEE Trans. Control Syst. Technol. **20**(1), 84–97 (2011)
11. W. Dressler, S. Ernst, Start and ignition assist systems, in *Handbook of Diesel Engines*, (Springer, Berlin, 2010), pp. 377–386
12. W. Egler, R.J. Giersch, F. Boecking, J. Hammer, J. Hlousek, P. Mattes, U. Projahn, W. Urner, B. Janetzky, Fuel injection systems, in *Handbook of Diesel Engines* (Springer, Berlin, 2010), pp. 127–174
13. L. Eriksson, Spark advance for optimal efficiency, in *SAE Transactions* (1999), pp. 789–800
14. L. Eriksson, L. Nielsen, *Modeling and control of engines and drivelines* (Wiley, 2014)
15. E. Feru, F. Willems, B. de Jager, M. Steinbuch, Model predictive control of a waste heat recovery system for automotive diesel engines, in *International Conference on System Theory, Control and Computing* (2014), pp. 658–663
16. E. Feru, F. Willems, B. De Jager, M. Steinbuch, Modeling and control of a parallel waste heat recovery system for EURO-VI heavy-duty diesel engines. Energies **7**(10), 6571–6592 (2014)
17. J. Gerhardt, H. Hönninger, H. Bischof, A new approach to functional and software structure for engine management systems – BOSCH ME7, in *SAE Transactions* (1998), pp. 1173–1184
18. L. Guzzella, C.H. Onder, *Introduction to Modeling and Control of Internal Combustion Engine Systems* (Springer, 2010)
19. R. Hedinger, P. Elbert, C. Onder, Optimal cold-start control of a gasoline engine. Energies **10**(10), 1548–1564 (2017)
20. J.B. Heywood, *Internal Combustion Engine Fundamentals* (McGraw-Hill Education, 2018)
21. R. Isermann, *Engine Modeling and Control* (Springer, 2014)
22. L. Johannesson, N. Murgovski, E. Jonasson, J. Hellgren, B. Egardt, Predictive energy management of hybrid long-haul trucks. Control Eng. Pract. **41**, 83–97 (2015)
23. B. Johansson, Fuels and combustion, in *Biofuels from Lignocellulosic Biomass: Innovations Beyond Bioethanol* (Wiley, 2016)
24. M. Keller, M. Neumann, K. Eichler, S. Pischinger, D. Abel, T. Albin, Model predictive control for an organic rankine cycle system applied to a heavy-duty diesel engine, in *IEEE Conference on Control Technology and Applications* (2020)
25. K. Kim, K. Choi, K. Lee, K. Lee, Active coolant control strategies in automotive engines. Int. J. Autom. Technol. **11**(6), 767–772 (2010)
26. P. Kiwitz, C. Onder, L. Guzzella, Control-oriented modeling of a three-way catalytic converter with observation of the relative oxygen level profile. J. Process Control **22**(6), 984–994 (2012)
27. O. Leufvén, L. Eriksson, A surge and choke capable compressor flow model - validation and extrapolation capability. Control Eng. Pract. **21**(12), 1871–1883 (2013)
28. G. Mancini, J. Asprion, N. Cavina, C. Onder, L. Guzzella, Dynamic feedforward control of a diesel engine based on optimal transient compensation maps. Energies **7**(8), 5400–5424 (2014)
29. U. Projahn, H. Randoll, E. Biermann, J. Brückner, K. Funk, T. Küttner, W. Lehle, J. Zuern, Fuel injection system control systems, in *Handbook of Diesel Engines* (Springer, 2010), pp. 175–191
30. C.M. Schär, C.H. Onder, H.P. Geering, Control of an SCR catalytic converter system for a mobile heavy-duty application. IEEE Trans. Control Syst. Technol. **14**(4), 641–653 (2006)
31. C.M. Schär, C.H. Onder, H.P. Geering, M. Elsener, Control-oriented model of an SCR catalytic converter system, in *SAE Technical Paper* (2004)
32. D. Selmanaj, G. Panzani, S. van Dooren, J. Rosgren, C. Onder, Adaptive and unconventional strategies for engine knock control. IEEE Trans. Control Syst. Technol. **27**(4), 1838–1845 (2019)
33. J. Sowman, D.S. Laila, A.J. Cruden, P. Fussey, A. Truscott, A predictive control approach to diesel selective catalytic reduction, in *European Control Conference* (2015), pp. 3073–3078
34. A. Thomasson, O. Leufvén, I. Criscuolo, L. Eriksson, Modeling and validation of a boost pressure actuation system, for a series sequentially turbocharged SI engine. Control Eng. Pract. **21**(12), 1860–1870 (2013)

35. P. Tona, J. Peralez, A. Sciarretta, Supervision and control prototyping for an engine exhaust gas heat recovery system based on a steam rankine cycle, in *IEEE/ASME International Conference on Advanced Intelligent Mechatronics (AIM)* (2012), pp. 695–701
36. S. Trimboli, S. Di Cairano, A. Bemporad, I.V. Kolmanovsky, Model predictive control for automotive time-delay processes: an application to air-to-fuel ratio control. IFAC Proc. Vol. **42**(14), 90–95 (2009)
37. H. Tschöke, A. Graf, J. Stein, M. Krüger, J. Schaller, N. Breuer, K. Engeljehringer, W. Schindler, Diesel engine exhaust emissions, in *Handbook of Diesel Engines* (Springer, 2010), pp. 417–485
38. G. Vagnoni, M. Eisenbarth, J. Andert, G. Sammito, J. Schaub, M. Reke, M. Kiausch, Smart rule-based diesel engine control strategies by means of predictive driving information. Int. J. Engine Res. **20**(10), 1047–1058 (2019)
39. C. Vermillion, K. Butts, K. Reidy, Model predictive engine torque control with real-time driver-in-the-loop simulation results, in *American Control Conference* (2010), pp. 1459–1464
40. D. von Wissel, A. Husson, V. Talon, L. Lansky, D. Pachner, M. Uchanski, Reducing engine calibration time and cost with model predictive control, in *IAV Automotive Powertrain Control Systems Conference* (2014)
41. A. Walker, Controlling particulate emissions from diesel vehicles. Topics in Catal. **28**(1), 165–170 (2004)
42. T.T. Wang, J. Wagner, Advanced automotive thermal management-nonlinear radiator fan matrix control. Control Eng. Pract. **41**, 113–123 (2015)
43. F. Willems, R. Cloudt, Experimental demonstration of a new model-based SCR control strategy for cleaner heavy-duty diesel engines. IEEE Trans. Control Syst. Technol. **19**(5), 1305–1313 (2010)
44. F. Willems, R. Cloudt, E. Van Den Eijnden, M. Van Genderen, R. Verbeek, B. de Jager, W. Boomsma, I. van den Heuvel, Is closed-loop SCR control required to meet future emission targets? in*SAE Technical Paper*, no. 2007-01-1574 (2007)
45. F. Willems, P. van Gompel, X. Seykens, S. Wilkins,Robust real-world emissions by integrated ADF and powertrain control development, in *in: Control Strategies for Advanced Driver Assistance Systems and Autonomous Driving Functions* (Springer, 2019), pp. 29–45
46. X. Yuan, H. Liu, Y. Gao, Diesel engine SCR control: current development and future challenges. Emiss. Control Sci. Technol. **1**(2), 121–133 (2015)
47. X. Zhen, Y. Wang, S. Xu, Y. Zhu, C. Tao, T. Xu, M. Song, The engine knock analysis-an overview. Appl. Energy **92**, 628–636 (2012)

Chapter 8
Low-Temperature Combustion Engine Control

Abstract Besides conventional spark ignition (SI) and compression ignition (CI) engines, low-temperature combustion (LTC) engines are investigated as an alternative combustion concept for use in engine systems. They hold promise to lower pollutant emissions while achieving a high efficiency. The working principle is based on the autoignition of a highly premixed charge. These properties come along with increased demands for the process control. In this chapter, three possibilities to realize LTC combustion are detailed. They are gasoline controlled autoignition (GCAI), the diesel-based premixed charge compression ignition (PCCI), and the dual-fuel-based reactivity controlled compression ignition (RCCI). The chapter focuses on the requirements and challenges concerning the process control. They are specific for each concept. For instance, depending on the fuel reactivity, different measures are applied to realize LTC conditions. As a consequence, the system dynamics also depend strongly on the setup.

8.1 Introduction to LTC Engines

The use of LTC in engines holds the promise to combine the advantages of SI and CI engines. It can deliver a high efficiency and give rise to considerably reduced engine-out emissions. The application of LTC to engines is still in the phase of ongoing development. However, it gains increasing attention due to its advantages.

The various approaches to realize LTC are characterized by a large degree of homogenization of the air–fuel mixture and the autoignition of the charge. The combustion is initiated by almost simultaneous autoignition at several points in the combustion chamber, leading to a combustion process without a defined flame front. This combustion initiation is entirely different from that of SI engines where the charge is ignited via spark and from CI engines where the combustion is triggered by the (late) injection timing. Instead, a reactivity controlled reaction takes place, which is

T. Albin Rajasingham, *Nonlinear Model Predictive Control of Combustion Engines*, Advances in Industrial Control,
https://doi.org/10.1007/978-3-030-68010-7_8

initiated by the temperature increase in the compression stroke. As a consequence, the chemical reaction kinetics play an important role. The heat release is associated with the LTC kinetics, which has distinctive characteristics, such as the negative temperature coefficient (NTC) regime [2]. In this regime, the reactivity decreases even though the temperature in the cylinder increases. The LTC process is operated in a highly diluted fashion, which can be achieved by high amounts of recirculated exhaust gas and by an excess of air. Indeed, lean as well as stoichiometric conditions can be used for the LTC combustion.

The LTC engine has a different operating range concerning temperature and air-to-fuel ratio compared to a CI engine which significantly affects the pollutant formation [10]. At high local temperatures and at conditions with an excess of air, nitrogen oxides (NO_x) are produced. The formation of soot occurs at moderate temperatures with a low oxygen content. For a conventional CI engine, the injection of fuel occurs very late in the compression stroke. From the CI concept, a stratified charge follows with high local temperatures, which results in the formation of soot and NO_x emissions. In LTC, the regions of NO_x and soot formation are avoided. The highly diluted charge leads to reduced peak temperatures in the cylinder and, due to the large degree of homogenization, rich regions in the combustion chamber are avoided [57]. At the same time, a high efficiency can be achieved due to several reasons. These are the reduction of pumping losses due to a dethrottled operation, an almost constant-volume combustion, the possibility to apply high compression ratios, the reduced wall heat losses because of the reduced temperatures, and the higher ratio of specific heat due to the diluted charge (carbon dioxide CO_2 has a higher specific heat than oxygen O_2) [64]. Another reason for dilution with EGR is the slowdown of the combustion process. This reduces the pressure rise gradient which, due to the bulk autoignition, for LTC, is higher than that of conventional SI and CI engines.

The LTC combustion process is largely determined by the low-temperature chemical reaction kinetics, while conventional combustion processes are determined by the stabilizing mixture-controlled high-temperature reactions. As a consequence, the LTC regime is characterized by a high sensitivity to the global and local thermodynamic state in the system. Many parameters influence the sensitive thermodynamic properties. They include the temperature, the chemical mixture composition, and the turbulent flow field, which affects the local composition of the mixture and thus the stratification. Depending on this thermodynamic state, very early combustion with very high pressure rise gradients or very retarded combustion up to the point of unwanted misfires can result. Consequently, the sensitivities pose challenges concerning stability and undesired emissions. Since the resulting irregular combustion behavior and increased emissions cannot be tolerated, the mastering of these sensitivities is a major task for the practical application of LTC. Another challenge for practical application is the limited operating range. With recent approaches, the LTC combustion can only be applied for a small operating range, i.e. especially medium load operating regions [52].

Closed-loop control strategies are commonly used to respond to these sensitivities arising in LTC engines. For conventional SI and CI combustion control, feedforward control in combination with simple linear feedback concepts are usually sufficient. Several controllers are applied that individually deal with specific control tasks, such as combustion average control [56]. The LTC process is characterized by high sensitivity, the difficulty to measure all relevant parameters during operation, and strongly nonlinear system dynamics with strong interaction of the different in- and outputs. These characteristics preclude such linear control approaches and the separate treatment of individual phenomena. Instead, the strongly nonlinear MIMO dynamics as well as the limitations of the actuating elements must be considered by appropriate feedback control approaches [61]. In the most common approaches, the control algorithm relies on the in-cylinder pressure trace, which is directly measured by an in-cylinder pressure sensor [63].

For an appropriate control of the process, the entire MIMO system dynamics have to be considered due to the strong nonlinear cross-couplings. Decentralized control approaches, e.g. separate controllers for the indicated mean effective pressure (IMEP) and the combustion average, are not sufficient, especially when the entire operation envelope is considered. As control algorithms, a broad range of control concepts is evaluated. Initial control concepts relied on gain-scheduled PID controllers [55]. In recent approaches, model-based control strategies are used primarily. The model-based control strategies show promising advantages to handle the complex system dynamics. Various model-based control strategies are investigated, such as sliding mode control [4], adaptive control [6], reference governor control [29], switching control [40], and model predictive control (MPC) [13, 14, 44, 47]. Especially, the optimization-based concepts show a high potential since limitations on the actuators and on the outputs have to be considered and, in some cases, non-minimum phase behavior is present.

For the LTC combustion process, the fuel plays an important role. In general, LTC is applicable to a wide range of fuels, such as gasoline [32], diesel [5], hydrogen [1], bio-fuels [34], and natural gas [17]. In addition, much research is conducted to design specific fuels that are advantageous for LTC [39, 41]. From a control point of view, most differences exist especially with regard to three different categories:

1. applications with fuels that are characterized by a low reactivity such as gasoline,
2. applications of LTC to fuels with high reactivity such as diesel,
3. applications where the reactivity of the fuel is actively changed to realize LTC, especially by use of two different injection systems to gain an in-cylinder fuel blending.

All three applications necessitate closed-loop control due to the sensitivities discussed. However, the control challenges are quite diverse and depend on the specific setup. In the following, one setup of each category is discussed.

8.2 Gasoline-Based LTC: Gasoline Controlled Autoignition

In this subsection, the control challenges associated with the gasoline-fueled LTC process are to be discussed. Various names are used in literature for this concept, such as GCAI or homogenous charge compression ignition (HCCI). The abbreviation GCAI for this combustion process will be used in the remainder of this text. Compared to the conventional SI engine, the GCAI process allows for an engine efficiency that is up to 25% higher than that of a conventional SI engine [60]. The main reason for the efficiency increase lies in the possibility of an unthrottled operation, which reduces the pumping losses. Additionally, GCAI produces almost no measurable soot and NO_x emissions due to the homogenization and the dilution with recirculated exhaust gas [60].

Gasoline is a low-reactivity fuel, i.e. it is less reactive than diesel fuel. The gasoline fuel for SI engines is designed such that knocking, an unwanted autoignition phenomenon, is prevented [20]. However, for GCAI, conditions have to be created where autoignition of the gasoline fuel occurs in a controlled setting. There exist different measures such as intake air heating [62], internal recirculation of exhaust gas [30], external recirculation of exhaust gas [35], or the use of variable compression ratios [7] that are combined with an appropriate fuel injection strategy. In the following, one exemplary realization of LTC conditions with internal exhaust gas recirculation is described.

As the basic engine setup, a SI engine with a moderately increased compression ratio is used. The main difference from conventional SI engines is the considerable recirculation of hot internal exhaust gas. Specifically, the exhaust gas from the last cycle delivers the thermal energy to initiate the compression-induced autoignition in the current cycle. A major control task is the appropriate adjustment of this quantity depending on the current operating point, which is characterized by the load, speed, and ambient conditions for instance. Inappropriate amounts of internal exhaust gas or thermal energy can lead to advanced combustion with exceeding maximum pressure rise gradients, retarded incomplete combustion, or even misfire.

For the development of control systems, the amount of recirculated internal exhaust gas can be varied by means of a variable valve train (VVT). Various valve strategies can be applied to recirculate the internal exhaust gas with a VVT. A commonly used strategy is the so-called combustion chamber recirculation (CCR) where the exhaust gas is trapped in the combustion chamber. Figure 8.1 shows the resulting cylinder pressure trace along with the valve timings. Alternatively, the exhaust gas can be trapped in the intake or exhaust port and recirculated from there [36]. The strategies differ in terms of the maximum amount of exhaust gas that can be recirculated and the properties of the mixture, such as the temperature. Depending on the valve strategy used, differences result in terms of stability and operating range of the combustion process that can be realized [36].

As mentioned above, all LTC concepts exhibit high sensitivities to the local and global thermodynamic states, which generally requires feedback concepts. In the case of GCAI, the particularity of a strong coupling of consecutive cycles must be

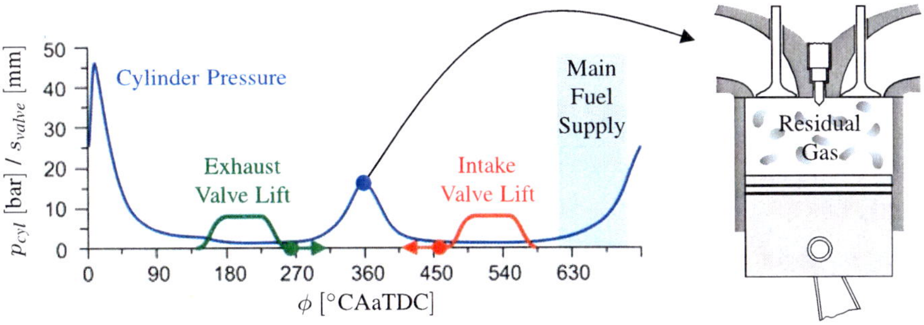

Fig. 8.1 Typical GCAI cylinder pressure trace along with actuated values

considered in the control algorithm. This strong coupling distinguishes the GCAI process from the diesel PCCI and the dual-fuel RCCI process to be described below. Due to the initiation of the combustion by the exhaust gas from the preceding cycle, the quality of combustion in the current cycle directly depends on the combustion process of the preceding cycle. This presents a considerable control problem at stationary conditions (especially at the limits of the operating range) and even more so for load transients.

The coupling of the combustion cycles is also called the memory effect. For instance, the disturbances arising from one cycle are propagated to the next cycles. Incomplete combustion in one cycle transfers HC molecules to the next cycle, which can lead to advanced combustion. This again affects the subsequent cycle. The strong coupling of consecutive cycles leads to a high tendency toward unstable process behavior [23].

In the case of load transients, these memory effects can be observed in terms of the dynamic behavior of the outputs, as depicted in Fig. 8.2. It shows a load transient that is conducted from one stable operating point to another stable operating point. The actuated values, i.e. exhaust valve closure (EVC) and the duration of injection (DOI), are changed in a step-wise manner. Directly after the application of the step input, the combustion phase is very retarded when increasing the load, leading to a temporarily incomplete combustion until a steady state is reached again. For steps toward lower loads, a very advanced combustion results with high pressure rise gradients. The dynamic behavior in load transients is due to the change in initial condition from a low temperature of the exhaust gas (at low loads) to higher temperatures of the exhaust gas (at high loads). If the load transients are bigger, even misfires can occur for a step-wise change in the actuated values. An appropriate control system has to take this dynamic behavior into account.

At the boundaries of the operating range, even in the case of stationary operation, combustion instabilities can be observed. In these operating points, the sensitivities mentioned are higher, and simultaneously the stochastic, cyclic variations increase, e.g. due to the turbulent flow field. The combination with the distinctive coupling of consecutive combustion cycles leads to bifurcative system dynamics such as those

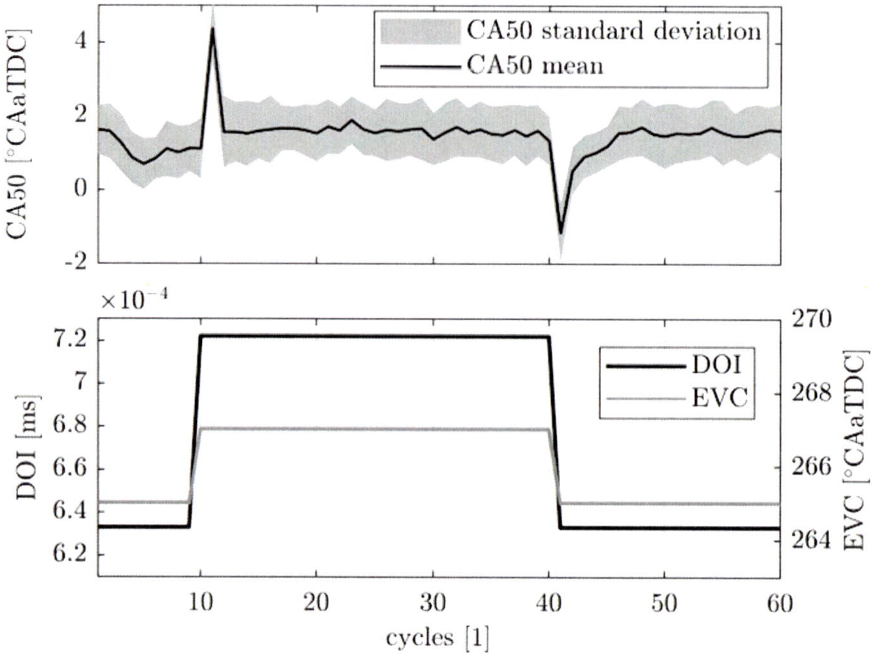

Fig. 8.2 Memory effect for load transients

known from chaotic systems [16, 22]. In Fig. 8.3, a schematic of such a bifurcative behavior is depicted by showing the return maps of the combustion average CA50. The return maps plot many measurements of the combustion average at cycle $k + 1$ in comparison to cycle k. The left-hand plot shows a stable GCAI operating point. For constant input values, a certain expected value results which is overlapped by a Gaussian-distributed white noise. The right-hand plot shows an unstable operating point which is located close to the limits of the operating range. Instead of a stationary behavior, various limit cycles are triggered by small disturbances. For an appropriate application of GCAI, a suitable closed-loop control system is required.

8.2.1 Cycle-to-Cycle-Based Control Strategies

The conventional approach to deal with these kinds of combustion problems is the use of a cycle-to-cycle-based control strategy, as shown in Fig. 8.4. It is characterized by two aspects; frequency of actuation and time resolution of the controlled variables. On the one hand, the actuated values are recalculated once every combustion cycle. On the other hand, they are based on integral surrogate parameters that characterize an individual engine cycle [15]. Important surrogate parameters include the following:

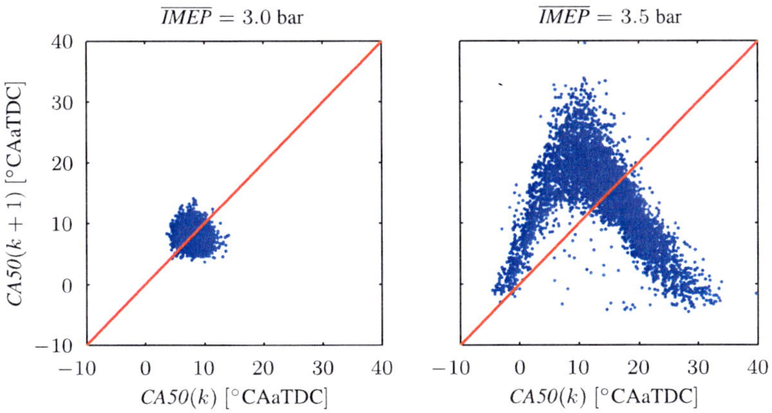

Fig. 8.3 Stable behavior (left-hand) and bifurcative behavior (right-hand) of GCAI combustion

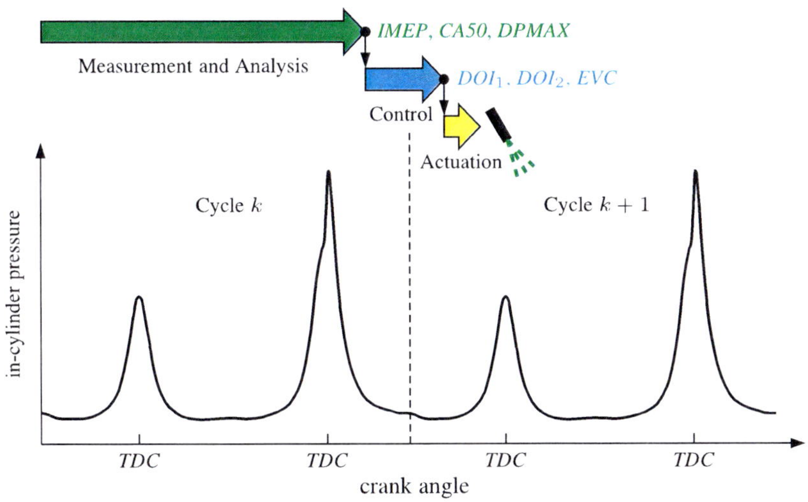

Fig. 8.4 Cycle-to-Cycle combustion control strategy

- the IMEP as a correlator for the load,
- the combustion average, i.e. the crank angle position where 50% of the heat is released (CA50) as an indicator for the efficiency, and
- the maximum pressure rise gradient (DPMAX) as an indicator for the noise.

These values are calculated based on sensor data obtained from the preceding cycle, as explained in more detail in Sect. 14.3. Typically, a cylinder pressure sensor is used. Based on these surrogate parameters, the control algorithm determines the actuated values for the subsequent cycle. In order to simultaneously affect the various controlled values (IMEP, CA50, DPMAX), a combination of fuel path parameters and valve timings is applied as actuated values. For the realization of CCR, the

VVT is used by an early closing of the exhaust valve, which leads to the desired trapping of a part of the exhaust gas of the last combustion cycle in the combustion chamber. By a symmetric adjustment of the timing of the closure of the exhaust valve and of the opening of the intake valve, the negative valve overlap can be changed, and thus, the amount of exhaust gas is adjusted. Additionally, a split injection of fuel can be applied in the case of direct injection. The first injection is applied in the recompression phase. This early injection leads to first chemical reactions. The second injection is the main injection. The timing of the injections is usually fixed and thus cannot be used for closed-loop control. The durations of the first injection (DOI1) and of the second injection (DOI2) are applied in the closed-loop setting [51]. Several control approaches have attempted to also include the deterministic coupling of subsequent cycles to reduce cyclic variations [21, 31].

A significant disadvantage of the GCAI process with cycle-to-cycle-based controllers is its limitation to a rather small range of operation in which the LTC can be stabilized [52]. Especially for the transient performance in practical applications, small parameter changes near the boundaries of the window of operation can lead to significant instabilities with adverse consequences for the combustion efficiency and the emissions. This holds especially true if external exhaust gas recirculation or turbocharging is used [37]. These are implemented for an extension of the operating range. However, they increase the fluctuations in the initial conditions of combustion. The major disadvantage of the cycle-to-cycle-based control approaches is the fact that only the system dynamics and disturbances that occur on a cycle-averaged time scale can be controlled. However, the delicate balance between chemistry and mixture formation in the LTC process occurs on smaller time scales. Consequently, the relevant physico-chemical processes determining the stability and emissions characteristics of LTC cannot be controlled sufficiently under all arising circumstances using a cycle-to-cycle-based controller with cycle-integral surrogate parameters.

8.2.2 Multi-scale Control Strategies

To overcome the drawbacks of cycle-to-cycle-based control strategies, multi-scale control algorithms are being investigated. They allow smaller time scales to be controlled than those possible with cycle-to-cycle-based control systems. Within multi-scale control approaches, an in-cycle controller is used in addition to the cycle-to-cycle controller. The in-cycle controller calculates feedback based on measurements of the same combustion cycle. The architecture of such a multi-scale control approach is shown in Fig. 8.5. An in-cycle controller can counteract disturbances arising within the same cycle and can thus substantially increase the stability in critical operation regions due to a faster feedback [59]. From a technical point of view, the in-cylinder pressure has to be evaluated fast. One possibility to do this is the use of a FPGA in order to determine the amount of unburnt fuel based on the pressure trace in the recompression phase, which significantly affects the subsequent main combustion [18, 60]. This information can be exploited for an in-cycle feedback that adjusts

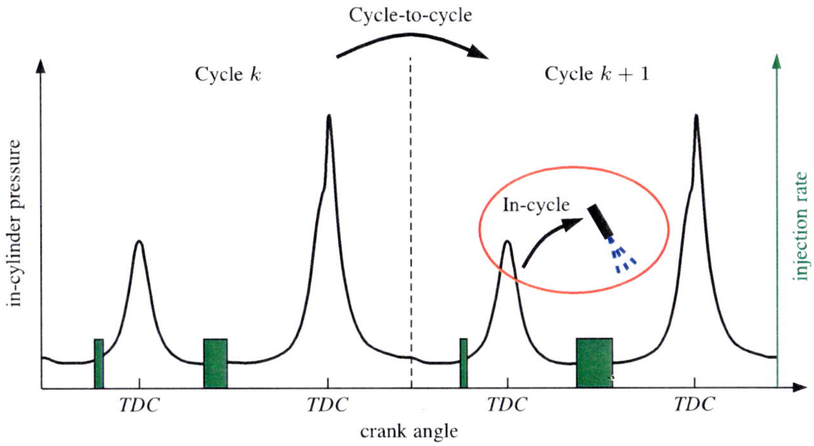

Fig. 8.5 Multi-scale control architecture for combustion control

the main injection. Other possibilities for in-cycle actuation are the adjustment of intake valve closure timing [38] or the use of direct water injection into the combustion chamber [59].

8.3 Diesel-Based LTC: Premixed Charge Compression Ignition

The LTC combustion can also be realized with diesel fuel, which is named PCCI or diesel HCCI. In the following, the abbreviation PCCI will be used for the diesel-based LTC process. In contrast to gasoline, diesel fuel is much more reactive. In a conventional CI engine, the diesel fuel is ignited via autoignition. Compared to conventional CI engines, the diesel-based LTC produces significantly less NO_x due to lower peak temperatures. It also produces lower soot emissions due to the high degree of homogenization, while the fuel efficiency is approximately equivalent to that of a conventional CI engine.

In order to achieve this high performance, the process has to fulfill several requirements. The combustion timing has to be set appropriately, which means that the combustion should not take place either too early or too late. At the same time, the formation of a suitable mixture has to be ensured. Ignition delay times that are too short can lead to the formation of an incomplete mixture. Ignition delay times that are too long can lead to high cyclic variations or even to complete misfires. A sufficient homogenization of the mixture thus has to be ensured, while simultaneously the unburned hydrocarbons (HC) and carbon monoxide (CO) emissions as well as the combustion noise and high efficiency must be considered [53]. Concerning the engine hardware, slightly modified conventional CI engines can be used. Modifications include the piston bowl geometry, for instance [42].

In order to establish the conditions discussed for PCCI, a lean mixture is applied. In addition, high amounts of cooled exhaust gas are recirculated to decrease peak temperatures and to increase the ignition delay times [48]. The high EGR rates also slow down the combustion process, which positively affects the maximum pressure rise gradient. As LTC combustion in general, the PCCI combustion is sensitive to the local and global thermodynamic states. Due to this sensitivity, the control of the air path becomes more important than it is in the conventional CI engine. Especially, the rate of recirculated exhaust gas and the intake pressure have to be controlled accurately and fast. The boundary condition of combustion is defined by the initial cylinder condition with respect to pressure, chemical composition, and temperature. Small deviations in these boundary conditions can lead to significant decreases in the performance of the engine. This is especially the case for transient operation when boundary conditions such as the EGR rate are changing quickly along with their influence on the ignition delay, for instance. The overall process control also has to take into account the cross-coupling of the air path and the combustion. A coordinated control approach of the air path and the combustion process thus is necessary. Various solutions exist, such as the use of a cascaded control law [12] or by taking into account the cross-couplings of the air path parameters on the combustion controller [24].

As the air path controlled variables are too slow for cycle-to-cycle-based control, the actuated variables from the combustion have to be used for this purpose. The combustion controller has to act very fast to ensure the appropriate PCCI conditions under all circumstances. The main actuator for combustion control is the fuel injector. Several possibilities exist to realize a high degree of homogenization. One possibility for a single injection is to inject the fuel very early. However, this approach allows suitable PCCI conditions only in a very narrow operating range. The early fuel injection in combination with high injection pressures leads to distinctive wall-wetting [3]. The initially low temperatures of the wall cause an incomplete combustion, resulting in high amounts of HC emissions that exceed the allowable range. Another challenge is the loss of the control authority as the injection timing only has a small effect on the combustion timing for early injections with (almost) homogeneous combustion [64].

Therefore, the application of a partially premixed charge is preferable for diesel-fueled LTC. Low pollutant emissions can be realized for a wider range of operation, while the combustion noise is decreased and a better controllability of the combustion timing is achieved [65–67]. A common approach is the use of a multi-pulse fuel injection strategy, which allows some portion of the fuel to be injected very early with a high degree of homogenization [9, 50]. The typical approach to control the IMEP and the combustion average CA50 is not sufficient for multi-pulse fuel injection. Figure 8.6 shows three different combustion traces. They are realized by the three injection strategies; single, double, and triple injection and still result in the same values for IMEP and CA50. Although these two typical performance parameters are

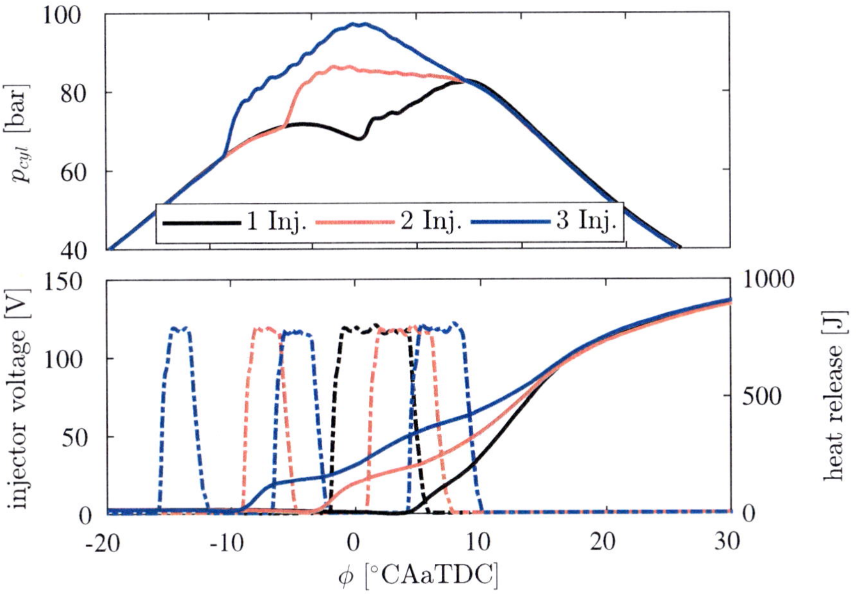

Fig. 8.6 Three different cylinder pressure traces (1 vs. 2 vs. 3 injections) with the same combustion average and load along with the injector voltage (dashed curve)

Table 8.1 Performance parameters for the three pressure traces depicted in Fig. 8.6

Number of Inj.	IMEP [bar]	CA50 [°CAaTDC]	η [%]	NO_x [$\frac{g}{kW}$]	DPMAX [bar/°CA]
1	9	15	37.05	1.66	3.21
2	9	15	37.12	1.84	5.87
3	9	15	36.81	2.49	7.46

similar, the performance varies. Table 8.1 lists the differences in NO_x, efficiency η, and in the maximum pressure rise gradient DPMAX for the three traces.

Thus, the reference tracking of the two cycle-integral surrogate parameters, CA50 and IMEP cannot guarantee achieving high-performance combustion in the entire operating range. Instead, controlled variables can be used that offer a higher resolution concerning the temporal development of the cylinder pressure or the temporal development of the heat release in the engine, respectively. They allow more in-cycle effects to be considered. This concept is called combustion rate shaping and can be applied for conventional CI as well as for PCCI conditions. The combustion rate shaping will be introduced in more detail in Sect. 14.4.

8.4 Dual-Fuel-Based LTC: Reactivity Controlled Compression Ignition

An alternative approach to realize LTC combustion is the use of a dual-fuel concept. Two different injection systems are applied to inject one fuel with a low reactivity and one with a high reactivity. Changing the ratio of the two fuel masses is used to appropriately blend the in-cylinder fuel. Thus, control-relevant parameters such as the ignition delay and the reactivity of the overall charge can be adapted. Depending on the conditions such as speed and required load, the fuel blending ratio can be adjusted for satisfying LTC conditions with a suitable mixture formation as well as appropriate combustion timings. As this ratio of the fuels can be adjusted on a cycle-to-cycle basis, it allows for a fast closed-loop combustion control. The control algorithm can be used for instance to reject disturbances and to allow for reference-tracking in load-transient operation. This combustion concept is known under the abbreviation RCCI [49]. The RCCI concept can be applied with a variety of different fuel combinations. Examples are the combination of gasoline with diesel [8], ethanol with diesel [54], and natural gas with diesel [43].

The mixing of natural gas with diesel offers several advantages. Natural gas consists mainly of methane (CH_4), which has a high hydrogen-to-carbon ratio and thus causes inherently lower CO_2 emissions to be formed during combustion. Methane produces CO_2 emissions of 55 $\frac{g}{MJ}$, whereas the long-chained hydrocarbon surrogate of diesel $C_{15}H_{28}$ emits 74.2 $\frac{g}{MJ}$ and the surrogate for gasoline C_7H_{15} emits 73.3 $\frac{g}{MJ}$ CO_2 emissions [33]. Thus, natural gas is able to lower CO_2 emissions levels by approximately 25% just due to its chemical structure. Also, methane can be easily stored, transported, and it can be produced synthetically by power-to-gas processes [19]. This feature allows surplus energy from fluctuating renewable energy carriers to be stored. Methane is even less reactive than gasoline, which has benefits in terms of the knock resistance, i.e. high compression ratios are feasible with advantages in terms of efficiency.

Various technologies exist to ensure combustion in a methane engine, such as the use of a spark plug. However, the use of a spark plug limits the applicability to a certain range of air-to-fuel ratios, thus prohibiting a very lean operation. Alternatively, methane can be ignited by the use of an additional diesel fuel injection. If diesel fuel is used as an ignition source, the operating range concerning the air-to-fuel ratio is much higher. The ignition energy of the diesel fuel can easily be adjusted by the diesel fuel mass injected, with the possibility to also operate at very lean conditions. A typical setup is the injection of methane into the intake port in order to gain a premixed charge of air and methane. Diesel as a high reactivity fuel is injected directly into the cylinder, as shown in Fig. 8.7.

The actuated values for closed-loop control are the duration of injection of methane, the duration of injection of diesel, and the time of injection of the diesel fuel. With these values, the load and the combustion timing are controlled. Compared to combustion control in conventional diesel engines, one surplus degree of freedom

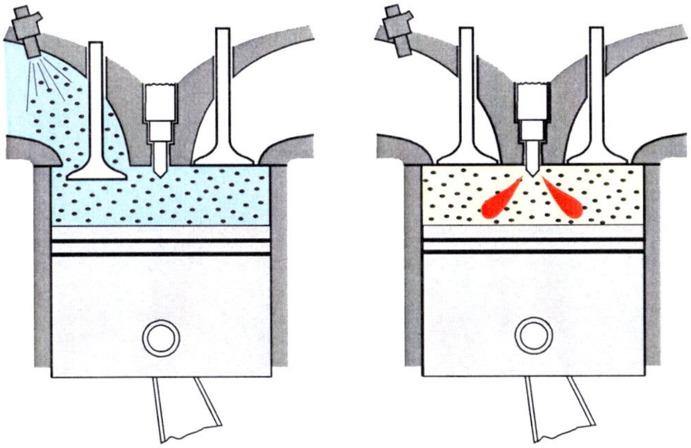

Fig. 8.7 Engine setup with dual-fuel combustion – Reproduced from [26], originally published open access under a Creative Commons CC BY 4.0 license, https://doi.org/10.3390/en10101450

remains. This degree of freedom can be optimized to achieve low emissions while maintaining a high efficiency and specific ratios of methane-to-diesel fuel mass.

Depending on the timing of diesel injection, different combustion concepts result, namely the diesel-ignited dual-fuel (DDF), the RCCI, and the diesel-minimum operation. Figure 8.8 shows the static behavior of the MIMO system. The combustion average is plotted for various diesel injection timings and various substitutions rates SR, i.e. the ratio $SR = \frac{m_{gas}}{m_{diesel}+m_{gas}}$. For sufficiently late diesel injections, a DDF combustion results. The homogeneous methane–air mixture is burnt by flame propagation resulting from the diesel pilot injection. This pilot ignition leads to mixing-controlled combustion comparable to a conventional SI combustion. On the other hand, the diesel can be injected very early, which leads to an RCCI combustion. The autoignition of a highly premixed charge of methane, diesel, and air results. The ignition delay is much longer than that of a DDF, such that enough time remains for the mixture formation. In fact, the same combustion average can be achieved with RCCI and DDF, although the timing of the diesel injection is very different. The third principle possibility is the diesel-minimum operation. The minimum diesel fuel is achieved at the minima of the parabolic curves in Fig. 8.8. Thus, for diesel-minimum operation, the surplus degree of freedom is exploited to achieve a certain combustion average and load with minimal substitution rate [68].

In all three combustion regions, DDF, diesel-minimum, and RCCI, different levels of pollutant emissions, efficiency, and CO_2 emissions result. The choice of the most appropriate combustion regime depends on the specific application. For a reduction of the NO_x emissions, for instance, the RCCI region is most appropriate [11]. However, if the goal is to lower the emissions levels of CO_2, the diesel-minimum operating regime should be applied. For closed-loop control, it is important to take the non-linearity of the process into account. This concerns especially the cross-coupling of

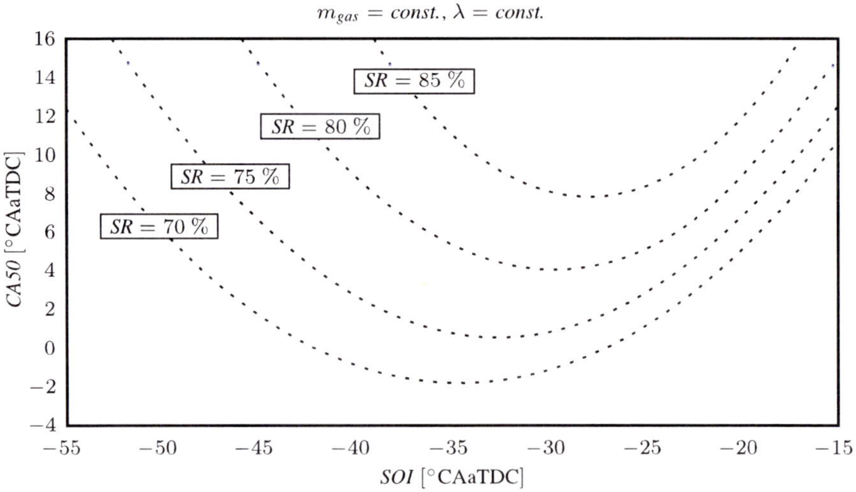

Fig. 8.8 Influence of diesel start of injection on combustion average

the various actuated values and also the reverse of the sign in the combustion timing [28, 45].

In addition to the MIMO system dynamics described, further control challenges arise in an RCCI operation. As in the case of diesel PCCI, the parameters of the air path, especially the intake pressure and the recirculated exhaust gas, have a strong effect on the amounts of pollutants emitted. Thus, it is important to control these parameters accurately and fast. Additionally, the strong interaction between the air path and the combustion has to be accounted for [58]. Another challenge arises as the operating range of a dual-fuel engine is limited to medium and high loads [26]. At low loads, the dual-fuel combustion cannot be used. Instead, a pure diesel combustion has to be applied. The engine controller thus has to be capable also of operating in pure diesel mode. Additionally, for load-transient operations, the switching between operating modes (RCCI, DDF, pure diesel) has to be accomplished as well [25, 27]. The requirement is that switching between combustion modes can be achieved without any degradation in torque and while avoiding peaks in the pollutant emissions. The necessity to be able to operate the engine in dual-fuel and also in diesel mode affects the injector as well. In a dual-fuel mode, only very small quantities of diesel are injected. On the other hand, in a pure diesel mode, very high quantities can be injected. The injector thus has to handle the entire range from very small to very high fuel quantity injections. Especially, in the RCCI mode, where the combustion is sensitive to the injection parameters, very high requirements exist on the injection quality, such as a shot-to-shot reproducibility. In [46], an injector is presented where the fuel amount injected is closed-loop-controlled in order to fulfill the requirements arising. The closed-loop control of the injection quantity allows for drifts over a lifetime to be compensated.

References

1. J.G. Antunes, R. Mikalsen, A. Roskilly, An investigation of hydrogen-fuelled HCCI engine performance and operation. Int.J. Hydrog. Energy **33**(20), 5823–5828 (2008)
2. F. Battin-Leclerc, Detailed chemical kinetic models for the low-temperature combustion of hydrocarbons with application to gasoline and diesel fuel surrogates. Progr. Energy Combust. Sci. **34**(4), 440–498 (2008)
3. H. Bendu, S. Murugan, Homogeneous charge compression ignition (HCCI) combustion: mixture preparation and control strategies in diesel engines. Renew. Sustain. Energy Rev. **38**, 732–746 (2014)
4. M. Bidarvatan, M. Shahbakhti, S. Jazayeri, C. Koch, Cycle-to-cycle modeling and sliding mode control of blended-fuel HCCI engine. Control Eng. Pract. **24**, 79–91 (2014)
5. A. Carlucci, D. Laforgia, S. Motz, R. Saracino, S. Wenzel, Advanced closed loop combustion control of a LTC diesel engine based on in-cylinder pressure signals. Energy Convers. Manag. **77**, 193–207 (2014)
6. C.-J. Chiang, C.-C. Chou, J.-H. Lin, Adaptive control of homogeneous charge compression ignition (HCCI) engines, in *American Control Conference* (2012), pp. 2066–2071
7. M. Christensen, A. Hultqvist, B. Johansson, Demonstrating the multi fuel capability of a homogeneous charge compression ignition engine with variable compression ratio. SAE Trans. 2099–2113 (1999)
8. S. Curran, R. Hanson, R. Wagner, R.D. Reitz, Efficiency and emissions mapping of RCCI in a light-duty diesel engine, SAE Technical Paper, no. 2013-01-0289 (2013)
9. W. de Ojeda, P. Zoldak, R. Espinosa, R. Kumar, Development of a fuel injection strategy for partially premixed compression ignition combustion. SAE Int. J. Eng. **2**(1), 1473–1488 (2009)
10. A.B. Dempsey, S.J. Curran, R.M. Wagner, A perspective on the range of gasoline compression ignition combustion strategies for high engine efficiency and low NOx and soot emissions: effects of in-cylinder fuel stratification. Int. J. Eng. Res. **17**(8), 897–917 (2016)
11. E. Doosje, F. Willems, R. Baert, Experimental demonstration of RCCI in heavy-duty engines using diesel and natural gas, SAE Technical Paper (2014)
12. P. Drews, T. Albin, F.-J. Hesseler, N. Peters, D. Abel, Fuel-efficient model-based optimal MIMO control for PCCI engines. IFAC Proc. Vol. **44**(1), 12 998–13 003 (2011)
13. K. Ebrahimi, C.B. Koch, Real-time control of HCCI engine using model predictive control, in *American Control Conference* (2018), pp. 1622–1628
14. S.M. Erlien, A.F. Jungkunz, J.C. Gerdes, Multicylinder HCCI control with coupled valve actuation using model predictive control. J. Dyn. Syst., Meas., Control **135** (2013)
15. M. Fathi, O. Jahanian, M. Shahbakhti, Modeling and controller design architecture for cycle-by-cycle combustion control of homogeneous charge compression ignition (HCCI) engines - a comprehensive review. Energy Convers. Manag. **139**, 1–19 (2017)
16. C.E.A. Finney, B.C. Kaul, C.S. Daw, R.M. Wagner, K.D. Edwards, J.B. Green Jr., A review of deterministic effects in cyclic variability of internal combustion engines. Int. J. Eng. Res. **16**(3), 366–378 (2015)
17. D. Flowers, S. Aceves, C. Westbrook, J. Smith, R. Dibble, Detailed chemical kinetic simulation of natural gas HCCI combustion: gas composition effects and investigation of control strategies. J. Eng. Gas Turb. Power **123**(2), 433–439 (2001)
18. D. Gordon, C. Wouters, M. Wick, F. Xia, B. Lehrheuer, J. Andert, C.R. Koch, S. Pischinger, Development and experimental validation of a real-time capable field programmable gate array-based gas exchange model for negative valve overlap. Int. J. Eng. Res. **21**(3), 421–436 (2018)
19. M. Götz, J. Lefebvre, F. Mörs, A.M. Koch, F. Graf, S. Bajohr, R. Reimert, T. Kolb, Renewable power-to-gas: a technological and economic review. Renew. Energy **85**, 1371–1390 (2016)
20. L. Guzzella, C.H. Onder, *Introduction to Modeling and Control of Internal Combustion Engine Systems* (Springer, 2010)
21. E. Hellström, J. Larimore, S. Jade, A.G. Stefanopoulou, L. Jiang, Reducing cyclic variability while regulating combustion phasing in a four-cylinder HCCI engine. IEEE Trans. Control Syst. Technol. **22**(3), 1190–1197 (2014)

22. E. Hellström, A. Stefanopoulou, J. Vavra, A. Babajimopoulos, D. Assanis, L. Jiang, H. Yilmaz, Understanding the dynamic evolution of cyclic variability at the operating limits of HCCI engines with negative valve overlap, SAE Technical Paper (2012)
23. E. Hellström, A.G. Stefanopoulou, L. Jiang, Cyclic variability and dynamical instabilities in autoignition engines with high residuals. IEEE Trans. Control Syst. Technol. **21**(5), 1527–1536 (2013)
24. M. Hillion, J. Chauvin, N. Petit, Combustion control of an HCCI diesel engine with cool flame phenomenon, in *European Control Conference* (2009), pp. 3827–3832
25. R. Hutter, S. Hänggi, T. Albin, C.H. Onder, Optimal transition control between combustion modes in a diesel-ignited gas engine. Control Eng. Pract. **97**, 104327 (2020)
26. R. Hutter, J. Ritzmann, P. Elbert, C. Onder, Low-load limit in a diesel-ignited gas engine. Energies **10**(10), 1450–1476 (2017)
27. A. Indrajuana, C. Bekdemir, E. Feru, F. Willems, Towards model-based control of RCCI-CDF mode-switching in dual fuel engines, *SAE Technical Paper* (2018)
28. A. Indrajuana, C. Bekdemir, X. Luo, F. Willems, Robust multivariable feedback control of natural gas-diesel RCCI combustion. IFAC-PapersOnLine **49**(11), 217–222 (2016)
29. S. Jade, E. Hellström, J. Larimore, A.G. Stefanopoulou, L. Jiang, Reference governor for load control in a multicylinder recompression HCCI engine. IEEE Trans. Control Syst. Technol. **22**(4), 1408–1421 (2014)
30. S. Jade, J. Larimore, E. Hellström, A.G. Stefanopoulou, L. Jiang, Controlled load and speed transitions in a multicylinder recompression HCCI engine. IEEE Trans. Control Syst. Technol. **23**(3), 868–881 (2015)
31. A.F. Jungkunz, N. Ravi, H.-H. Liao, S.M. Erlien, J.C. Gerdes, An analytical method for reducing combustion instability in homogeneous charge compression ignition engines through cycle-to-cycle control. Int. J. Eng. Res. **16**(3), 485–500 (2014)
32. G.T. Kalghatgi, P. Risberg, H.-E. Angström, Advantages of fuels with high resistance to auto-ignition in late-injection, low-temperature, compression ignition combustion, SAE Technical Paper (2006)
33. P.E. Kapus, A. Fuerhapter, H. Fuchs, G.K. Fraidl, Ethanol direct injection on turbocharged SI engines – potential and challenges, SAE Technical Paper (2007)
34. N. Komninos, C. Rakopoulos, Modeling HCCI combustion of biofuels: a review. Renew. Sustain. Energy Rev. **16**(3), 1588–1610 (2012)
35. Y. Kugimachi, Y. Nakamura, N. Iida, Model-based combustion control of a HCCI engine using external EGR and the exhaust rebreathed, SAE Technical Paper (2014)
36. O. Lang, W. Salber, J. Hahn, S. Pischinger, K. Hortmann, C. Bücker, Thermodynamical and mechanical approach towards a variable valve train for the controlled auto ignition combustion process. SAE Trans. 722–734 (2005)
37. B. Lehrheuer, B. Morcinkowski, S. Pischinger, M. Nijs, Low temperature gasoline combustion – potential, challenges, process modeling and control, in *Active Flow and Combustion Control* (Springer, 2015)
38. B. Lehrheuer, S. Pischinger, M. Wick, J. Andert, D. Berneck, D. Ritter, T. Albin, M. Thewes, A study on in-cycle combustion control for gasoline controlled autoignition, SAE Technical Paper (2016)
39. W. Leitner, J. Klankermayer, S. Pischinger, H. Pitsch, K. Kohse-Höinghaus, Advanced biofuels and beyond: chemistry solutions for propulsion and production. Angewandte Chemie International Edition **56**(20), 5412–5452 (2017)
40. H.H. Liao, A. Widd, N. Ravi, A.F. Jungkunz, J.M. Kang, J.C. Gerdes, Control of recompression HCCI with a three region switching controller. Control Eng. Pract. **21**(2), 135–145 (2013)
41. X. Lu, D. Han, Z. Huang, Fuel design and management for the control of advanced compression-ignition combustion modes. Progr. Energy Combust. Sci. **37**(6), 741–783 (2011)
42. M.P. Musculus, P.C. Miles, L.M. Pickett, Conceptual models for partially premixed low-temperature diesel combustion. Progr. Energy Combust. Sci. **39**(2–3), 246–283 (2013)
43. D.E. Nieman, A.B. Dempsey, R.D. Reitz, Heavy-duty RCCI operation using natural gas and diesel. SAE Int. J. Eng. **5**(2), 270–285 (2012)

44. E. Nuss, M. Wick, J. Andert, J. De Schutter, M. Diehl, D. Abel, T. Albin, Nonlinear model predictive control of a discrete-cycle gasoline-controlled auto ignition engine model: simulative analysis. Int. J. Eng. Res. **20**(10), 1025–1036 (2019)
45. T. Ott, F. Zurbriggen, C. Onder, L. Guzzella, Cylinder individual feedback control of combustion in a dual fuel engine. IFAC Proc. Vol. **46**(21), 600–605 (2013)
46. E. Rauer, M. Rotter, M. Willmann, C. Senghaas, Intelligent injectors - digitization in the injection technology of large engines. MTZ Worldwide **80**(6), 58–65 (2019)
47. N. Ravi, H.H. Liao, A.F. Jungkunz, A. Widd, J.C. Gerdes, Model predictive control of HCCI using variable valve actuation and fuel injection. Control Eng. Pract. **20**(4), 421–430 (2012)
48. R.D. Reitz, Directions in internal combustion engine research. Combust. Flame **160**(1), 1–8 (2013)
49. R.D. Reitz, G. Duraisamy, Review of high efficiency and clean reactivity controlled compression ignition (RCCI) combustion in internal combustion engines. Progr. Energy Combust. Sci. **46**, 12–71 (2015)
50. S.Z. Rezaei, F. Zhang, H. Xu, A. Ghafourian, J.M. Herreros, S. Shuai, Investigation of two-stage split-injection strategies for a dieseline fuelled PPCI engine. Fuel **107**, 299–308 (2013)
51. D. Ritter, J. Andert, D. Abel, T. Albin, Model-based control of gasoline-controlled auto-ignition. Int. J. Eng. Res. **19**(2), 189–201 (2018)
52. S. Saxena, I.D. Bedoya, Fundamental phenomena affecting low temperature combustion and HCCI engines, high load limits and strategies for extending these limits. Progr. Energy Combust. Sci. **39**(5), 457–488 (2013)
53. L. Shi, Y. Cui, K. Deng, H. Peng, Y. Chen, Study of low emission homogeneous charge compression ignition (HCCI) engine using combined internal and external exhaust gas recirculation (EGR). Energy **31**(14), 2665–2676 (2006)
54. D. Splitter, R. Hanson, S. Kokjohn, R.D. Reitz, Reactivity controlled compression ignition (RCCI) heavy-duty engine operation at mid- and high-loads with conventional and alternative fuels, SAE Technical Paper (2011)
55. P. Strandh, J. Bengtsson, R. Johansson, P. Tunestal, B. Johansson, Cycle-to-cycle control of a dual-fuel HCCI engine. SAE Trans. 589–598 (2004)
56. M. Thor, B. Egardt, T. McKelvey, I. Andersson, Closed-loop diesel engine combustion phasing control based on crankshaft torque measurements. Control Eng. Pract. **33**, 115–124 (2014)
57. H. Tschöke, K. Mollenhauer, R. Maier, *Handbuch Dieselmotoren* (Springer, 2017)
58. R. van der Weijst, T. van Keulen, F. Willems, Constrained multivariable extremum-seeking for online fuel-efficiency optimization of diesel engines, in *Control Engineering Practice* (2019), pp. 133–144
59. M. Wick, J. Bedei, D. Gordon, C. Wouters, B. Lehrheuer, E. Nuss, J. Andert, C.R. Koch, In-cycle control for stabilization of homogeneous charge compression ignition combustion using direct water injection. Appl. Energy **240**, 1061–1074 (2019)
60. M. Wick, B. Lehrheuer, T. Albin, J. Andert, S. Pischinger, Decoupling of consecutive gasoline controlled auto-ignition combustion cycles by field programmable gate array based real-time cylinder pressure analysis. Int. J. Eng. Res. **19**(2), 153–167 (2018)
61. A. Widd, H. Liao, H. Gerdes, J.C. Tunestal, R. Johansson, Hybrid model predictive control of exhaust recompression HCCI. Asian J. Control **2**, 370–381 (2014)
62. A. Widd, K. Ekholm, P. Tunestal, R. Johansson, Physics-based model predictive control of HCCI combustion phasing using fast thermal management and VVA. IEEE Trans. Control Syst. Technol. **20**(3), 688–699 (2011)
63. F. Willems, Is cylinder pressure-based control required to meet future HD legislation? IFAC-PapersOnLine **51**(31), 111–118 (2018)
64. M. Yao, Z. Zheng, H. Liu, Progress and recent trends in homogeneous charge compression ignition (HCCI) engines. Progr. Energy Combust. Sci. **35**(5), 398–437 (2009)
65. L. Yin, M. Lundgren, Z. Wang, P. Stamatoglou, M. Richter, Ö. Andersson, P. Tunestål, High efficient internal combustion engine using partially premixed combustion with multiple injections. Appl. Energy **233**, 516–523 (2019)

66. L. Yin, G. Turesson, P. Tunestål, R. Johansson, Evaluation and transient control of an advanced multi-cylinder engine based on partially premixed combustion. Appl. Energy **233**, 1015–1026 (2019)
67. L. Yin, G. Turesson, T. Yang, R. Johansson, P. Tunestål, Partially premixed combustion (PPC) stratification control to achieve high engine efficiency. IFAC-PapersOnLine **51**(31), 694–699 (2018)
68. F. Zurbriggen, R. Hutter, C. Onder, Diesel-minimal combustion control of a natural gas-diesel engine. Energies **9**(1), 58 (2016)

Part III
In-Depth Case Studies: Air Path Control

Chapter 9
Fundamentals of Control-Oriented Air Path Modeling

Abstract This chapter outlines the fundamentals of air path modeling for the purpose of optimization-based control. First, an overview of the requirements is given. To realize appropriate real-time optimization, a suitable trade-off is needed between the accuracy of a model and its computational complexity. Additionally, if gradient-based methods are applied for a numerical solution of the optimization problem, models are required that are continuously differentiable. Gray-box modeling is proposed as a suitable solution. Wherever it is computationally feasible, models based on first principles are applied. When such first-principle models become very complex, data-driven approaches are investigated instead. Based on these considerations, the various components of the air path are reviewed and suitable models are proposed. Among the components investigated are volumes, flow restrictions, and turbochargers.

9.1 Introduction to Control-Oriented Air Path Modeling

The air path of an engine can be modeled using various approaches. Depending on the purpose of the model, one or another approach is preferable. In the context of model predictive control (MPC), the main modeling challenge is the trade-off required. While the accuracy of the model determines the quality of the closed-loop control algorithm, the model still has to be simple enough to be useable in a real-time optimization control setting. A common way to achieve real-time-feasibility, especially when considering nonlinear models, is the use of powerful gradient-based optimization methods. In order to make use of them, the model has to consist of continuously differentiable functions. Additionally, the model has to be robust against outliers. For instance, an outlier in a state estimation should not produce a division by zero or the square root of a negative value during the calculation of the gradient.

T. Albin Rajasingham, *Nonlinear Model Predictive Control of Combustion Engines*, Advances in Industrial Control,
https://doi.org/10.1007/978-3-030-68010-7_9

In the following, the requirements on the control-oriented air path models are summarized. Subsequently, the requirement of continuous differentiability is detailed. Based on these fundamentals, common components of the air path are revisited in order to highlight reasonable models for use within MPC. The purpose is twofold. On the one hand, an overview of model equations is presented which have been proven to work well in an MPC context. On the other hand, the goal is to demonstrate that even with strong simplifications, models can be developed that are suitable for a good reproduction of the system dynamics while allowing for real-time-feasible optimization algorithms.

For use within MPC, nonlinear state-space models offer a suitable description of the air path. The submodels have to be combined into a single model that describes the entire air path. The various components are connected via their in- and outputs. Most commonly, two different types of components are present. The first type can be classified as a volume. Its internal states are pressure and temperature, and it requires at least the in- and outflowing masses as model inputs. The other type consists of the class of flow restriction models. These take the pressure states of the connecting volumes as inputs and compute at the very least a mass flow through the restriction. In most cases, the flow restriction models are connected to volumes and vice versa.

9.1.1 Requirements on Control-Oriented Air Path Models

In order to find applicable models, two important considerations are the timescale and the spatial resolution to be modeled. If the goal is the geometrical design of certain components such as the intake manifold or the piping, detailed simulations should be conducted such as 3D CFD simulations. They can resolve processes with small timescales. For instance, the dynamics resulting from the opening and closing of the intake or exhaust valves are resolved. These are non-steady processes occurring during the strokes, which introduce oscillations in the air path. However, if the target application is the closed-loop control of the engine air path, these processes on a very small timescale do not need to be considered. The dominant dynamics are given by the dynamics of the turbochargers and the volumes. For quite some time, so-called mean-value models thus have been the quasi-standard for control applications. A good overview on this topic is given in [3, 5].

All mean-value models are characterized by only resolving the dynamics on timescales that are relevant for control. Processes occurring on very small timescales are not modeled, e.g. the dynamics mentioned resulting from the valve opening or closing are one order of magnitude smaller than the turbocharger dynamics. If these dynamics were considered as well, very stiff models would result that will introduce unnecessary complexity. Furthermore, the mean-value models are always simplified such that spatial uniformity results. Instead of spatially resolving the dynamics, developers thus lump together entire volumes. The modeling of a pipe results in a 0D modeling approach, for instance. From a system-theoretic point of view, the models are described by ordinary differential equations (ODE) instead of partial differential equations (PDE). The models resulting from a simplification of spatially distributed physical systems into discrete entities are called lumped-parameter models.

In order to find a reasonable trade-off between accuracy and complexity, gray-box models are typically used. They represent the combination between physically motivated white-box models and data-driven black-box models; see [9]. The white-box models usually have better properties concerning their ability to be generalized. In the best cases, a suitable process model is obtained by determining just a few parameters, e.g. by system identification. The resulting models usually are strongly nonlinear and represent multiple physical domains. For some physical processes, simple white-box models do exist. However, for other processes, a physics-based approach can result in a model that is too complex to be used within optimization-based control systems.

One advantage of black-box models is the fact that they allow for the simulation of complex nonlinear system dynamics with very low computational effort. Even processes which result in complex physics-based models can be reproduced in a computationally efficient manner with black-box models. Many black-box models even are universal function approximators. If enough basis functions are used, they can approximate any continuous function arbitrarily well. Various methods exist to develop black-box models. Depending on the specific purpose, a tailor-made black-box modeling approach should be used. Some of the examples are polynomial models, Gaussian process models, neural networks, and local linear model trees (LoLiMoT) [12]. The drawback of black-box models is their decreased capability for extrapolation. Additionally, enough data is needed to reproduce the system behavior in an accurate manner. Those have to be obtained by measurements or via simulations of complex models.

9.1.2 Continuous Differentiability

An important requirement is the continuous differentiability of the air path model. This allows to use the powerful gradient-based optimization methods. For conventional air path models, this requirement is often not that important. Hence, many available air path models are not continuously differentiable. For instance, they often contain "if-else" conditions and look-up tables.

Typically these expressions can be approximated by a continuously differentiable function. Two examples are detailed in the following. First, "min" and "max" functions are considered. These often occur when certain states are supposed to be above or below a certain limit. For instance, limiting values of y_{lim} to be greater than zero can be expressed by

$$y_{lim}(x) = \max(x, 0) \tag{9.1}$$

A continuously differentiable alternative with ϵ chosen to be positive and close to zero is given by

$$y_{lim,approx}(x) = \frac{1}{2}\sqrt{x^2 + \epsilon} + \frac{x}{2} \tag{9.2}$$

This approximation can also be used to prevent divisions by zero. If the replacement function given in (9.2) is applied to the entire denominator, the minimum value is always greater than zero up to numerical accuracy. A more generally approximation can be set up, for the case that x is supposed to be greater than a value a.

$$y_{lim}(x) = \max(x, a) \tag{9.3}$$

A continuously differentiable alternative with ϵ chosen to be positive and close to zero is given by

$$y_{lim,approx}(x) = \frac{1}{2}\sqrt{(x - a)^2 + \epsilon} + \frac{(x + a)}{2} \tag{9.4}$$

Equivalently, the case can be handled that x is supposed to be smaller than a value a.

$$y_{lim}(x) = \min(x, a) \tag{9.5}$$

A continuously differentiable alternative with ϵ chosen to be positive and close to zero is given by

$$y_{lim,approx}(x) = -\frac{1}{2}\sqrt{(x - a)^2 + \epsilon} + \frac{(x + a)}{2} \tag{9.6}$$

Besides enabling a continuous differentiability, these approximation functions can also be used to increase the robustness of the model against outliers. It can be ensured that a certain expression stays within a specific region.

Furthermore, piecewise functions exist in conventional air path models. They typically occur when case distinctions are necessary, e.g. by if-else statements or look-up tables. It is important to ensure a smooth transition between the various regions. Sigmoid functions can be used to enable if-else statements in a closed-form state-space description. The if-else statement can be approximated by a continuous differentiable function by adding sigmoid functions. The sigmoid function is given by

$$y_r(x) = \left(1 + \exp\left(-\frac{x - \gamma}{\delta}\right)\right)^{-1} \tag{9.7}$$

In the state equation, all the cases are summed up with a weighting of either $y_r(x)$ or $(1 - y_r(x))$, depending on whether they belong to the if or the else statement. The variable γ can be adapted to shift the sigmoid function to the left or the right. A proper choice of γ allows the location of the case distinction to be reproduced. The slope of the sigmoid function can be adjusted by the variable δ, which thus controls the smoothness of the transition. Smaller values of δ lead to more steep transitions. The continuous differentiable function replacement is demonstrated with one example.

Assume, the following if-else statement is given:

$$g(x) = \begin{cases} a(x), & \text{for } x > x_{lim} \\ b(x), & \text{for } x \leq x_{lim} \end{cases} \tag{9.8}$$

The piecewise function (9.8) can be replaced by the following continuously differentiable function, where $\delta = 1$ is chosen:

$$g_c(x) = \frac{1}{1 + \exp(-(x - x_{lim}))} a(x) + \left(1 - \frac{1}{1 + \exp(-(x - x_{lim}))}\right) b(x) \tag{9.9}$$

9.2 Volume

Important components of the air path model are the volumes. Examples in an engine system which can be described by this module are the intake and exhaust manifolds. For the volumes, the potential variables are given by the mass and the energy. The volume exchanges the flow variables, specifically the in- and the outflow of mass and enthalpy with the connected components.

In order to develop a reduced-order model, several assumptions are made. The spatial distribution inside the volumes is neglected, thus resulting in a lumped-parameter model. The gas inside the volume is assumed to have constant gas properties. This concerns especially c_p and c_v, which represent the specific heats at constant pressure and constant volume, respectively. Additionally, the gas is assumed to be ideal, meaning that it can be well approximated by the ideal gas law:

$$p(t)V = m(t)R\vartheta(t) \tag{9.10}$$

The volume is given by V, the pressure is denoted by p, the mass by m, and the temperature by ϑ. The specific gas constant $R = \frac{R_u}{M}$ is calculated by dividing the universal gas constant R_u by the molar mass M. The relationship between R and the specific heats can be expressed by $R = c_p - c_v$.

The dynamics of the volumes can be derived by the conservation laws of mass and energy, which are shown in (9.11) and (9.12).

$$\frac{\mathrm{d}}{\mathrm{d}t} m(t) = \dot{m}_{in}(t) - \dot{m}_{out}(t) \tag{9.11}$$

$$\frac{\mathrm{d}}{\mathrm{d}t} U(t) = \dot{H}_{in}(t) - \dot{H}_{out}(t) - \dot{Q}(t) \tag{9.12}$$

Here, $\dot{m}_{in}(t)$ and $\dot{m}_{out}(t)$ correspond to the in- and outflowing masses, $U(t)$ is the internal energy, while $\dot{H}_{in}(t)$ and $\dot{H}_{out}(t)$ describe the enthalpy flows. The heat transfer out of the volume is given by $\dot{Q}(t)$.

Pressure and Temperature

From a technical point of view, the pressure $p(t)$ and the temperature $\vartheta(t)$ are the most relevant parameters, as they can be directly measured. In the following reduced-order models are derived for their calculations. The internal energy can be expressed by the pressure and the temperature as follows:

$$U(t) = m(t)c_v\vartheta(t) = \frac{1}{\kappa - 1}p(t)V \tag{9.13}$$

The ratio of specific heats is given by κ, i.e. $\kappa = c_p/c_v$. The enthalpy flows can be expressed with respect to the temperature, as shown in (9.14)–(9.15). The temperature of the outflowing gas is set equal to that inside the volume.

$$\dot{H}_{in}(t) = \dot{m}_{in}(t)c_p\vartheta_{in}(t) \tag{9.14}$$

$$\dot{H}_{out}(t) = \dot{m}_{out}(t)c_p\vartheta(t) \tag{9.15}$$

Substituting (9.13)–(9.15) in (9.12) yields

$$\frac{\mathrm{d}}{\mathrm{d}t}\left(\frac{1}{\kappa - 1}p(t)V\right) = \dot{m}_{in}(t)c_p\vartheta_{in}(t) - \dot{m}_{out}(t)c_p\vartheta(t) - \dot{Q}(t) \tag{9.16}$$

Finally, for obtaining an expression describing the pressure dynamics, this equation can be rearranged as follows:

$$\frac{\mathrm{d}}{\mathrm{d}t}p(t) = \frac{\kappa - 1}{V}\left(\dot{m}_{in}(t)c_p\vartheta_{in}(t) - \dot{m}_{out}(t)c_p\vartheta(t) - \dot{Q}(t)\right) \tag{9.17}$$

When deriving the equations for the temperature dynamics, the internal energy shown in (9.13) has to be differentiated with respect to time.

$$\frac{\mathrm{d}}{\mathrm{d}t}U(t) = c_v\vartheta(t)\frac{\mathrm{d}}{\mathrm{d}t}m(t) + c_v m(t)\frac{\mathrm{d}}{\mathrm{d}t}\vartheta(t) \tag{9.18}$$

The substitution in (9.12) yields

$$c_v\vartheta(t)\frac{\mathrm{d}}{\mathrm{d}t}m(t) + c_v m(t)\frac{\mathrm{d}}{\mathrm{d}t}\vartheta(t) = \dot{m}_{in}(t)c_p\vartheta_{in}(t) - \dot{m}_{out}(t)c_p\vartheta(t) - \dot{Q}(t) \tag{9.19}$$

Using the ideal gas law, the equation governing for temperature can be derived.

$$\frac{\mathrm{d}}{\mathrm{d}t}\vartheta(t) = \frac{R\vartheta(t)}{c_v p(t)V}\left(\dot{m}_{in}(t)c_p\vartheta_{in}(t) - \dot{m}_{out}(t)c_p\vartheta(t) - \dot{Q}(t) - c_v\vartheta\left(\dot{m}_{in}(t) - \dot{m}_{out}(t)\right)\right) \tag{9.20}$$

Heat Transfer

In some cases, the heat transfer $\dot{Q}(t)$ has to be accounted for, e.g. due to considerable wall heat losses. A common approach to model the heat transfer is given by Newton's law of heat transfer [7].

$$\dot{Q}(t) = k_{vol} A_{vol} \Big(\vartheta(t) - \vartheta_{amb}(t) \Big) \tag{9.21}$$

The parameter k_{vol} represents the heat transmission coefficient, A_{vol} the surface area of the volume, and ϑ_{amb} the ambient temperature. However, often simplifications are made regarding the heat transfer, as described below.

Adiabatic Model Simplification
A common assumption for volumes is the adiabatic model simplification. In this case, the heat transfer is assumed to be zero, i.e. $\dot{Q} = 0$. The result is the adiabatic model, where the two following equations result:

$$\frac{\mathrm{d}}{\mathrm{d}t}\vartheta(t) = \frac{R\vartheta(t)}{c_v p(t) V} \Big(\dot{m}_{in}(t) c_p \vartheta_{in}(t) - \dot{m}_{out}(t) c_p \vartheta(t) - c_v \vartheta \big(\dot{m}_{in}(t) - \dot{m}_{out}(t) \big) \Big) \tag{9.22}$$

$$\frac{\mathrm{d}}{\mathrm{d}t} p(t) = \frac{\kappa R}{V} \Big(\dot{m}_{in}(t) \vartheta_{in}(t) - \dot{m}_{out}(t) \vartheta(t) \Big) \tag{9.23}$$

Isothermal Model Simplification
Another common assumption is the isothermal model simplification. In this case, the temperatures are assumed to be constant, i.e. $\vartheta_{in}(t) = \vartheta_{amb}(t) = \vartheta(t)$. The equations governing can be directly derived from the ideal gas law. Due to the isothermal assumption, the pressure dynamics can be simplified to

$$\begin{aligned} \frac{\mathrm{d}}{\mathrm{d}t} p(t) &= \frac{R\vartheta}{V} \frac{\mathrm{d}}{\mathrm{d}t} m(t) \\ &= \frac{R\vartheta}{V} \Big(\dot{m}_{in}(t) - \dot{m}_{out}(t) \Big) \end{aligned} \tag{9.24}$$

Gas Composition
In the case of exhaust gas recirculation, not only the pressure and the temperature inside the volume are of interest, but also the gas composition in the volume. The concentration of a species x is denoted as F_x. The term F_x is defined as the fraction of the mass of species x denoted as m_x of the total mass m.

$$F_x = \frac{m_x}{m} \tag{9.25}$$

The change of mass $m_x(t)$ in the volume can be expressed in relation to the concentration $F_{x,in}(t)$ of the inflowing and the concentration $F_x(t)$ of the outflowing mass flow. The concentration $F_x(t)$ of the outflowing mass flow is set equal to that inside the volume.

$$\frac{\mathrm{d}}{\mathrm{d}t} m_x(t) = F_{x,in}(t) \dot{m}_{in}(t) - F_x(t) \dot{m}_{out}(t) \tag{9.26}$$

In order to calculate the change in the fraction $F_x(t)$, the expression (9.25) needs to be differentiated with respect to time:

$$\frac{\mathrm{d}}{\mathrm{d}t}F_x(t) = \frac{m(t)\frac{\mathrm{d}}{\mathrm{d}t}m_x(t) - m_x(t)\frac{\mathrm{d}}{\mathrm{d}t}m(t)}{m^2(t)} \tag{9.27}$$

Substituting (9.11) and (9.26) in this equation leads to

$$\frac{\mathrm{d}}{\mathrm{d}t}F_x(t) = \frac{m(t)\Big(F_{x,in}(t)\dot{m}_{in}(t) - F_x(t)\dot{m}_{out}(t)\Big) - m_x(t)\Big(\dot{m}_{in}(t) - \dot{m}_{out}(t)\Big)}{m^2(t)} \tag{9.28}$$

The equation governing for $F_x(t)$ can be obtained, by inserting the relation $m_x(t) = F_x(t)m(t)$ and the use of the ideal gas law to replace $m(t)$.

$$\frac{\mathrm{d}}{\mathrm{d}t}F_x(t) = \frac{R\vartheta(t)}{p(t)V}\Big(F_{x,in}(t) - F_x(t)\Big)\dot{m}_{in}(t) \tag{9.29}$$

9.3 Engine

In the context of optimization-based air path control, the engine does not have to be modeled in the detail level that the crank-angle-based dynamics are resolved; a mean-value model is sufficient. A common approach is to model the engine as a volumetric pump that enforces a continuous average volume flow [5]. Especially the following quantities are of interest for control: the in- and out-mass flows of the total gas, the gas composition of the mass flows, and the engine-out temperature.

Total Gas Mass Flow

Once per engine cycle, the displaced volume is aspirated from the intake manifold into the cylinders. If the entire volume were displaced, the average mass flow $\dot{m}_{asp,th}$ given in (9.30) would result.

$$\dot{m}_{asp,th}(t) = \rho_{im}(t)\dot{V}(t) = \frac{p_{im}(t)}{R\vartheta_{im}(t)}\frac{n_{eng}(t)}{N}V_{dis} \tag{9.30}$$

It involves the pressure in the intake manifold p_{im}, the specific gas constant R, and the temperature in the intake manifold ϑ_{im}. The displacement volume of all the cylinders combined is given by V_{dis}, which is divided by the number of revolutions per cycle N with $N = 1$ for a two-stroke engine and with $N = 2$ for a four-stroke engine. The engine speed n_{eng} is given in revolutions per second. The first term represents the density of the gas inside the intake manifold:

$$\rho_{im}(t) = \frac{p_{im}(t)}{R\vartheta_{im}(t)} \tag{9.31}$$

However, in reality, the volume is not displaced entirely due to effects such as pressure oscillations in the intake and exhaust port, the internal exhaust gas recirculation, or the influences among the cylinders. To account for these losses, the volumetric efficiency η_v is introduced, which describes the effectiveness of the gas aspiration.

$$\dot{m}_{asp}(t) = \eta_v \rho_{im}(t)\dot{V}(t) = \eta_v \frac{p_{im}(t)}{R\vartheta_{im}(t)} \frac{n_{eng}(t)}{N} V_{dis} \tag{9.32}$$

Modeling the volumetric efficiency η_v in a physical manner is very difficult due to numerous underlying processes. Data-driven approaches thus are usually applied instead. For the parameter identification of η_v, steady-state measurements are conducted at various conditions. The parameter η_v can be assumed to be constant. Alternatively, for a more accurate modeling, it can be modeled as a function that depends on parameters such as the engine speed n_{eng}, the intake manifold pressure p_{im}, the intake manifold gas density ρ_{im}, the exhaust manifold pressure p_{em}, or the pressure ratio over the engine $\frac{p_{em}}{p_{im}}$. For a function, low-order polynomials yield good approximations [3], such as

$$\eta_v = f\left(n_{eng}, \rho_{im}, p_{im}, p_{em}\right) \tag{9.33}$$

In the case of *port-fuel injection*, it is reasonable to assume that the aspirated mass flow calculated by (9.32) consists of the entire fuel mass injected $\dot{m}_{fuel}$ as well as fresh air and potentially recirculated exhaust gas, consisting of air and burnt gas. The engine-out mass flow $\dot{m}_{eo}$ is equal to this aspirated mass flow $\dot{m}_{asp}$.

$$\dot{m}_{asp} = \dot{m}_{air,asp} + \dot{m}_{bg,asp} + \dot{m}_{fuel} \tag{9.34}$$

$$\dot{m}_{eo} = \dot{m}_{asp} \tag{9.35}$$

For the case of a *direct-injection engine*, the aspirated mass flow only consists of fresh air and perhaps some recirculated exhaust gas. The engine-out mass flow is given by the aspirated mass flow $\dot{m}_{asp}$ and the fuel mass injected into the cylinder $\dot{m}_{fuel}$:

$$\dot{m}_{asp} = \dot{m}_{air,asp} + \dot{m}_{bg,asp} \tag{9.36}$$

$$\dot{m}_{eo} = \dot{m}_{asp} + \dot{m}_{fuel} \tag{9.37}$$

Gas Composition of the Mass Flow

When exhaust gas recirculation is applied, not only the total gas flow is of relevance but also its composition. In the case of a *stoichiometric* operation, a complete combustion with an air-to-fuel equivalence ratio of $\lambda = 1$ can be assumed. Thus, all exhaust gas is burnt gas. The burnt gas is defined to be the products from combustion, i.e. mainly CO_2, H_2O, and N_2. As a consequence, the following engine-out mass flow concentrations result:

$$F_{O_2,eo} = F_{air,eo} = 0 \tag{9.38}$$

$$F_{bg,eo} = 1 \tag{9.39}$$

In fact, for stoichiometric operation, controlling the exhaust gas fraction and the oxygen fraction in the intake manifold is equivalent. Consequently, either of the two can be tracked.

For controlling the combustion process of a *lean burning* engine, such as a CI engine, it is advantageous to not only know the amount of recirculated exhaust gas but also its chemical composition. Due to the lean operation, the recirculated exhaust gas consists of burnt gas, which does not contain oxygen, as well as unburnt air, which does contain oxygen. The varying amounts of oxygen in the recirculated exhaust gas can participate in the subsequent combustion process. As a result, it is advisable to track either the oxygen fraction or the burnt gas fraction in the various volumes rather than the exhaust gas fraction. The two values, oxygen fraction F_{O_2} and the burnt gas fraction F_{bg}, can be converted into each other by taking into account the mass fraction of oxygen in the air $F_{O_2,air}$, which for typical ambient conditions is $F_{O_2,air} = 0.23$:

$$F_{bg} = 1 - \frac{F_{O_2}}{F_{O_2,air}} \tag{9.40}$$

As the lambda sensor can measure the oxygen content, the use of oxygen fraction or burnt gas fraction has slight advantages compared to the tracking of the air fraction. The total engine-out exhaust gas can be split up into the engine-out burnt gas fraction $F_{bg,eo}$ and the engine-out air fraction $F_{air,eo}$.

$$F_{eg} = F_{bg,eo} + F_{air,eo} = 1 \tag{9.41}$$

The air fraction F_{air} can be converted to the oxygen fraction F_{O_2} by

$$F_{O_2} = F_{air} F_{O_2,air} \tag{9.42}$$

The engine-out air fraction $F_{air,eo}$ and the engine-out air mass flow $\dot{m}_{air,eo}$ directly depend on the air-to-fuel equivalence ratio λ of the combustion process. In order to calculate these values, the air remaining after combustion needs to be calculated with AFR_s being the stoichiometric air-to-fuel ratio. For direct-injection engines, it results:

$$\dot{m}_{air,eo} = \dot{m}_{asp} F_{air,im} - \dot{m}_{fuel} AFR_s = \dot{m}_{asp} F_{air,im} \left(1 - \frac{1}{\lambda}\right) \tag{9.43}$$

This yields the oxygen fraction flowing out of the cylinder:

$$F_{air,eo} = \frac{\dot{m}_{asp} F_{air,im} - \dot{m}_{fuel} AFR_s}{\dot{m}_{eg}} \tag{9.44}$$

Correspondingly, the engine-out oxygen fraction can be calculated:

$$F_{O_2,eo} = \frac{\dot{m}_{asp} F_{O_2,im} - \dot{m}_{fuel} AFR_s F_{O_2,air}}{\dot{m}_{eg}} \tag{9.45}$$

Engine-Out Temperature

The temperature of the exhaust gas flow out of the engine ϑ_{eo} depends on numerous parameters, such as the engine speed, the engine load, the air-to-fuel ratio, the amount of recirculated exhaust gas, and the spark timing in SI engines or the injection timing in CI engines. Due to the many processes involved, the physical modeling is quite complex. The data-driven approach is also quite demanding due to the measurement situation characterized by the hot temperatures and the unsteady flow conditions. Hence, simple models for ϑ_{eo} are used often within MPC.

Typically, data-driven approaches are applied where the temperature only depends on a few parameters. One simple formulation is given in [3] for the case of a standard SI engine which is run at optimal spark timing and with $\lambda = 1$. The engine-out temperature is modeled to be linearly dependent on the engine-out mass flow. In the model, the values of $\vartheta_{eo,0}$ and the proportionality constant k are fitted to data obtained on a testbench.

$$\vartheta_{eo} = \vartheta_{eo,0} + k\, \dot{m}_{eo} \tag{9.46}$$

Another approach to model the engine-out temperature is the use of a polynomial function f that depends on the engine speed n_{eng} and the fuel mass flow $\dot{m}_{fuel}$:

$$\vartheta_{eo} = f\left(n_{eng}, \dot{m}_{fuel}\right) \tag{9.47}$$

9.4 Flow Restriction Model

In the following, models are described for components that restrict the airflow. The driving potential for the mass flow is the difference between the upstream pressure p_{us} and the downstream pressure p_{ds}. There are many components in the air path which can be described by this module, e.g. passive devices such as air filters and intercoolers. Also, there are components among these that are used for actively controlling the gas flow, such as valves and throttles. In general, there exist two types of flow restriction models, and they are distinguished by whether the gas is assumed to be incompressible or compressible. Within the engine, this differentiation mainly depends on the speed of the gas within the component [3]. The incompressible flow restriction model is used when small gas velocities are present, i.e.the speed of the gas through the component is well below 30% of the speed of sound; see [5]. The compressible flow restriction model is used when this limit is surpassed.

9.4.1 Incompressible Flow Restriction Model

The components which can be represented well by the incompressible flow restriction model are characterized by small gas velocities within the component. Typically, a small pressure ratio or a large cross-section area is present. Several components exist in the air path which can be described by this model, such as air filters, intercoolers, and the exhaust gas aftertreatment systems. For deriving the equations, various assumptions are made such as no friction in the flow, zero-dimensional flow phenomena, and isolated conditions. With these assumptions, Bernoulli's law can be applied to derive (9.48); see [6] for details.

$$\dot{m}(t) = c_d A \sqrt{\frac{2}{R\vartheta_{us}(t)}} p_{us}(t) \psi\Big(\Pi(t)\Big) \tag{9.48}$$

The discharge coefficient is denoted by c_d, and A is the reference flow area. Although the discharge coefficient c_d generally is not constant because it depends e.g. on the pressure ratio, the variations are usually very small, such that it can be assumed to be constant. Generally, the factor $c_d A$ is used as a parameter which is identified to fit the measurement data. For incompressible flow restriction models, the so-called flow function ψ depends on the pressure ratio Π over the component. The pressure ratio is defined as

$$\Pi = \frac{p_{ds}}{p_{us}} \tag{9.49}$$

The flow function ψ can be determined by

$$\psi = \sqrt{1 - \Pi} \tag{9.50}$$

This relation cannot be directly used within the nonlinear MPC (NMPC) concept. The flow function presented is not defined for values of $\Pi > 1$, and for $\Pi = 1$, the derivative approaches infinity, i.e. $p_{us} = p_{ds}$. The function $\psi(\Pi)$ thus does not fulfill the Lipschitz condition, which is needed for the existence and uniqueness of the solution. In order to use the function within NMPC, it needs to be slightly adapted such that continuous differentiability and the Lipschitz condition are fulfilled. One suitable approximation of the flow function is given by

$$\psi_{approx} = \sqrt{f_{lim}\,(1 - \Pi)} \tag{9.51}$$

$$f_{lim}(x) = \frac{1}{2}\sqrt{x^2 + \epsilon} + \frac{x}{2} \tag{9.52}$$

The changes in the potential and kinetic energy are sufficiently small that they can be neglected. Thus, there is no temperature change in the flow restriction model [3].

$$\vartheta_{ds} = \vartheta_{us} \tag{9.53}$$

9.4.2 Compressible Flow Restriction Model

Compressible flow restriction models are well suited for components with large gas velocities. Typically, large pressure ratios or small cross-section areas are present in this case. Examples where these models are commonly used are throttles, valves, turbine wastegates, and compressor bypasses. In these components, the gas velocity can go up to the speed of sound for a nearly closed position of the actuator. The effects of compressibility in the gas thus become significant when the mass flow is considered. The mass flow of the compressible flow restriction model can be described by the following model, as derived in detail in [6]. As in the case of the incompressible flow restriction model, the changes in the temperature can be neglected.

$$\vartheta_{ds} = \vartheta_{us} \tag{9.54}$$

The mass flow is calculated by

$$\dot{m}(t) = c_d A \frac{p_{us}(t)}{\sqrt{R\vartheta_{us}(t)}} \psi\left(\Pi(t)\right) \tag{9.55}$$

The factor c_d is assumed to be constant and is used as a parameter which is identified to fit the measurement data. The model is often applied for components which allow for an active control of the gas flow by changing the effective opening area. Thus, the opening area is usually a function of the actuated value $A(t) = f(u(t))$, which is set by the controller.

For calculation of the flow function, two cases are distinguished, i.e. pressure ratios smaller and bigger than the critical pressure ratio. The critical pressure ratio Π_{cr} is defined to be

$$\Pi_{cr} = \left(\frac{2}{\kappa + 1}\right)^{\frac{\kappa}{\kappa-1}} \tag{9.56}$$

For pressure ratios Π smaller than Π_{cr}, the flow can be considered to be choked. In this case, the gas velocity is limited by the speed of sound in the gas. The flow function $\psi(\Pi)$ for the compressible flow restriction model is given by

$$\psi\left(\Pi\right) = \begin{cases} \sqrt{\kappa \cdot \left(\frac{2}{\kappa+1}\right)^{\frac{\kappa+1}{\kappa-1}}}\,, & \Pi \leq \left(\frac{2}{\kappa+1}\right)^{\frac{\kappa}{\kappa-1}} \\ \Pi^{\frac{1}{\kappa}} \sqrt{\frac{2\kappa}{\kappa-1} \cdot \left(1 - \Pi^{\frac{\kappa-1}{\kappa}}\right)}\,, & \Pi > \left(\frac{2}{\kappa+1}\right)^{\frac{\kappa}{\kappa-1}} \end{cases} \tag{9.57}$$

This compressible flow restriction model cannot be used directly within the NMPC approach. It is defined in a piecewise manner. On top, it is not defined for $\Pi > 1$, and the Lipschitz condition is not fulfilled. To overcome these issues, the function needs to be adapted. The formulation in (9.57) can be transformed to the following equivalent relation:

$$\psi\left(\Pi\right)=\Pi_{a}^{\frac{1}{\kappa}}\sqrt{\frac{2\kappa}{\kappa-1}\cdot\left(1-\Pi_{a}^{\frac{\kappa-1}{\kappa}}\right)}\ ,\ \Pi_{a}=\max\left(\Pi_{cr},\Pi\right) \tag{9.58}$$

This allows the derivation of a suitable approximation, which is usable within NMPC:

$$\begin{aligned}\psi_{approx}\left(\Pi\right)&=\Pi_{a,lim}^{\frac{1}{\kappa}}\sqrt{\frac{2\kappa}{\kappa-1}f_{lim}\left(1-\Pi_{a,lim}^{\frac{\kappa-1}{\kappa}}\right)}\\ \Pi_{a,lim}&=\frac{1}{2}\sqrt{\left(\Pi-\Pi_{cr}\right)^{2}+\epsilon_{1}}+\frac{\Pi+\Pi_{cr}}{2}\\ f_{lim}(x)&=\frac{1}{2}\sqrt{x^{2}+\epsilon_{2}}+\frac{x}{2}\end{aligned} \tag{9.59}$$

9.5 Turbocharger

As described in Sect. 7.2, a turbocharger consists of a compressor and a turbine which are connected by a common shaft. In the following, control-oriented models for turbine and compressor are investigated.

9.5.1 *Compressor*

Historically, turbocharging components often have been studied by looking at dimensionless quantities [13]. A remainder of that view are the corrections used to normalize the mass flow and the rotational speed. Both quantities are normalized with respect to the inlet temperature ϑ_{us} and pressure p_{us}, such that the process behavior can be compared across various operating conditions. Thus, the complexity of the models and the amount of measurement data needed to characterize their behavior can be reduced. The corrections are shown in (9.60) and (9.61). In these equations, the reference values for temperature and pressure are defined by the inlet conditions at which the tests were conducted. For compressors, these typically are $\vartheta_{ref}=20\%$ and $p_{ref}=1\text{bar}$.

$$\dot{m}_{c,corr}(t)=\dot{m}_{c}(t)\sqrt{\frac{\vartheta_{us}(t)}{\vartheta_{ref}(t)}\frac{p_{ref}}{p_{us}(t)}} \tag{9.60}$$

$$n_{tc,corr}(t)=n_{tc}(t)\sqrt{\frac{\vartheta_{ref}}{\vartheta_{us}(t)}} \tag{9.61}$$

The pressure ratio over the compressor Π_c is described by (9.62). In normal operation, the pressure ratio Π_c is greater than one.

$$\Pi_c(t) = \frac{p_{ds}(t)}{p_{us}(t)} \tag{9.62}$$

Compressing the intake gas requires energy that is delivered to the compressor by a rotational momentum. Therefore, the model of the compressor has to compute the power consumed during the compression process as well as the gas conditions downstream of the compressor. When modeling the compression process, an isentropic process is typically used as a reference. The value η_c expresses the efficiency of the compressor to represent further losses compared to the isentropic reference. Based on the equation for power $P_c = \dot{m}_c \Delta h$ and the standard enthalpy difference equation $\Delta h = c_p \Delta \vartheta$, the equations for temperature ϑ_{ds} and power consumption P_c can be derived:

$$\vartheta_{ds}(t) = \vartheta_{us}(t) + \frac{\vartheta_{us}(t)}{\eta_c} \cdot \left(\left(\frac{p_{ds}(t)}{p_{us}(t)} \right)^{\frac{\kappa-1}{\kappa}} - 1 \right) \tag{9.63}$$

$$P_c(t) = c_p \dot{m}_c(t) \frac{\vartheta_{us}(t)}{\eta_c} \cdot \left(\Pi_c^{\frac{\kappa-1}{\kappa}}(t) - 1 \right) \tag{9.64}$$

The compression is characterized by the interplay between the mass flow through the compressor $\dot{m}_c$, the pressure ratio Π_c, the turbocharger speed n_{tc}, and the efficiency of operation η_c. A compressor map graphically represents the operation of a turbocharger described by these four main quantities. Figure 9.1 shows a typical compressor map. The pressure ratio is shown on the y axis. The x axis shows the corrected mass flow through the compressor. The dotted lines indicate operating points with the same corrected turbocharger speed, while the efficiency of the compression process is shown by the solid isolines. The data for generating these compressor maps can typically be obtained from the component supplier.

Compressors have limited operating regions that depend on their design. In the compressor map shown, the operating region is bounded by several limits. The right dashed line presents the *choke limit*. It is characterized by the maximum mass flow the compressor can manage. The velocity of the fluid reaches the speed of sound in the narrowest passage of the compressor and thus cannot be further increased. The left dashed line presents the *surge limit*. Operating points to the left of this limit are unstable, and the compressor has to transport the intake gas against a pressure ratio that is too high. If the turbocharger speed is insufficient to internally maintain a pressure as high as the back pressure, the direction of the flow is reversed and the gas flows backward through the compressor until the flow is stabilized again. Such a cyclic behavior can cause serious damage to the compressor. A detailed study of the surge behavior in compressors is described in [4]. There are also limits concerning the rotational speed of the turbocharger. Operation above the maximum speed limit leads to mechanical damage. At zero rotational speed, the compressor is blocking and acts as a flow restriction.

Modeling a compressor is difficult because the four quantities efficiency, pressure ratio, mass flow, and turbocharger speed that describe the operating point are coupled

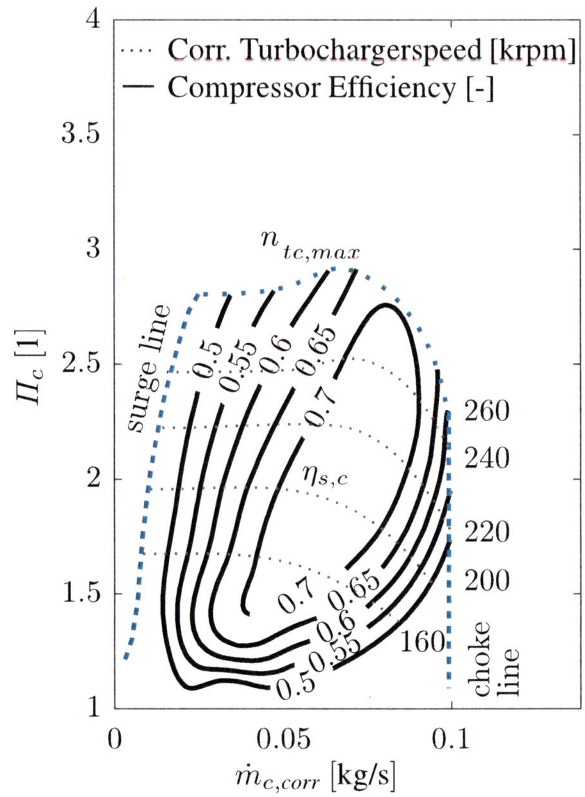

Fig. 9.1 Compressor map – Reproduced from [2], originally published open access under a Creative Commons CC BY 4.0 license, https://doi.org/10.3390/en9070530

in a nonlinear fashion. Data-based models are used typically to represent this process behavior. In the following models for the efficiency, the compressor mass flow and the direct calculation of the pressure ratio are revisited.

Efficiency

The authors of [5] develop a simplified model to describe the compressor efficiency η_c in dependence of the corrected compressor mass flow and the pressure ratio. For normal operation, the functions shown in (9.65) and (9.66) are proposed. Within this model, the matrix $\boldsymbol{Q} \in \mathbb{R}^{2x2}$ is used for the parametrization of the model. In total, seven parameters are contained, namely the matrix $\boldsymbol{Q}$, and the parameters at optimal operation $\eta_{c,opt}$, $\Pi_{c,opt}$, and $\dot{m}_{c,corr,opt}$.

$$\eta_c(\Pi_c(t), \dot{m}_{c,corr}(t)) = \eta_{c,opt} - X^{\mathsf{T}} Q X \tag{9.65}$$

$$X^{\mathsf{T}} = [\Pi_c(t) - \Pi_{c,opt}, \dot{m}_{c,corr}(t) - \dot{m}_{c,corr,opt}] \tag{9.66}$$

The seven parameters have to be fitted to either measurement data gained during engine operation or to the compressor map delivered by the supplier.

Mass Flow

The compressor mass flow can be modeled as a function in dependence of the turbocharger speed and the pressure ratio. Within the overall model, the compressor module is connected to a volume on the inlet and the outlet, such that the pressure ratio serves as an input for the calculation. Various data-driven models have been proposed such as neural networks and multivariate polynomials [10]. To achieve a low-order polynomial model, the corrected mass flow can be calculated in dependence of the corrected turbocharger speed and the pressure ratio:

$$\dot{m}_{c,corr}(t) = f(\Pi_c(t), n_{tc,corr}(t)) \tag{9.67}$$

The parameters of the multivariate polynomial have to be fitted to either measurement data or to the compressor map. Attention has to be paid when using the compressor maps. They are typically measured at a gas test bench. For several reasons, they do not precisely reflect the behavior of the turbocharger in an engine installation, e.g. due to different conditions of operation [8].

Approximation of the Compressor Pressure Ratio

Alternatively to calculating the mass flow through the compressor, in some applications, the compressor pressure ratio can be directly calculated. In [11], the authors identify a strong linear relationship between the pressure ratio Π_c and the squared corrected turbocharger speed $n^2_{tc,corr}$, which itself correlates to the kinetic energy. This model is extended in [1] by taking into account the engine speed n_{eng}. The resulting model is given by

$$\Pi_c(t) = \frac{n^2_{tc,corr}(t) - b\, n_{eng}(t) - d}{a\, n_{eng}(t) + c} \tag{9.68}$$

In some applications, this can be simplified to

$$\Pi_{c,simp}(t) = \frac{n^2_{tc,corr}(t) - b_{simp}}{a_{simp}} \tag{9.69}$$

The parameters a, b, c, d resp. a_{simp}, b_{simp} are determined by identification with measurement data. Figure 9.2 shows such a correlation for two different compressors. The measured behavior of the compressor (a) can be reproduced very well with the model given in (9.68). For the compressor (b), the engine speed dependency can be neglected, such that (9.69) results in a very good correlation.

The reasoning for the quality of the fit is detailed in Fig. 9.3 where the compressor map is shown along with engine operating points for the compressor (a) in Fig. 9.2. For low mass flows $\dot{m}_{c,corr}$, the lines of constant turbocharger speed have a slope close to zero. Thus, in this operating range, one turbocharger speed corresponds to a single pressure ratio. As a consequence, one linear fit is sufficient, independent of the engine speed. For larger mass flows, the lines of constant turbocharger speed do have

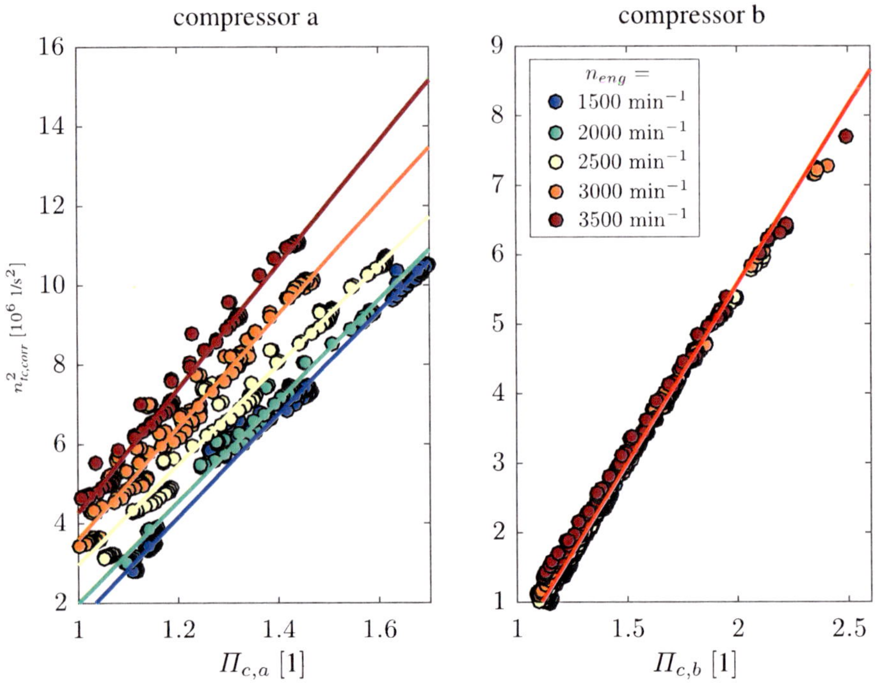

Fig. 9.2 Measured and simulated data for the correlation between pressure ratio and squared corrected turbocharger speed for two compressors – © [2015] IEEE. Reprinted, with permission, from [1]

a negative slope. Therefore, an additional offset for the relation between the pressure ratio and the corrected mass flow results which depends on the engine speed. The operating points of the compressor (a) are in this region.

The result can be utilized to compute the pressure ratio directly given the corrected turbocharger speed and the engine speed. This approach contradicts the common modeling philosophy of independent components, which are connected solely via in- and output flow variables. The main advantage is that it enables a very lean model structure and thereby a computationally rather efficient model. At the same time, the simulation is numerically very robust. The correlations with slopes that are close to zero, which are numerically demanding, are not used in this approach. For instance, the relation between mass flow and pressure ratio at constant turbocharger speed for low mass flows has a slope close to zero. However, typically differential–algebraic equations (DAE) result, such that implicit integration schemes have to be applied.

To calculate the power consumed during the compression process, the compressor mass flow is still needed. It is approximated by the mass flow of the next downstream component, which may be a throttle, another compressor, or the engine directly. For instance, when the next downstream component is the engine, it is assumed that the mass flow through the compressor is equal to the aspirated mass flow of the engine:

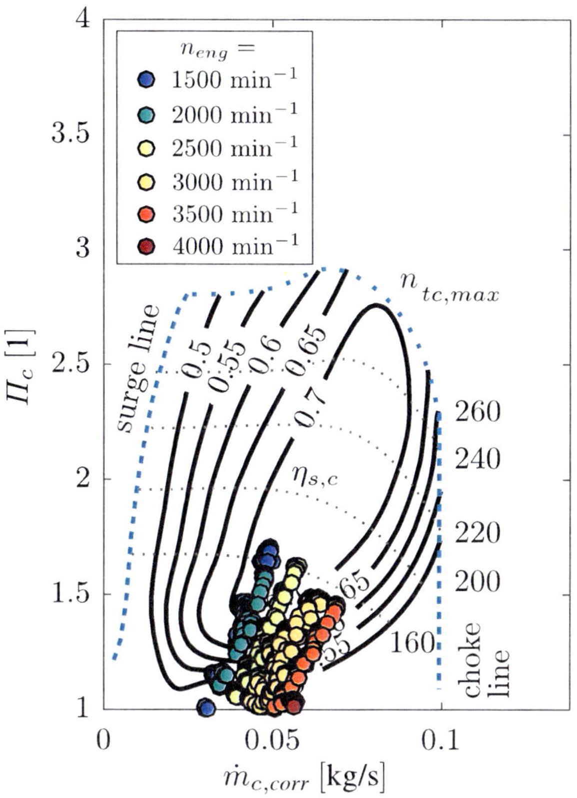

Fig. 9.3 Engine speed-dependent operating points in the compressor map – Reproduced from [2], originally published open access under a Creative Commons CC BY 4.0 license, https://doi.org/10.3390/en9070530

$$\dot{m}_c(t) = \dot{m}_{asp}(t) \tag{9.70}$$

This implies that the pressure dynamics in the volume between the compressor and downstream component are negligible. Thus, the pressure immediately goes to a constant state, such that the in- and outflowing masses in the volume are equal.

9.5.2 Turbine

Turbines are the counterparts of compressors. They extract enthalpy from the hot exhaust gases and deliver it to the compressor via the turbocharger shaft. For turbines, corrected quantities are applied as well, which originate from dimensional analysis. These are used in order to make different inlet conditions comparable. The spread of inlet conditions is bigger for turbines than for compressors. Thus, the corrections actually play a greater role for turbines. The mass flow and the turbocharger speed are modified in the same way as described for compressors:

$$\dot{m}_{t,corr}(t) = \dot{m}_t(t)\sqrt{\frac{\vartheta_{us}(t)}{\vartheta_{ref}}\frac{p_{ref}}{p_{us}(t)}} \tag{9.71}$$

$$n_{tc,corr}(t) = n_{tc}(t)\sqrt{\frac{\vartheta_{ref}}{\vartheta_{us}(t)}} \tag{9.72}$$

For turbines, the reference temperature typically is $\vartheta_{ref} = 600\,^\circ\mathrm{C}$, while the reference pressure is at $p_{ref} = 1\,\mathrm{bar}$ [3]. The expansion ratio Π_t is not defined unanimously in the literature. Both the form shown in (9.73) as well as the reciprocal are used. Within the book, the following definition is used:

$$\Pi_t(t) = \frac{p_{us}(t)}{p_{ds}(t)} \tag{9.73}$$

The expansion ratio is $\Pi_t > 1$ for normal operation. The fluid mechanics in turbines are fundamentally different from those in compressors. Instead of transporting the fluid against a high downstream pressure, the fluid is expanded. Therefore, stall cannot happen inside a turbine. The only operation limit that has to be considered in few cases is the turbine inlet temperature; a too high temperature can damage the turbine.

For modeling the expansion process, an isentropic process is typically used as a reference. The value η_t expresses the efficiency of the turbine to represent further losses compared to the isentropic process. Based on the equation for power $P_t = \dot{m}_t \Delta h$ and the standard enthalpy difference equation $\Delta h = c_p \Delta\vartheta$, the equations for the temperature ϑ_{ds} and the power generated P_t can be derived:

$$\vartheta_{ds} = \vartheta_{us} + \vartheta_{us}\eta_t \cdot \left(\left(\frac{p_{ds}}{p_{us}}\right)^{\frac{\kappa-1}{\kappa}} - 1\right) \tag{9.74}$$

$$P_t(t) = c_p\dot{m}_t(t)\eta_t(t)\vartheta_{us,t}(t) \cdot \left(1 - \Pi_t^{\frac{1-\kappa}{\kappa}}(t)\right) \tag{9.75}$$

In the following, models for the mass flow through the turbine $\dot{m}_t$ as well as for the turbine efficiency η_t are revisited.

Mass Flow

The mass flow through the turbine $\dot{m}_t$ depends mainly on the pressure ratio Π_t. The turbocharger speed only has a minor impact on the mass flow. Thus, the dependency on the turbocharger speed is often neglected for modeling. To characterize the mass flow through a turbine, simple formulations are commonly applied. The turbine mass flow is often modeled as an orifice, as for example in [5]. The standard orifice equations presented in (9.55) and (9.59) can be applied. Also, modified versions of the orifice equation exist to model the turbine mass flow. For instance, a model based on [11] is given by

$$\dot{m}_t(t) = c_d A \frac{p_{us}(t)}{\sqrt{R\vartheta_{us}(t)}} \psi(\Pi_t(t)) \tag{9.76}$$

$$\psi(\Pi_t(t)) = \Pi_t^{\frac{\kappa-2}{2\kappa}} \sqrt{\frac{2\kappa}{\kappa-1} f_{lim}\left(1 - \Pi_t^{\frac{-\kappa+1}{\kappa}}\right)} \tag{9.77}$$

$$f_{lim}(x) = \frac{1}{2}\sqrt{x^2 + \epsilon_2} + \frac{x}{2} \tag{9.78}$$

The actuators of the turbocharger belong to the main actuators for the air path control. Two types of actuators typically exist to control the turbocharger, namely a wastegate on the turbine side and a variable turbine geometry (VTG). The wastegate simply enables a part of the mass flow to bypass the turbine. The model equations for the turbine stay the same even when the turbine is equipped with an additional wastegate. The VTG actuation affects the inlet geometry of the turbine and thereby the angle of incidence for the rotor. This changes the mass flow behavior of the turbine. Within the orifice equation, this can be taken into account by modeling the factor $c_d A$ by a data-driven function that depends on the VTG position:

$$c_d A(t) = f\left(u_{vtg}(t)\right) \tag{9.79}$$

An alternative to the use of orifice-like equations to model the mass flow through the turbine is the application of polynomial models. The parameters in the polynomial model are identified for a best fit of the measurement data of the corrected turbine mass flow:

$$\dot{m}_{t,corr}(t) = f(\Pi(t)) \tag{9.80}$$

If the turbine is equipped with a VTG, a multivariate polynomial model can be used:

$$\dot{m}_{t,corr}(t) = f\left(\Pi(t), u_{vtg}(t)\right) \tag{9.81}$$

Efficiency

A very common approach to model the efficiency is based on the dimensionless blade speed ratio (BSR) [3]. The BSR characterizes the ratio between the speed of the blade tip and the fluid velocity. When the blade speed is much lower than that of the fluid velocity, the kinetic energy of the fluid is lost, while a very fast blade is slowed down by the fluid. There exists a sweet spot in between these two situations where the maximum efficiency is reached. The efficiency is often approximated as a quadratic function of BSR with a unique maximum. The calculation of the BSR is shown in (9.82), whereas (9.83) can be used to compute the turbine efficiency. The radius of the blade is given by r_t; the angular speed of the turbocharger is denoted by ω_{tc}. The tuning parameters $\eta_{t,max}$ and BSR_{opt} can be used to shape the parabola. These values are estimated to either fit measurement data or a turbine map. These formulations were proposed in [3]; however, similar approaches exist that employ a different parametrization for the parabola, as in [5], for example.

$$\mathrm{BSR}(t) = \frac{\omega_{tc}(t) r_t}{\sqrt{2 c_p \vartheta_{us}(t) \left(1 - \Pi_t^{(1-\kappa)/\kappa}\right)}} \tag{9.82}$$

$$\eta_t\left(\mathrm{BSR}(t)\right) = \eta_{t,max} \cdot \left(1 - \left(\frac{\mathrm{BSR}(t) - \mathrm{BSR}_{opt}}{\mathrm{BSR}_{opt}}\right)^2\right) \tag{9.83}$$

In case of VTG actuation, the parameter $\eta_{t,max}$ can be modeled as a polynomial function in dependence of the VTG position:

$$\eta_{t,max} = f(u_{vtg}) \tag{9.84}$$

9.5.3 *Turbocharger Dynamics*

The turbocharger dynamics are given by the interaction between the turbine on the exhaust gas side and the compressor on the inlet side. In transients, the momenta of the turbine as well as that of the compressor and the friction are not in balance causing an acceleration or a deceleration of the turbocharger shaft until steady-state conditions are reached where the momenta are in balance again. Newton's second law can be used to derive the equations of motion:

$$\frac{\mathrm{d}}{\mathrm{d}t}\omega_{tc}(t) = \frac{1}{J_{tc}\omega_{tc}(t)}\left(P_t(t) - P_c(t) - P_{fric}(t)\right) \tag{9.85}$$

The turbine delivers the power P_t by the extraction of enthalpy from the hot exhaust gases, while the compressor consumes the power P_c in order to compress the intake gas to a higher pressure, thus increasing the gas density. For the description of the air path dynamics, the moment of inertia of the turbocharger J_{tc} plays a significant role. The value of J_{tc} mainly depends on the sizing of the turbocharger and can vary in the order of one magnitude. The angular velocity of the turbocharger is represented by ω_{tc}. Additionally, a friction power P_{fric} is considered to describe the losses occurring in the bearings. Typically, either P_{fric} is assumed to be constant, $P_{fric} = const.$, or it can be modeled as a function that depends on the angular speed of the turbocharger $P_{fric} = f(\omega_{tc})$.

An alternative approach to consider the friction losses is the introduction of the mechanical efficiency η_{tc} of the turbocharger which can be accounted for the turbine:

$$\frac{\mathrm{d}}{\mathrm{d}t}\omega_{tc}(t) = \frac{1}{J_{tc}\omega_{tc}(t)}\left(\eta_{tc} P_t(t) - P_c(t)\right) \tag{9.86}$$

The equations are not defined for $\omega_{tc}(t) = 0$. In normal operation, it holds that $\omega_{tc}(t) > 0$. If situations occur with $\omega_{tc}(t) \leq 0$, limiting functions can be introduced to increase numerical robustness. For instance, the following model can be used:

$$\begin{aligned}\frac{\mathrm{d}}{\mathrm{d}t}\omega_{tc}(t) &= \frac{1}{J_{tc}\omega_{tc,lim}(t)}\left(\eta_{tc}P_t(t) - P_c(t)\right)\\ \omega_{tc,lim}(t) &= \frac{1}{2}\sqrt{\omega_{tc}^2(t)+\epsilon} + \frac{\omega_{tc}(t)}{2}\end{aligned} \tag{9.87}$$

References

1. T. Albin, D. Ritter, D. Abel, N. Liberda, R. Quirynen, M. Diehl, Nonlinear MPC for a two-stage turbocharged gasoline engine airpath, in *IEEE Conference on Decision and Control* (2015), pp. 849–856
2. T. Albin, D. Ritter, N. Liberda, D. Abel, Boost pressure control strategy to account for transient behavior and pumping losses in a two-stage turbocharged air path concept. Energies **9**(7), 530–545 (2016)
3. L. Eriksson, L. Nielsen, *Modeling and Control of Engines and Drivelines* (Wiley, New Jersey, 2014)
4. E.M. Greitzer, Surge and rotating stall in axial flow compressors - part I: theoretical compression system model. J. Eng. Power **98**(2), 190–198 (1976)
5. L. Guzzella, C.H. Onder, *Introduction to Modeling and Control of Internal Combustion Engine Systems* (Springer, Berlin, 2010)
6. J.B. Heywood, *Internal Combustion Engine Fundamentals* (McGraw-Hill Education, New York, 2018)
7. R. Isermann, *Engine Modeling and Control* (Springer, Berlin, 2014)
8. O. Leufvén, L. Eriksson, Engine test bench turbo mapping, SAE Technical Paper (2010)
9. L. Ljung, Perspectives on system identification. Ann. Rev. Control **34**(1), 1–12 (2010)
10. P. Moraal, I. Kolmanovsky, Turbocharger modeling for automotive control applications, SAE Technical Paper (1999)
11. P. Moulin, J. Chauvin, Modelling and control of the air system of a turbocharged gasoline engine, in *Control Engineering Practice* (2011), pp. 287 – 297
12. O. Nelles, *Nonlinear System Identification: From Classical Approaches to Neural Networks and Fuzzy Models* (Springer, Berlin, 2013)
13. N. Watson, M.S. Janota, *Turbocharging the Internal Combustion Engine* (Macmillan Press Ltd., 1982)

Chapter 10
Combined Exhaust Gas Recirculation and VTG: Modeling and Analysis

Abstract In this case study, a very common control task within the air path of a compression ignition (CI) engine is discussed. The engine setup investigated uses exhaust gas recirculation (EGR) in combination with a turbocharger that is equipped with variable turbine geometry (VTG). The topic of investigation is the simultaneous control of the burnt gas ratio and the pressure in the intake manifold. The present chapter outlines the process modeling and an analysis of the system dynamics. A state-space model based on ordinary differential equations (ODE) is derived which consists of five dynamic states. The process model is validated for the static and the dynamic behavior by the use of measurement data from an engine test bench. The model shows a good correlation with the measurement data while it is still computationally efficient. This validated model is subsequently used to analyze the system dynamics of the process. Control-relevant characteristics, such as pole-zero location and relative gain array (RGA) matrix are discussed. The air path exhibits complex system dynamics including nonlinearity and sign reversal.

10.1 System Setup

To achieve low tailpipe pollutant emissions for a CI engine, two approaches are applied. First, the engine is operated such that low engine-out emissions are formed. Second, to further lower the emissions to the limits set by legislation, aftertreatment systems are used. One common measure to obtain already low levels of engine-out emissions while still achieving a high efficiency is the use of EGR in combination with a VTG turbocharger. For the appropriate reduction of the pollutant emissions with these devices, the air path control plays a crucial role. Section 7.3.2 describes details on this system and on the physical effects. In the present case study, the complex process control problem of this setup is investigated. Figure 10.1 shows the CI engine setup investigated.

T. Albin Rajasingham, *Nonlinear Model Predictive Control of Combustion Engines*, Advances in Industrial Control,
https://doi.org/10.1007/978-3-030-68010-7_10

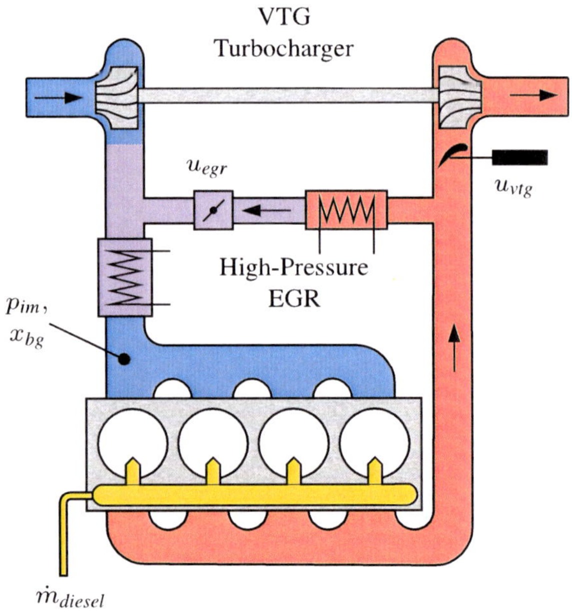

Fig. 10.1 Investigated CI engine setup

The EGR valve position u_{egr} and the guide vane position u_{vtg} of the VTG are used as actuated variables. They allow to control the intake manifold pressure p_{im} and the burnt gas ratio x_{bg} in the intake manifold. The value p_{im} is measured by the use of a pressure sensor. For the determination of x_{bg}, an oxygen sensor in the intake manifold is applied. The sensor measures the fraction of oxygen mass in the intake manifold denoted by F_{im}. The variable $F_{O_2,air}$ denotes the oxygen mass fraction in the air which is assumed constant, i.e. $F_{O_2,air} = 0.23$. The burnt gas ratio x_{bg} can be correlated to these two values by a simple algebraic relation:

$$x_{bg}(t) = 1 - \frac{F_{im}(t)}{F_{O_2,air}} \tag{10.1}$$

The system is characterized by nonlinear dynamics and strong cross-couplings between EGR and VTG. The control of the EGR system in combination with VTG turbocharging has been widely covered in scientific publications due to its importance in emissions reduction. Various kinds of control concepts were applied for the SI as well as for the CI engine. While there exist publications for the SI case such as [4, 5, 11], the majority of research for this system setup has been conducted for the CI case. Especially initial publications investigated the application of classical control algorithms such as gain-scheduled PI/PID-based control [1, 10], sliding mode control [8], control Lyapunov functions [9], and flatness-based control [6]. More recent publications especially demonstrate the advantages of nonlinear model predictive control (NMPC) over the classical control approaches, such as [2, 3, 7]. The following case study details the benefits of NMPC for this application.

In the remainder of this chapter, a system model is derived with a special focus on the trade-off between the accuracy of the model and its computational demands. Subsequently, a model-based analysis of the process is conducted. In Chap. 11, an NMPC method is developed for controlling this system. The results then are compared to PID and linear MPC (LMPC) controllers.

10.2 Control-Oriented Process Model

As a process model, a nonlinear state-space model is developed which takes the following form:

$$\begin{aligned}\dot{\boldsymbol{x}}(t) &= \boldsymbol{f}(\boldsymbol{x}(t), \boldsymbol{u}(t)) \\ \boldsymbol{y}(t) &= \boldsymbol{g}(\boldsymbol{x}(t))\end{aligned} \tag{10.2}$$

Two actuated variables are available for the control purpose, the guide vane position of the VTG u_{vtg} and the EGR valve position u_{egr}, which are summarized by $\boldsymbol{u}(t)$.

$$\boldsymbol{u}(t) = \begin{bmatrix} u_{vtg}(t) \\ u_{egr}(t) \end{bmatrix} \begin{matrix} \text{position VTG} \\ \text{position EGR valve} \end{matrix} \tag{10.3}$$

These two actuated variables allow the burnt gas ratio and the intake manifold pressure to be controlled which are summarized by $\boldsymbol{y}(t)$.

$$\boldsymbol{y}(t) = \begin{bmatrix} p_{im}(t) \\ x_{bg}(t) \end{bmatrix} \begin{matrix} \text{intake manifold pressure} \\ \text{burnt gas ratio} \end{matrix} \tag{10.4}$$

All in all, the system is described by five system states which are summed up as $\boldsymbol{x}(t)$. Two gas volumes are present, i.e. the intake manifold and the exhaust manifold. Each gas volume is described by its pressure $p(t)$ and its oxygen mass fraction $F(t)$. The angular speed of the turbocharger $\omega_{tc}(t)$ acts as an additional state variable.

$$\boldsymbol{x}(t) = \begin{bmatrix} \omega_{tc}(t) \\ p_{im}(t) \\ p_{em}(t) \\ F_{im}(t) \\ F_{em}(t) \end{bmatrix} \begin{matrix} \text{angular speed of turbocharger} \\ \text{pressure at intake manifold} \\ \text{pressure at exhaust manifold} \\ \text{oxygen mass fraction in intake manifold} \\ \text{oxygen mass fraction in exhaust manifold} \end{matrix} \tag{10.5}$$

The system state $p_{im}(t)$ is the first controlled variable of the system. The second controlled variable $x_{bg}(t)$ can be calculated from the system state $F_{im}(t)$ by (10.1).

The differential equations characterizing the five system states are developed according to the descriptions in Chap. 9. For determining the turbocharger speed, Newton's second law is used. Within the equation of motion (10.6), the power generated by the turbine P_t, the power consumed by the compressor P_c, and the friction losses P_f are considered. The friction losses P_f are assumed to be constant. The value J_{tc} describes the moment of inertia of the turbocharger. Both values P_f and J_{tc} are determined by system identification for best fit with measurement data. For the two gas volumes, the pressure dynamics are defined using the isothermal model simplification, i.e. the inlet temperature is equal to the outlet temperature. The dynamics of oxygen fractions in the intake $\dot{F}_{im}$ and the exhaust manifold $\dot{F}_{em}$ are well described by (10.9) and (10.10). For the thermodynamic calculations, the gas properties of air are used, i.e. the specific gas constant is chosen as $R = 287\frac{\text{J}}{\text{kgK}}$, the specific heat capacity as $c_p = 1005\frac{\text{J}}{\text{kgK}}$, and the heat capacity ratio as $\kappa = 1.4$. Due to the lean operation of the engine, the air is a dominant content within the entire air path.

$$\dot{\omega}_{tc}(t) = \frac{1}{\omega_{tc}(t) J_{tc}} \Big(P_t(t) - P_c(t) - P_f \Big) \tag{10.6}$$

$$\dot{p}_{im}(t) = \frac{R\vartheta_{im}}{V_1} \Big(\dot{m}_c(t) + \dot{m}_{egr}(t) - \dot{m}_{asp}(t) \Big) \tag{10.7}$$

$$\dot{p}_{em}(t) = \frac{R\vartheta_{em}}{V_2} \Big(\dot{m}_{eo}(t) - \dot{m}_{egr}(t) - \dot{m}_t(t) \Big) \tag{10.8}$$

$$\dot{F}_{im}(t) = \frac{R\vartheta_{im}}{V_3 p_{im}(t)} \Big(\Big(F_{O_2,air} - F_{im}(t) \Big) \dot{m}_c(t) + \Big(F_{em}(t) - F_{im}(t) \Big) \dot{m}_{egr}(t) \Big) \tag{10.9}$$

$$\dot{F}_{em}(t) = \frac{R\vartheta_{em}}{V_4 p_{em}(t)} \Big(\Big(F_{eo}(t) - F_{em}(t) \Big) \dot{m}_{eo}(t) \Big) \tag{10.10}$$

All terms that have no explicit dependence on time, i.e. J_{tc}, P_f, R, ϑ_{im}, ϑ_{em}, V_1, V_2, V_3, V_4, and $F_{O_2,air}$, are assumed to be constant. The conditions upstream of the compressor and downstream of the turbine are assumed to be identical to constant ambient conditions with $p_{amb} = 1\text{bar}$, $\vartheta_{amb} = 293\text{K}$, and $F_{O_2,air} = 0.23$. Alternatively, sensors may be employed to continuously measure the ambient conditions. For ϑ_{im} and ϑ_{em}, the numerical values are chosen to represent medium values of their value range. This avoids the inclusion of additional temperature relations which would result in more accurate models, but increases its complexity. The values V_1,... V_4 are chosen to reflect the dynamic behavior.

The terms $P_t(t)$, $P_c(t)$, $\dot{m}_c(t)$, $\dot{m}_{egr}(t)$, $\dot{m}_{asp}(t)$, $\dot{m}_{eo}(t)$, $\dot{m}_t(t)$, and $F_{eo}(t)$ are determined by algebraic relations as detailed in the following. Many quantities depend on pressure ratios, which are denoted by Π. More specifically, the following pressure ratios over the engine, the turbine, and the compressor are used:

$$\Pi_{eng} = \frac{p_{em}}{p_{im}} \tag{10.11}$$

$$\Pi_t = \frac{p_{em}}{p_{amb}} \tag{10.12}$$

$$\Pi_c = \frac{p_{im}}{p_{amb}} \tag{10.13}$$

Three different *engine mass flows* are considered, i.e. the mass flow $\dot{m}_{asp}$ aspirated into the engine from the intake manifold, the mass flow of fuel injected directly into the cylinders $\dot{m}_{fuel}$, and the engine-out mass flow exhausted from the engine $\dot{m}_{eo}$. Taking the direct injection of the fuel into account, the mass flow $\dot{m}_{eo}$ can be calculated by

$$\dot{m}_{eo} = \dot{m}_{asp} + \dot{m}_{fuel} \tag{10.14}$$

The value for $\dot{m}_{fuel}$ is available in the engine control unit as it is an actuated value. The aspirated mass flow into the engine $\dot{m}_{asp}$ is determined by a data-driven approach. The following function that depends on the exhaust manifold pressure p_{em}, the angular engine speed ω_{eng}, and the pressure ratio over the engine Π_{eng} is used. The constants $i_0, \ldots, i_3$ are identified to fit the measurement data.

$$\dot{m}_{asp} = i_0 + i_1 p_{im} + i_2 \Pi_{eng} + i_3 \omega_{eng} \tag{10.15}$$

Based on the engine mass flows, the relation for the *engine-out oxygen mass fraction* is calculated by (10.16). The relation takes the stoichiometric air-to-fuel ratio AFR_s into account, i.e. for diesel fuel, it holds $AFR_s F_{O2,air} = 3.4$.

$$F_{eo} = \frac{1}{\dot{m}_{eo}} \cdot \left(F_{im} \dot{m}_{asp} - \dot{m}_{fuel} AFR_s F_{O2,air} \right) \tag{10.16}$$

The *mass flow through the EGR valve* is determined using a data-driven model. The function depends on the position of the EGR valve u_{egr} and the pressure ratio over the engine Π_{eng}. The constants $h_0, \ldots, h_8$ are fitted to the measurement data.

$$\dot{m}_{egr} = h_0 \frac{1 - \exp\left(- \left(\frac{u_{egr}(t) + h_1}{h_2} \right)^{h_3} \right)}{h_4 \exp\left(- h_5 \Pi_{eng} \right) + h_6} + f_{lim}(h_7 u_{egr}(t) - h_8) \tag{10.17}$$

The limit function f_{lim} is defined as

$$f_{\text{lim}}(x) = \frac{1}{2}\sqrt{x^2 + \varepsilon} + \frac{1}{2}x \tag{10.18}$$

The *turbocharger* is described by the following equations. The power generated by the turbine P_t and the power consumed by the compressor P_c are calculated by

$$P_t = c_p \eta_t \vartheta_{em} \dot{m}_t \left(1 - \Pi_t^{\frac{1-\kappa}{\kappa}} \right) \tag{10.19}$$

$$P_c = c_p \vartheta_{amb} \frac{\dot{m}_c}{\eta_c} \left(\Pi_c^{\frac{\kappa-1}{\kappa}} - 1 \right) \tag{10.20}$$

For η_c and η_t, constant values are assumed. They are determined by system identification with measurement data.

The *mass flow through the compressor* $\dot{m}_c$ is determined by a polynomial model. The constants involved l_{ij} are fitted to reproduce the measurement data. The correction term, as given in (9.60), does not have a big influence on the compressor mass flow in this case, as the compressor is operated at ambient conditions that are very similar to the reference conditions. Thus, no correction is conducted. The polynomial function is given by

$$\dot{m}_c = l_{02}\omega_{tc}^2 + l_{20}\Pi_c^2 + \sum_{i=0}^{1}\sum_{j=0}^{1} l_{ij}\Pi_c^i\omega_{tc}^j \tag{10.21}$$

For calculation of the *mass flow through the turbine* $\dot{m}_t$ as well a polynomial model is used. More precisely, the polynomial model calculates the corrected mass flow through the turbine. The constants of the polynomial model t_{ij} are fitted to reproduce the measurement data. For the correction term, the reference pressure is assumed to be equal to the ambient pressure $p_{ref} = p_{amb}$. The following relations result:

$$\dot{m}_c = l_{02}\omega_{tc}^2 + l_{20}\Pi_c^2 + \sum_{i=0}^{1}\sum_{j=0}^{1} l_{ij}\Pi_c^i\omega_{tc}^j \tag{10.22}$$

$$\dot{m}_{t,corr} = \sum_{i=0}^{4}\sum_{j=0}^{3} t_{ij}\Pi_t^i u_{vtg}^j \tag{10.23}$$

$$\dot{m}_t = \dot{m}_{t,corr}\Pi_t\sqrt{\frac{\vartheta_{ref}}{\vartheta_{em}}} \tag{10.24}$$

There is also a considerable dead-time arising in the process. The dead-time is mainly acting on the system output, i.e. by sensor delay. As a consequence, the model lumps together all arising dead-times as one that is acting on each output. Both dead-times are assumed to be constant.

$$y_{p_{im}} = p_{im}(t - T_{D,1}) \tag{10.25}$$

$$y_{x_{bg}} = x_{bg}(t - T_{D,2}) \tag{10.26}$$

The dead-times are identified via system identification.

10.3 Validation of the Process Model

To validate the process model derived, it is compared to data gathered from an engine test bench. The engine is a 2-liter production-type diesel engine with a compression ratio of 16.5:1 and four cylinders. All validations are conducted for a constant engine

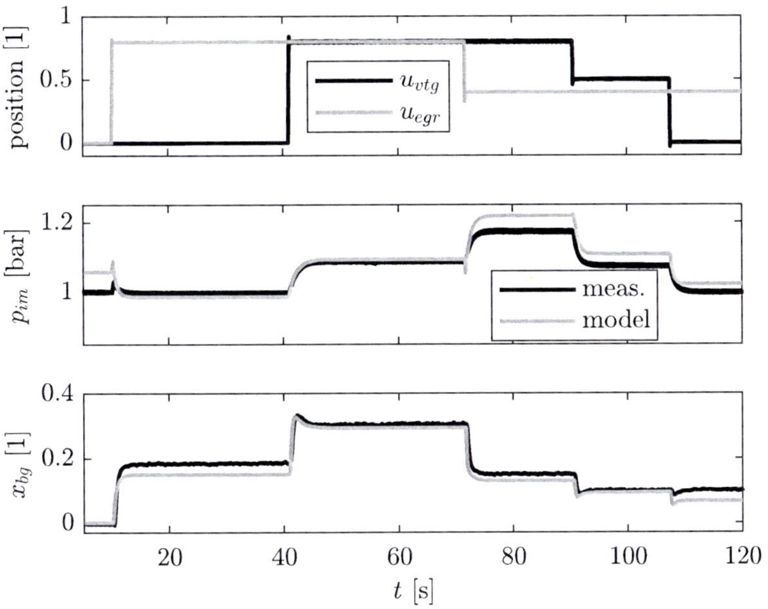

Fig. 10.2 Comparison of measurement data and simulative data of the process model

speed and for a constant diesel fuel mass injected. Figure 10.2 shows the transient behavior gained from experimental measurements and from simulations with the process model. For validation purposes, steps with various heights are applied on u_{vtg} and on u_{egr}. The experimental data exhibits overshoots as well as non-minimum phase behavior. Both are reproduced by the process model. The model accurately captures the transient dynamics. In certain areas, a slight offset in the steady-state behavior can be observed. For the intake manifold pressure $y_{p_{im}}$, the dead-time is identified to be $T_{D,1} = 0.1$s and for the burnt gas ratio $y_{x_{bg}}$, the value $T_{D,2} = 0.4$s is used. With these values, the dead-time can be captured very well.

Figures 10.3 and 10.4 detail the steady-state behavior of the burnt gas ratio x_{bg} and the intake manifold pressure p_{im} in dependence of the two actuated values u_{egr} and u_{vtg} at a constant engine speed. The measured values as well as the ones resulting from simulation of the process model are depicted. The process model is able to capture the steady-state behavior sufficiently well over the entire operating range. Quantitatively, the maximum error is acceptable for both values. Qualitatively, the process model is able to capture the nonlinearities of the system behavior. The process model can be further improved by the inclusion of additional submodels. Submodels e.g. for capturing the efficiencies of the turbine and the compressor or for capturing the temperature dynamics can be added. However, such submodels cause a higher model complexity.

In summary, the process model quantitatively and qualitatively shows a good reproduction behavior. Both the steady-state and the transient behavior can be

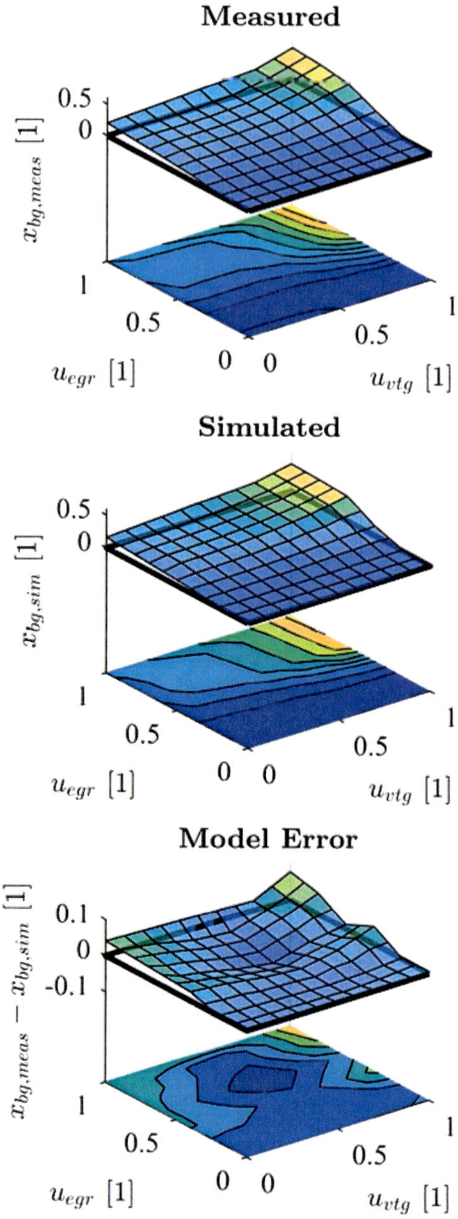

Fig. 10.3 Steady-state values of burnt gas ratio

captured. The model is a low-order nonlinear model that represents a good compromise between accuracy and complexity. Furthermore, it fulfills the requirement to be continuously differentiable.

Fig. 10.4 Steady-state values of intake manifold pressure

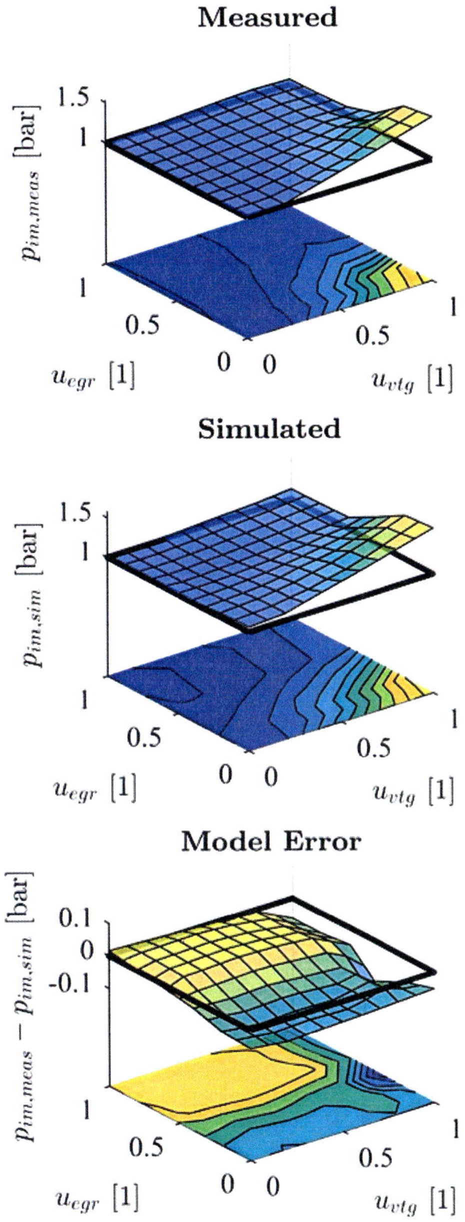

10.4 Analysis of the System Dynamics

First, an analysis of the system dynamics of the air path system has to be conducted in order to develop suitable model-based control algorithms. In the following, various characteristics such as pole-zero locations as well as the RGA matrix are investigated based on the process model derived.

10.4.1 Nonlinearity

The nonlinear behavior of the plant is readily visible in the steady-state maps shown in Figs. 10.3 and 10.4. To further quantify these nonlinearities, the sensitivities at steady-state operation are investigated. These are calculated for all input–output combinations by determining the gradients at stationary conditions.

$$\left.\frac{\partial \boldsymbol{y}_i}{\partial \boldsymbol{u}_j}\right|_{\boldsymbol{x}_0,\,\boldsymbol{u}_0} \quad \text{with} \quad \boldsymbol{f}(\boldsymbol{x}_0, \boldsymbol{u}_0) = \dot{\boldsymbol{x}} = \boldsymbol{0} \tag{10.27}$$

Figure 10.5 shows the sensitivity of each of the two actuated values to each of the two controlled variables. A linear plant has a constant sensitivity over the entire operating range, i.e. the DC gain. In contrast, the nonlinear EGR VTG system has very variable sensitivities, i.e. the sensitivities of the system are strongly dependent on the operating point. Additionally, in some operating regions, a sign reversal occurs. When $u_{vtg} = 1$ is kept constant and u_{egr} is reduced from 1 to 0, then the sensitivity for p_{im} to u_{egr} is at first positive, while for values $u_{egr} < 0.4$, the sign subsequently switches to negative values. This behavior is known as sign reversal and generally is very difficult to control.

To illustrate the nonlinearities in the transient behavior, a simulation is conducted with a linearized system model. The model is linearized around the steady-state operating point resulting from $u_{egr} = u_{vtg} = 0.5$. Figure 10.6 shows the simulation differences between the linearized and the nonlinear system. While a moderate change in the VTG position can be captured quite well with the linearized model, considerable discrepancies occur at a moderate step input of the EGR position.

10.4.2 Pole-Zero Locations

Crucial insights of the system behavior can be gained by an inspection of the pole-zero locations of the linearized plant. For the analysis, the linearizations are calculated over the entire operating range of the actuated values – all at steady-state conditions. Figure 10.7 shows the location of the pole with the highest value of the linearized MIMO transfer function in dependence of the steady-state actuated values. Numerical values for three exemplary operating points are highlighted. Clearly, at any point

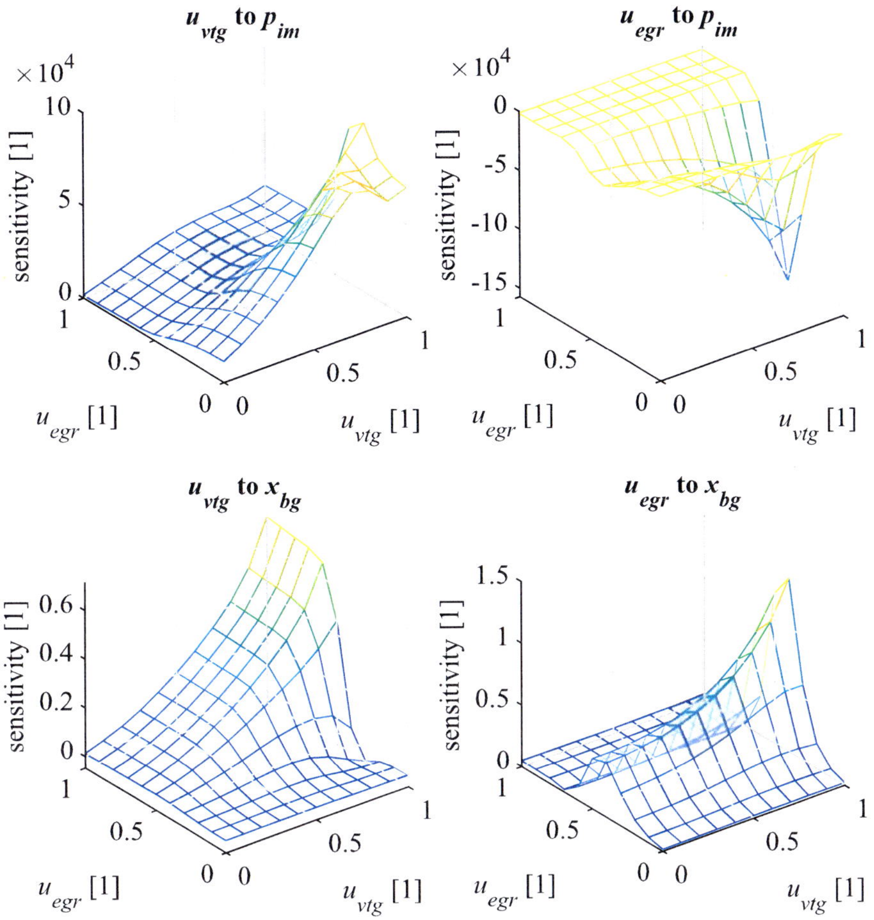

Fig. 10.5 Sensitivities of the input–output behavior at steady-state conditions

of the linearization, all poles are negative. The linearized plants thus are asymptotically stable. As a result, all possible equilibrium points of the nonlinear system are asymptotically stable within the operating range investigated.

Figure 10.8 shows the location of the zero with the highest value of the MIMO transfer function in dependence of the point of linearization. The numerical values for three exemplary linearization points are highlighted. In the entire operation range, the linear plant always includes a non-minimum phase zero. For the design of a conventional controller such as a PID controller, the non-minimum phase zeros limit the achievable crossover frequency of a controller and thus its possible performance. In fact, the nonlinear system also exhibits this typical non-minimum phase behavior, as the step response in Fig. 10.2 shows.

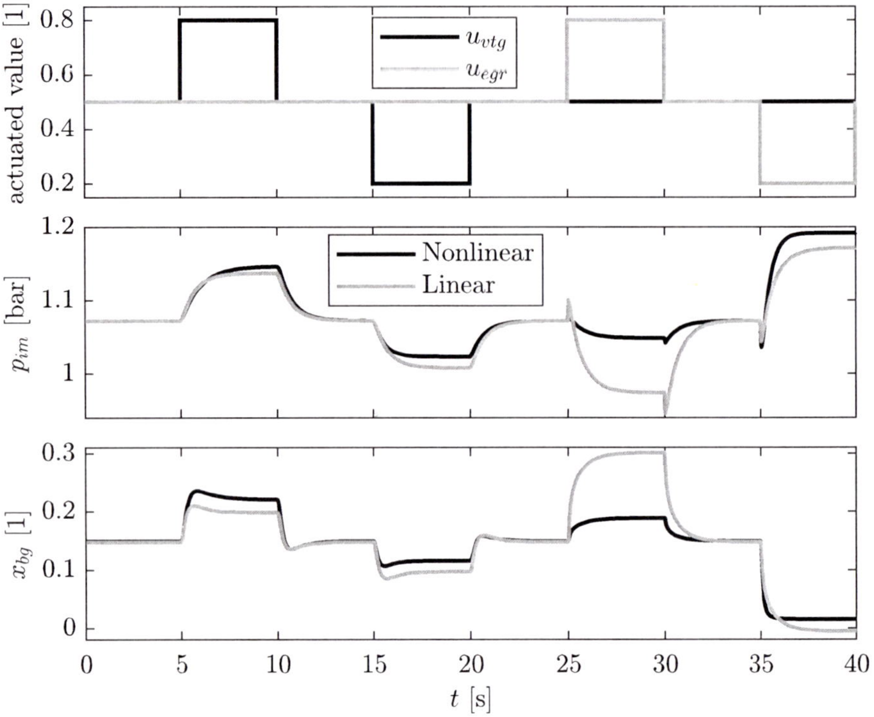

Fig. 10.6 Comparison of the transient behaviors of the linearized and the nonlinear model

Fig. 10.7 Maps with maximum poles of the linearized MIMO transfer function

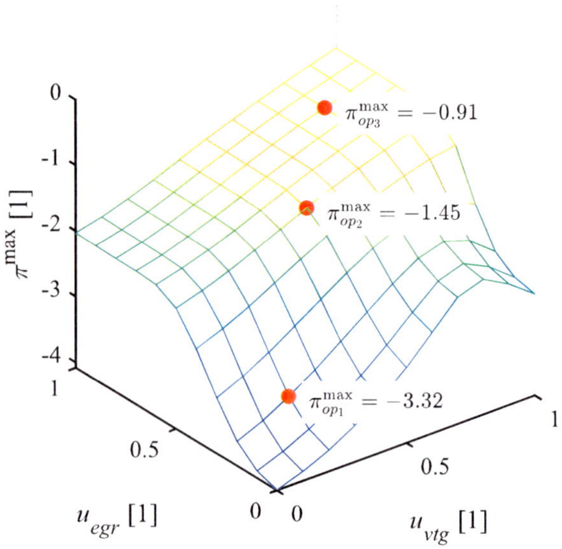

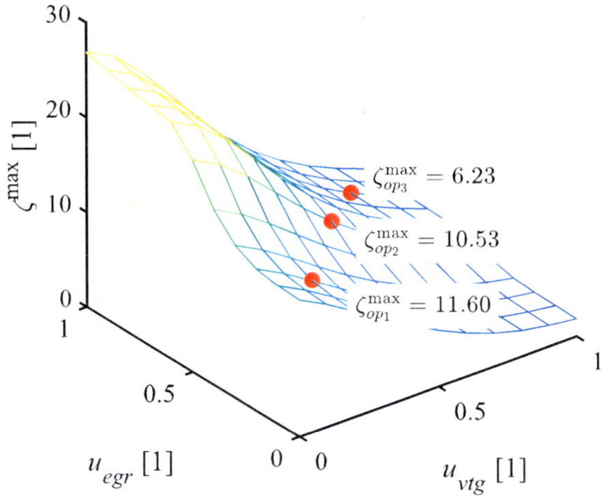

Fig. 10.8 Maps with maximum zeros of the linearized MIMO transfer function

10.4.3 Relative Gain Array

In Sect. 2.1.2, the RGA matrix $\boldsymbol{\Lambda}(s)$ is introduced as a measure to quantify the cross-couplings in MIMO systems. Figure 10.9 shows the values of the RGA matrix of the linearized system for the various linearization points. The RGA matrix is evaluated at the steady state $s = 0$. Specifically, the first values of $\boldsymbol{\Lambda}(0)$, i.e. $\Lambda_{11}(0)$, are illustrated, which correspond to the channel $u_{vtg} \rightarrow p_{im}$. In the case of 2×2 MIMO systems, one element of $\boldsymbol{\Lambda}(s)$ always is sufficient to define the entire matrix, since all columns and rows of $\boldsymbol{\Lambda}(s)$ must add up to 1. The numerical values for three exemplary linearization points are highlighted. The effects of cross-couplings strongly depend on the operating point.

- OP_1: With $_{11}(0) = 0.97$, the plant is diagonally dominant, i.e. using two SISO controllers for the channels $u_1 \rightarrow y_1$ (here $u_{vtg} \rightarrow p_{im}$) and $u_2 \rightarrow y_2$ (here $u_{egr} \rightarrow x_{bg}$) should work satisfactorily.
- OP_2: Here, the entry $_{11}(0) = 0.66$ clearly differs from 1; therefore, a MIMO approach should be pursued for controlling the plant in this operating point.
- OP_3: As $_{11}(0) = 0.48$, the off-diagonal entries are even closer to 1. In fact, the RGA suggests that it is better to control p_{im} with u_{egr} and x_{bg} with u_{vtg}.

In summary, in certain operating conditions, the diagonal entries differ significantly from unity. Hence, the RGA matrix reveals that achieving a satisfactory control performance with a pure SISO control approach is unlikely. Instead, a decoupling matrix should be used for conventional control approaches.

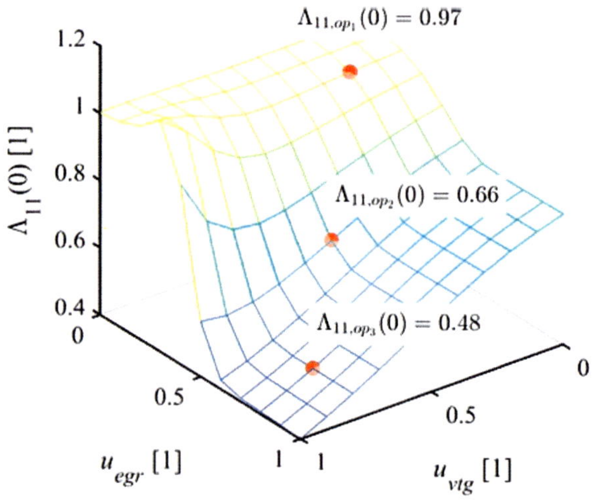

Fig. 10.9 Steady-state relative gain array of the linearized MIMO transfer function

References

1. M. Ammann, N.P. Fekete, L. Guzzella, A.H. Glattfelder, Model-based control of the vgt and egr in a turbocharged common-rail diesel engine: theory and passenger car implementation. SAE Trans. 527–538 (2003)
2. D. Gagliardi, T. Othsuka, L. del Re, Direct c/gmres control of the air path of a diesel engine. IFAC Proc. Vol. **47**(3), 3000–3005 (2014)
3. M. Huang, H. Nakada, K. Butts, I. Kolmanovsky, Nonlinear model predictive control of a diesel engine air path: a comparison of constraint handling and computational strategies. IFAC-PapersOnLine **48**(23), 372–379 (2015)
4. M. Keller, S. Geiger, D. Abel, T. Albin, Physics-based modeling and mpc for the air path of a two-stage turbocharged si engine with low pressure egr, in *European Control Conference* (2020)
5. M. Keller, S. Geiger, M. Günther, S. Pischinger, D. Abel, T. Albin, Model predictive air path control for a two-stage turbocharged spark-ignition engine with low pressure exhaust gas recirculation. Int. J. Engine Res. **21**(10), 1835–1845 (2020)
6. P. Kotman, A. Kugi et al., Flatness-based feedforward control of a diesel engine air system with egr. IFAC Proc. Vol. **43**(7), 598–603 (2010)
7. A. Murilo, M. Alamir, D. Alberer, A general NMPC framework for a diesel engine air path. Int. J. Control **87**(10), 2194–2207 (2014)
8. V. Utkin, H.-C. Chang, I. Kolmanovsky, J.A. Cook, Sliding mode control for variable geometry turbocharged diesel engines. Am. Control Conf. **1**(6), 584–588 (2000)
9. M. Van Nieuwstadt, I. Kolmanovsky, P. Moraal, A. Stefanopoulou, M. Jankovic, Egr-vgt control schemes: experimental comparison for a high-speed diesel engine. IEEE Control Syst. Mag. **20**(3), 63–79 (2000)
10. J. Wahlstrom, L. Eriksson, L. Nielsen, Egr-vgt control and tuning for pumping work minimization and emission control. IEEE Trans. Control Syst. Technol. **18**(4), 993–1003 (2009)
11. A.P. Wiese, A.G. Stefanopoulou, A.Y. Karnik, J.H. Buckland, Model predictive control for low pressure exhaust gas recirculation with scavenging, in *American Control Conference* (2017), pp. 3638–3643

Chapter 11
Combined Exhaust Gas Recirculation and VTG: Control

Abstract In this chapter, various control algorithms for the combined control of burnt gas ratio and intake manifold pressure are investigated. Based on the process model derived, three model-based controllers are designed. The most elaborate one of them is a nonlinear model predictive control (NMPC) algorithm. The design aspects which have been covered in the book, such as offset-free control, are implemented. The choice of several parameters, such as the number of SQP steps or the length of the control horizon, is investigated in detail for the NMPC algorithm. For comparison, a linear MPC (LMPC) algorithm is developed as well, which relies on the linearization of the nonlinear model. Additionally, a decentralized control approach based on PI controllers is investigated. For an appropriate decentralized control synthesis, several modules have to be included, such as decoupling control and anti-reset windup. Finally, the three concepts are compared, including the validation of the NMPC algorithm on an engine test bench.

11.1 Nonlinear MPC Algorithm

An NMPC approach is presented for the control of the air path of a CI engine. The investigated CI engine setup is depicted in Fig. 10.1. The objective is to simultaneously control the burnt gas ratio and the pressure in the intake manifold. First, the controller-internal model is discussed. Subsequently, the parametrization of the control algorithm is investigated.

11.1.1 Controller-Internal Model

The controller-internal model used within the NMPC algorithm has to be continuously differentiable and of low complexity in order to be implemented for a gradient-based real-time optimization scheme. The base model used for this case study is presented in Sect. 10.2 and consists of two inputs $\boldsymbol{u}(t)$, two outputs $\boldsymbol{y}(t)$, and five system states $\boldsymbol{x}(t)$. The model is defined by an ODE system.

T. Albin Rajasingham, *Nonlinear Model Predictive Control of Combustion Engines*, Advances in Industrial Control,
https://doi.org/10.1007/978-3-030-68010-7_11

Within the NMPC algorithm, a discretized model is applied. The continuous-time model is discretized using an integration scheme. For comparison, various integration schemes are investigated, i.e. the Runge–Kutta 4 scheme and the implicit Euler method. The continuous-time dead-time T_D is discretized using the sampling time T_s to the integer value $k_D = \left\lfloor \frac{T_D}{T_s} \right\rfloor$. The general discretized model has the following form, where k is the discrete sampling instant and $\boldsymbol{f}_{dis}$ represents the discretized dynamics:

$$\begin{aligned} \boldsymbol{x}(k+1) &= \boldsymbol{f}_{dis}(\boldsymbol{x}(k), \boldsymbol{u}(k)) \\ \boldsymbol{y}(k) &= \boldsymbol{g}(\boldsymbol{x}(k-k_D)) \end{aligned} \tag{11.1}$$

One of the goals of the closed-loop controller is to achieve an offset-free tracking even in the presence of disturbances and a model–plant mismatch. More precisely, the goal is to track the reference of both outputs, namely the intake manifold pressure p_{im} and the burnt gas ratio x_{bg}, without offset for references and disturbances that are asymptotically constant. Since the plant is stable, observable and the noise is negligible, an approach using a deadbeat observer is applied as discussed in Sect. 6.2.2.

For this purpose, a model for disturbance estimation is needed. A pure output disturbance model is used. The nominal outputs are thus augmented by additional disturbance states $\boldsymbol{d}(k) \in \mathbb{R}^2$. The following model for disturbance estimation results:

$$\begin{aligned} \boldsymbol{x}(k+1) &= \boldsymbol{f}_{dis}(\boldsymbol{x}(k), \boldsymbol{u}(k)) \\ \boldsymbol{y}_{aug}(k) &= \boldsymbol{g}(\boldsymbol{x}(k-k_D)) + \boldsymbol{d}(k) \end{aligned} \tag{11.2}$$

The states $\boldsymbol{x}(k)$ are calculated by an open-loop estimation. A deadbeat observer is used to estimate the disturbances $\hat{\boldsymbol{d}}(k+1)$ and thus to achieve an offset-free control algorithm. The disturbance estimates $\hat{\boldsymbol{d}}(k+1)$ are calculated with the recent measurement of the outputs $\boldsymbol{y}_m(k)$ by

$$\hat{\boldsymbol{d}}(k+1) = \boldsymbol{y}_m(k) - \boldsymbol{g}(\boldsymbol{x}(k-k_D)) \tag{11.3}$$

The disturbance states are assumed to be constant over the prediction horizon.

$$\hat{\boldsymbol{d}}(\cdot|k) = \hat{\boldsymbol{d}}(k+1) \tag{11.4}$$

11.1.2 Formulation of the Optimization Problem

The task of the controller is to drive the values $\boldsymbol{y}(k)$, i.e. the intake manifold pressure and the burnt gas ratio to its reference values $\boldsymbol{r}(k)$. The reference values $\boldsymbol{r}(k)$ are assumed to be constant over the prediction horizon. For this purpose, a reference tracking problem formulation is used. The cost function uses quadratic costs on the reference tracking term and on the change of actuated values.

$$J = \sum_{i=1}^{N} \left|\left| g(\boldsymbol{x}(k+i|k)) + \hat{\boldsymbol{d}}(k+1) - \boldsymbol{r}(k) \right|\right|^2_{\boldsymbol{Q}_i} + \sum_{i=0}^{N-1} ||\boldsymbol{u}(k+i|k) - \boldsymbol{u}(k+i-1|k)||^2_{\boldsymbol{R}_i} \tag{11.5}$$

For the controller-internal model that is used for optimization, the dead-time can be neglected as the dead-time is present solely on the outputs and the reference in each time step is assumed to be constant over the prediction horizon. In the given case, the dead-time only has to be considered for disturbance estimation (11.3). The discrete-time model enters the optimization problem as given for multiple shooting. The optimization problem also considers the initial state, the last actuated value applied, and the box constraints on the actuated values. The resulting NLP looks as follows:

$$\begin{aligned} \min_{\boldsymbol{u}(\cdot|k),\, \boldsymbol{x}(\cdot|k)} \quad & J \\ \text{s.t.} \quad & \boldsymbol{x}(k+i+1|k) = \boldsymbol{f}_{dis}\big(\boldsymbol{x}(k+i|k), \boldsymbol{u}(k+i|k)\big), \quad i = 0, \dots, N-1, \\ & \boldsymbol{x}(k|k) = \boldsymbol{x}(k), \\ & \boldsymbol{u}(k-1|k) = \boldsymbol{u}(k-1), \\ & \boldsymbol{0} \leq \boldsymbol{u}(k+i|k) \leq \boldsymbol{1}, \quad i = 0, \dots, N-1 \end{aligned} \tag{11.6}$$

11.1.3 Parametrization of the Numerical Solver

In this case study, an SQP algorithm is used to solve the resulting NLP given by (11.6). The design of the numerical solver significantly influences various important performance parameters. There is especially one trade-off that dominates the choice of parameters: the trade-off between the computational effort, measured in CPU time, and the control performance, measured for example in the sum of tracking error and the change of actuated values. This trade-off concerns most of the parameters discussed. Any gain in one of the metrics is "paid for" by a decrease in the other.

Important choices that affect these performance parameters are the calculation of the Hessian matrix and the numerical integration scheme. Either the exact Hessian can be used or an approximation can be applied instead. As a reference tracking problem is used within the NMPC algorithm, the Gauss–Newton scheme is investigated as an approximation. The system exhibits stiff system dynamics, i.e. fast as well as slow dynamics are present. Thus, the use of implicit integration schemes for numerical simulation becomes relevant. To evaluate the influence of these choices, the following three setups are compared.

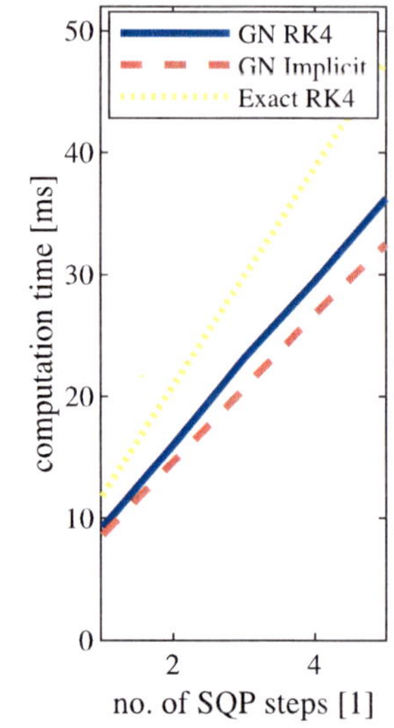

Fig. 11.1 Comparison of the three numerical solvers settings in terms of computation time

- Exact Hessian calculation with five Runge–Kutta 4 integration steps per sample time
- Gauss–Newton Hessian calculation with five Runge–Kutta 4 integration steps per sample time
- Gauss–Newton Hessian calculation with one implicit Euler integration step

To study the influence of these three setups, various simulations are carried out. All other parameters, such as sampling time T_s and number of prediction steps N, are set equal. Figure 11.1 shows a comparison of the computation times of the three setups in dependence of the number of SQP steps. The average computation time of one NMPC step is depicted. The computation time grows linearly with the number of SQP steps. Compared to the exact Hessian, the calculation of the Hessian matrix with the Gauss–Newton approach considerably reduces the computation time. The replacement of five Runge–Kutta 4 steps by a single implicit Euler integration scheme further reduces the computational effort.

In addition to the computation time, the control quality is the essential performance parameter. Closed-loop control simulations are carried out for comparing the three setups. A step in the intake manifold pressure as well as the burnt gas ratio at $t = 1$ s is used as a reference. Figure 11.2 shows the closed-loop control result for the three design choices, each with one SQP step. All choices result in a similar tracking behavior. In theory, the exact Hessian has a better convergence rate than the Gauss–Newton calculation. Still, there is no visible difference. Besides the reduced computation time, the Gauss–Newton Hessian calculation has the advantage that the approximated Hessian matrix calculated is always positive semidefinite.

The number of SQP steps performed in each sampling step to solve the resulting NLP has a direct effect on how close the actuated value applied is to that of the fully

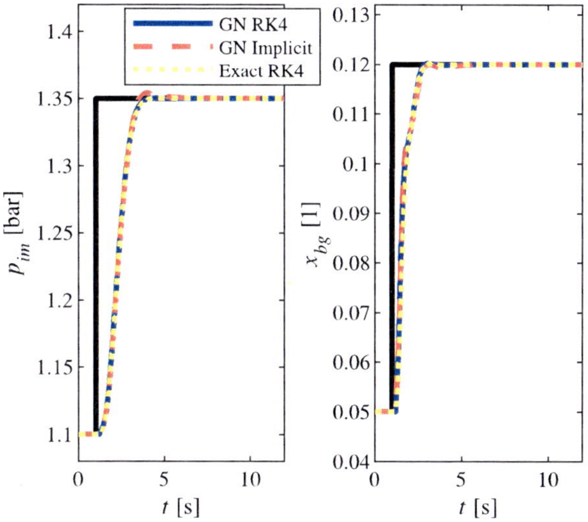

Fig. 11.2 Comparison of the three numerical solvers settings in terms of control performance

converged solution. Possible choices are to only perform one step, as proposed in the real-time iteration (RTI) scheme, perform multiple SQP steps, or even let the algorithm run until it converges. Figure 11.3 shows a comparison between using 50 SQP steps where a converged solution is gained in most of the sampling steps and using one SQP step. The solution found with one SQP step is almost as good as the one that is nearly converged. The reason lies in the combination of the receding-horizon principle, multiple shooting discretization, and the SQP method. This combination allows open-loop predictions to be initialized with the solution of the preceding step. Thus, each solution is continuously improved in every time step. After a few time steps, the initialization is already close to the optimal solution.

For a further analysis of the air path control problem, the following choices are made. As the difference in the closed-loop control performance is small, the RTI scheme is used, which consists of a single SQP step $N_{sqp} = 1$. As all three setups yield comparable results for the tracking performance, the fastest NLP formulation is used for the further analysis, which is the Gauss–Newton approximation for the Hessian matrix with an implicit Euler integration scheme.

11.1.4 Parametrization of the Cost Function

In the following, the design parameters of the cost function are analyzed. First, the influence of the number of prediction steps N is investigated. Figure 11.4 shows its influence on the NMPC tracking cost and on the computation time. All simulations use a fixed shooting interval of $T_{sh} = 0.1$ s. Hence, the increase in the number of

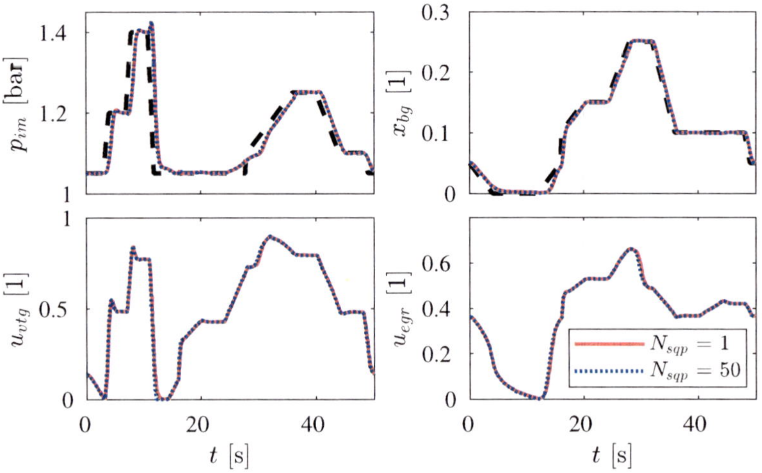

Fig. 11.3 Effect of the number of SQP iterations on the closed-loop solution

prediction steps also leads to an increase in the time horizon that is predicted, i.e. the prediction horizon H. The prediction horizon H is determined by the number of prediction steps N and the length of the shooting interval T_{sh} by $H = T_{sh}N$. Figure 11.4 shows that the tracking cost significantly decreases with an increasing number of prediction steps. However, there is a threshold where any further increase of prediction steps has no effect. Figure 11.4 shows that the computation time increases exponentially with the number of prediction steps. With an increasing number of prediction steps, the number of optimization variables increases too, which is the reason for the exponential increase in computation time.

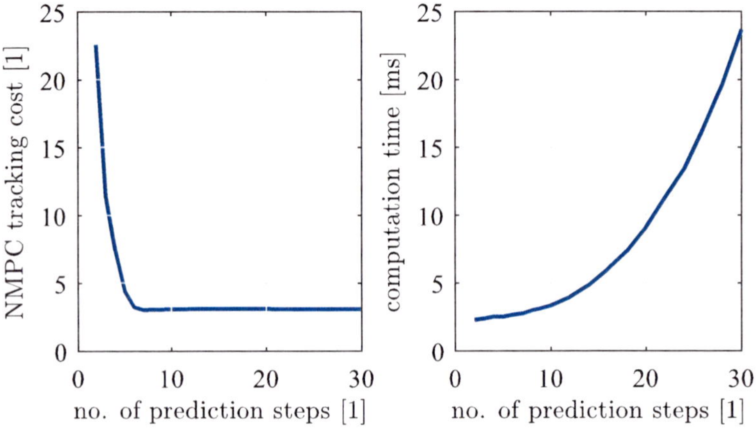

Fig. 11.4 Effect of number of prediction steps on tracking cost and computation time

Another design choice within NMPC is the length of the shooting interval T_{sh}. Typically, it is chosen to be equal to the sampling time T_s. However, it can also be chosen to be bigger than the sampling time T_s. The length of the shooting interval T_{sh} itself almost has no effect on the computation time, in contrast to the number of prediction steps N which directly correlates to the number of optimization variables. Thus, increasing the shooting interval T_{sh} allows for a long prediction horizon H while having a reduced number of optimization variables. However, the shooting interval T_{sh} affects the tracking cost. Figure 11.5 depicts the influence of the prediction horizon H and the number of prediction steps N on the tracking cost. For an appropriate control performance, the prediction horizon H has to be higher than a certain threshold, regardless of the number of prediction steps N. The number of prediction steps N has to be higher than a certain threshold as well. Any further increase in one of these two values does not improve the performance of the controller.

For determining suitable parameters, first, the maximum number of prediction steps N can be chosen such that an acceptable computation time results. In the given setup, an acceptable computation time is 10 ms, which allows the use of $N = 20$. The prediction horizon H can be chosen according to Fig. 11.5 for low tracking cost. A suitable shooting interval is $T_s = 0.1$ s, which is equivalent to the sampling time T_s. The prediction horizon results to be $H = 2$ s.

Once the various design choices are made, the tuning parameters such as Q and R can be determined. The tuning of the parameters can be conducted within a simulative environment. It is important to have a different model than the one used within the controller. For instance, a more complex system model can be used as a plant for the closed-loop simulation.

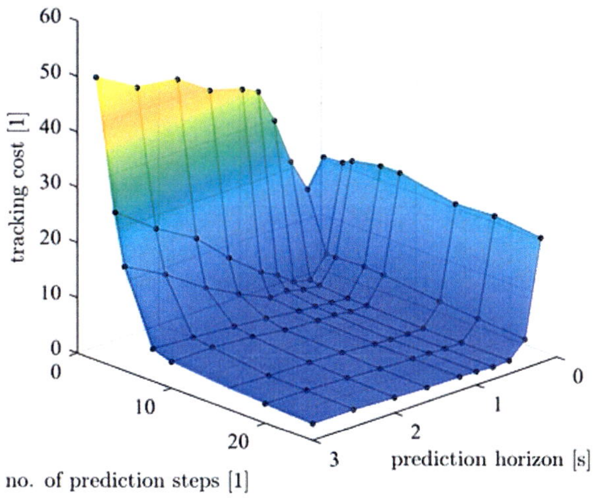

Fig. 11.5 Tracking cost in dependence of the prediction horizon and the number of prediction steps

11.2 Model-Based Synthesis of PI Controllers

PID-based controllers are state of the art for series-production engine control algorithms. The fundamentals for the design of the PID-based controllers for typical engine characteristics are discussed in Chap. 2. In the following, those methods are applied for the EGR VTG control problem, and their performance is analyzed subsequently.

11.2.1 Decentralized Synthesis of PI Controllers

The EGR VTG control problem is characterized by multiple-input multiple-output (MIMO) system dynamics. Thus, decentralized controllers are designed. The pairing of inputs to outputs is based on the RGA matrix analysis described in Sect. 10.4.3. In this case study, two single-input single-output (SISO) PI controllers are used. The controllers C_1 and C_2 are designed, where the first one controls the plant from u_{vtg} to p_{im}, while the second one controls from u_{egr} to x_{bg} according to the RGA matrix. This corresponds to the intuitive choice.

First, the system is treated as a decoupled system to design the controllers, thus ignoring the cross-couplings arising. The controllers C_1 and C_2 both are designed to achieve a certain bandwidth. For this reason, the crossover frequencies are set to $0.8\,\frac{\text{rad}}{\text{s}}$ for controller C_1 and $0.5\,\frac{\text{rad}}{\text{s}}$ for controller C_2. The sampling time is chosen to be $T_s = 0.05$ s. It is chosen to be smaller than the one used for NMPC to account for the reduced computation time compared to the NMPC controller. A summary of all parameters is shown in Table 11.1. For the design of the controllers, a linearized model is used. The linearization point is chosen to be the steady state $\boldsymbol{x}_{ss}(\boldsymbol{u}_{lin})$ for the input $\boldsymbol{u}_{lin} = [0.5\;\; 0.5]^\mathsf{T}$.

While for the design, the plant is assumed to consist of two perfectly decoupled linear SISO plants, in the real application, the plant is a coupled, nonlinear system. As the PI controllers are designed on the simplified plant, no guarantee can be given concerning their performance on the nonlinear plant or even on the full linear plant, i.e. including the cross-couplings. To evaluate the difference in control quality, simulations are conducted. Figure 11.6 shows a comparison of the closed-loop control results gained in simulation. The controllers are tested against the linear decoupled system (simplified), against the linear system including cross-couplings (linear), and against the nonlinear system model (nonlinear). While the controllers perform very

Table 11.1 Controller specification for the SISO controllers

Crossover frequency of C_1	0.8 rad/s
Crossover frequency of C_2	0.5 rad/s
Sampling time T_s	0.05 s

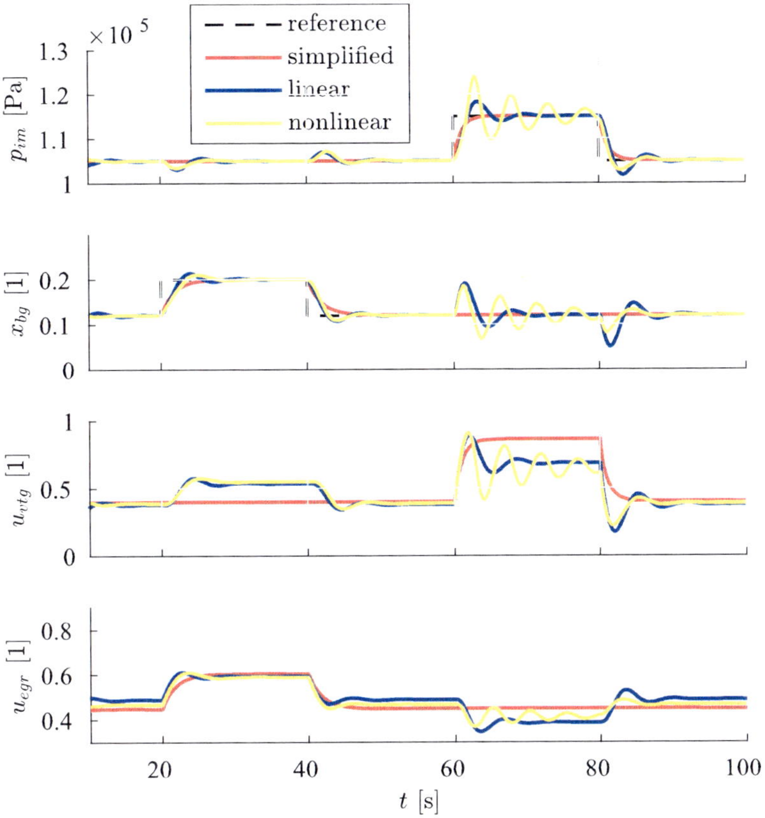

Fig. 11.6 Simulation of the decentralized SISO controllers on the decoupled linear plant, linear plant with cross-couplings, and the nonlinear plant

well on the simplified, decoupled plant, the controllers perform already significantly worse on the full linear plant due to the cross-couplings. As expected, the worst performance is observed in the simulation with the nonlinear process model. In the case of a nonlinear plant model, the control quality is unacceptable, as very high oscillations result. To take the cross-couplings into account and thus further improve the control result, a decoupling matrix can be added.

11.2.2 Decoupling Control Approach

To improve the control behavior of the SISO controllers for the air path system, a decoupling matrix can be added, as discussed in Sect. 2.1.3. With the goal to recover the diagonal entries of the original plant, the decoupling matrix can be calculated by

$$\begin{aligned} D_{11} &= \frac{G_{22}}{G_{11}G_{22} - G_{12}G_{21}} G_{11}, \quad D_{12} = -\frac{G_{12}}{G_{11}}, \\ D_{22} &= \frac{G_{11}}{G_{11}G_{22} - G_{12}G_{21}} G_{22}, \quad D_{21} = -\frac{G_{21}}{G_{22}}. \end{aligned} \tag{11.7}$$

Due to the system dynamics present, i.e. non-minimum phase zeros, exact decoupling leads to rather complex decoupling terms. A simple, approximate decoupling approach is used instead. One feasible approach is the use of static gains for decoupling. They result to be

$$\begin{aligned} D_{11}(s=0) &= 0.66 \qquad D_{12}(s=0) = 1.52 \\ D_{21}(s=0) &= -0.33 \qquad D_{22}(s=0) = 0.66 \end{aligned} \tag{11.8}$$

To evaluate the improvement in performance due to the decoupling matrix, simulations are conducted. The same decentralized SISO controllers C_1 and C_2 as designed in Sect. 11.2.1 are used. The simulations are conducted with the nonlinear plant. Figure 11.7 shows the results of the simulation with and without the decoupling matrix.

Even though only static decoupling is implemented, the decoupling matrix significantly improves the performance of the decentralized PI controllers. The overshoots can be drastically reduced, while the effects of the cross-couplings are considerably attenuated as well.

11.2.3 Anti-windup and Dead-Time

There exist constraints on the actuated values which have to be taken into account for control design. Both actuated variables are subject to box constraints which stem from physical constraints.

$$0 \leq u_{vtg} \leq 1 \tag{11.9}$$

$$0 \leq u_{egr} \leq 1 \tag{11.10}$$

Active constraints on actuated values may cause controllers with integrative behavior to windup. To prevent integrator windup, controllers with integrative behavior have to be enhanced with anti-windup measures. In this case, the back-calculation scheme, which is introduced in Sect. 2.2.1 is implemented.

In addition to constraints, the real air path system is also subject to dead-times. The closed-loop control simulations with the PI-based controllers in this subsection neglected this dead-time in the plants and the controller design so far. It is generally hard to deal with dead-times in MIMO systems with decentralized controllers. However, the handling is simplified in this specific case as the dead-times only arise on the outputs. The same controllers C_1 and C_2 can be used; however, to compensate for

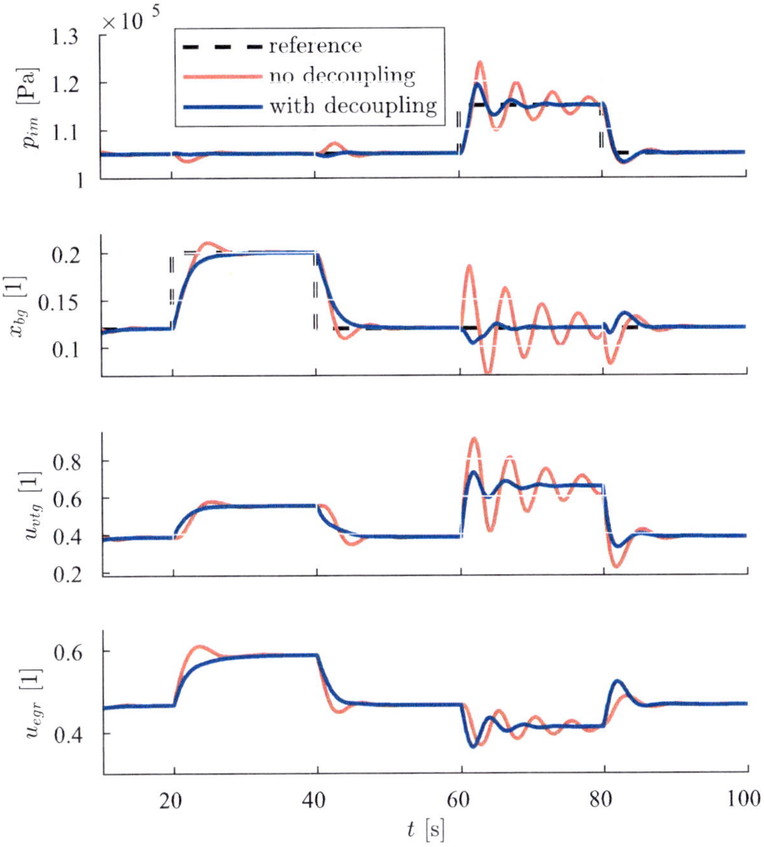

Fig. 11.7 Performance of the decentralized SISO controllers with and without decoupling with nonlinear plant

the dead-times, an additional Smith predictor is used. In this case, a Smith predictor can be applied for each output to compensate for the effect of each dead-time [1, 2].

11.3 Simulative Comparison of the Controllers

In the following, the closed-loop control results obtained by simulations of the designed PI and NMPC controllers are compared. For this comparison, an LMPC is evaluated as well. The LMPC is characterized by the use of a constant linear process model, given by

$$\begin{aligned} \boldsymbol{x}(k+1|k) &= \boldsymbol{A}\boldsymbol{x}(k|k) + \boldsymbol{B}\boldsymbol{u}(k|k) \\ \boldsymbol{y}(k|k) &= \boldsymbol{C}\boldsymbol{x}(k|k) \end{aligned} \tag{11.11}$$

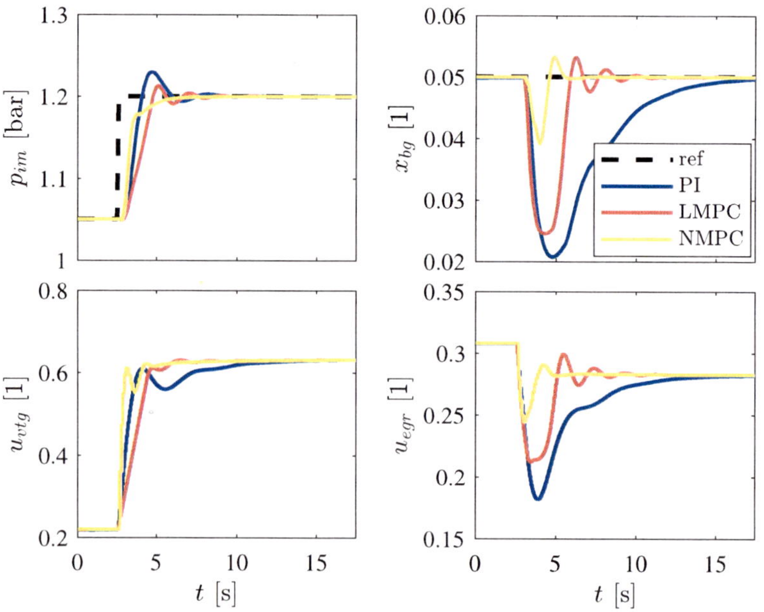

Fig. 11.8 Comparison of closed-loop control results for LMPC, NMPC, and PI for a step in the reference signal of the intake manifold pressure against a high-fidelity nonlinear model

The linear state-space matrices are derived by a linearization of the nonlinear model and a subsequent discretization. The same linearization point as the one used for the PI design is chosen. It is the steady-state value of $\boldsymbol{x}_{ss}(\boldsymbol{u}_{lin})$ for the input $\boldsymbol{u}_{lin} = [0.5\ \ 0.5]^\mathsf{T}$. The resulting state-space model is augmented by states that represent the dead-time. The dead-time is mapped into poles in $z = 0$. For the LMPC algorithm, a sampling time of $T_s = 0.05$ s is used to account for the reduced computation time compared to the NMPC controller. The value for the number of prediction steps is increased, i.e. $N = 50$.

In all simulations, the same plant model is used. It is a nonlinear model that is more complex, i.e. it contains more system states than the one presented in Chap. 10. Thus, it is also different from the controller-internal model of the NMPC controller.

Figure 11.8 shows the control results for a step input for the intake manifold pressure reference and a constant reference value for the burnt gas ratio. All control algorithms are able to track the references in an offset-free manner. However, the transient behavior differs. As expected, the NMPC controller performs best. It takes into account the main nonlinearities and allows for a suitable reference tracking. The LMPC performs slightly worse. It shows oscillatory behavior on the intake manifold, and the burnt gas ratio drops during the step input. The response time of the intake manifold pressure is significantly slower when using the PI control algorithm. Additionally, a considerable drop in the burnt gas ratio is visible. These results are according to the expectations. The control performance corresponds to the available model knowledge of the various controllers. While LMPC already allows to systematically take into account cross-couplings, dead-times, and constraints, its

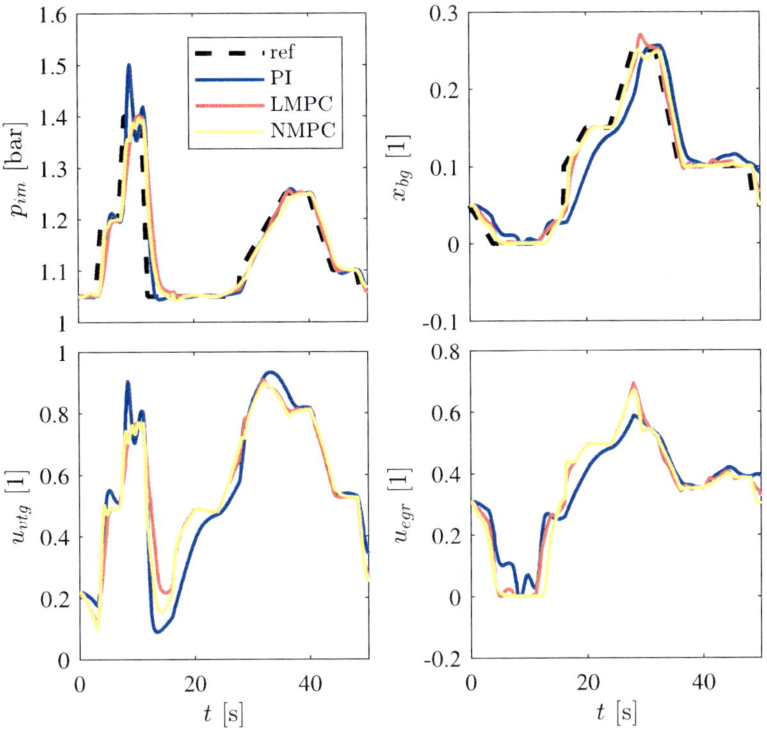

Fig. 11.9 Comparison of closed-loop control results for LMPC, NMPC, and PID against a high-fidelity nonlinear model

performance is especially appropriate close to the region of the state-space at which the model is linearized. The NMPC does not contain this restriction and offers a high performance in the entire state-space.

Figure 11.9 shows the closed-loop control result obtained in simulations of the three controllers for a more complex reference trajectory. The NMPC is able to reference-track the profiles for the intake manifold pressure and the burnt gas ratio with a very high control quality. The LMPC is able to reference-track the profiles as well. However, it is slower in the transients than the NMPC. The PI control algorithm is able to appropriately track the burnt gas ratio, but the tracking of the intake manifold pressure is rather slow.

11.4 Experimental Validation

In the simulations, the NMPC controller outperforms the LMPC and the PI controller for the air path control problem. The simulative environment makes compar-

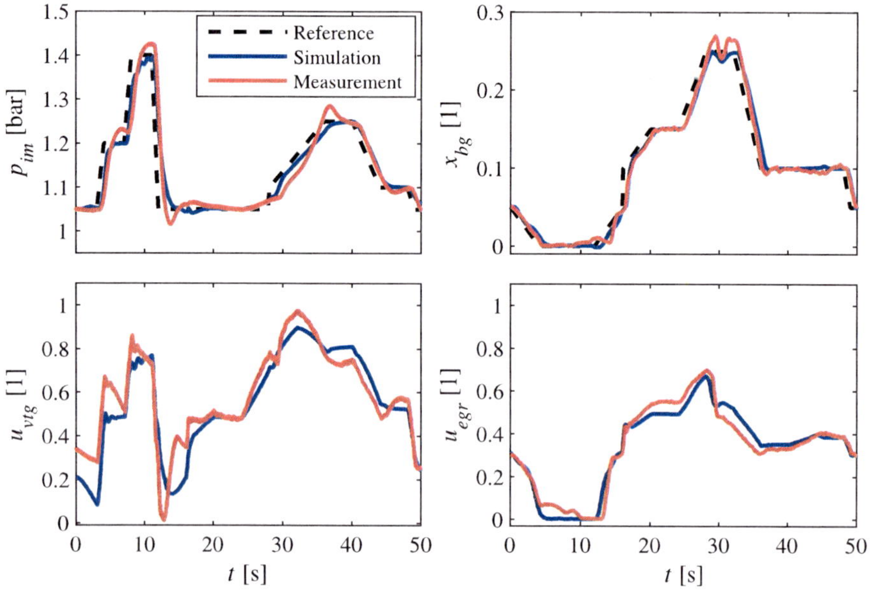

Fig. 11.10 Experimental validation of the NMPC controller, along with simulations of the closed-loop control behavior

isons possible, as the controllers can be tested with identical conditions. However, to validate the suitability of the NMPC algorithm for practical applications, an experimental validation is mandatory. The design of the NMPC controller still depends on the assumption that the nonlinear process model is a good approximation of the real engine behavior. Additionally, further requirements such as real-time feasibility have to be achieved.

Figure 11.10 shows the closed-loop control results of the NMPC algorithm at the engine test bench. They show that the NMPC algorithm is able to track the reference on a real engine with a high control quality. All requirements on the control algorithm, such as offset-free control and real-time feasibility, can be fulfilled. The trajectory of the simulated closed-loop control result is depicted together with the measurement data from the engine test bench. For both the simulation and the actual measurements, the same reference trajectory is used. The two trajectories match very well, with only slight offsets. This validates the suitability of the simulation.

References

1. N. Abe, K. Yamanaka, Smith predictor control and internal model control – a tutorial, in *SICE Conference* (2003), pp. 1383–1387
2. Q.-G. Wang, *Decoupling Control* (Springer, Berlin, 2002)

Chapter 12
Two-Stage Turbocharging: Modeling and Analysis

Abstract A second case study for a control system of the air path is examined. While the first one investigated the combined control of burnt gas ratio and VTG turbocharger for a CI engine, now, the control of the intake pressure for an SI engine is investigated. A setup is examined with two turbocharging units which are placed in series. This setup allows for a fast transient increase in load and at the same time for high specific power. In this chapter, the modeling of the two-stage turbocharging system is inspected. A suitable state-space model is derived which is computationally efficient and is still able to reproduce the system behavior with reasonable accuracy. Additionally, the system dynamics are analyzed, which becomes especially important for the design of suitable formulations of the optimization problem. This concerns for example the influence of the actuation on the pumping losses.

12.1 Introduction

Conventional turbocharger architectures consist of a single turbine and a single compressor which are connected by a common shaft. These single-stage turbocharging concepts exhibit inherent limitations on the simultaneous improvement of transient load performance and increase of the specific power which is advantageous to reduce fuel consumption. The sizing of the charging unit directly affects this trade-off. Large devices enhance the specific power output and reduce fuel consumption, especially at high engine speeds. However, they impair the transient load performance. Small devices, on the other side, provide a fast torque response and therefore, improve the driveability of the vehicle, especially at low engine speeds but cannot offer high specific power. To attain a combination of the best aspects of large and small devices, increasingly complex air path architectures are investigated which cope with the severe requirements. For SI engines, the two-stage turbocharging is one of the concepts analyzed. In this case study, the control system of a two-stage turbocharging unit with a nonlinear model predictive control (NMPC) algorithm is investigated. The chapter summarizes the results of earlier publications by the author [1–4].

The two-stage turbocharging concept typically consists of a small high-pressure turbocharger and a larger low-pressure turbocharger combined in series. Compared to

T. Albin Rajasingham, *Nonlinear Model Predictive Control of Combustion Engines*, Advances in Industrial Control,
https://doi.org/10.1007/978-3-030-68010-7_12

a single-stage concept, an improvement of the transient behavior is realized due to the small high-pressure stage, while the large low-pressure stage enables a higher power output. In order to fully exploit the system capabilities given by the hardware, the closed-loop control system has to manage the coordinated timing of the engagement and disengagement of the stages. The control task is to track the reference value for the charging pressure p_{char} while rejecting any disturbances acting on the system.

The requirements on the closed-loop control behavior result from various aspects. Due to the quantitative load control of a SI engine, a fixed air-to-fuel ratio is maintained for the combustion. As a consequence, there is a direct correlation between the charging pressure, the amount of air inducted in the cylinder, the fuel amount injected, and the torque generated. The reference value for the charging pressure should thus be reached as quickly as possible, as this determines the transient acceleration capability of the vehicle. However, in case of a step on the reference value, the output should additionally be achieved without any strong overshoots as they negatively influence the driving behavior. For CI engines, this requirement is not as pronounced. The working principle of a CI engine does not require a fixed air-to-fuel ratio. Instead, as an excess of air is present all the time, the torque is mainly based on the fuel amount injected. Therefore, in a CI engine, the torque and the charging pressure are decoupled such that oscillations can be tolerated up to a certain amount. Additionally, the control algorithm for the two-stage turbocharging system must respect the upper limit constraints for the high-pressure and low-pressure turbocharger speeds, as exceeding these limits might damage the turbocharger. This becomes especially challenging as the exhaust gas temperatures from an SI engine are very high, such that the limits can be reached relatively quickly and the turbocharger speed typically is not measured in a series-production configuration.

In summary, three substantial requirements have to be fulfilled at the same time, which are

- quick reference tracking,
- no oscillatory behavior for a step reference, and
- considering limits on turbocharger speeds.

These demands have to be fulfilled while the system to be controlled is very nonlinear, the system dynamic behavior is very dependent on disturbance signals such as the engine speed, and a considerable dead-time is present.

Various control concepts have been investigated for the control of turbochargers. Due to the complex system dynamics, especially model-based control techniques are applied. An exemplary approach is presented in [12, 13], where flatness-based control is used based on a model derived in [21]. The work published in [18, 19] use advanced internal model control (IMC) algorithms for the control task. Among the various model-based control techniques, especially the MPC algorithm shows advantages, due to the fact that it can consider constraints on actuated signals and on system states, for instance, constraints on the turbocharger speeds must be considered. Various MPC techniques have been investigated, and these concern online optimization techniques [6, 7, 11, 17, 20] as well as explicit MPC formulations such as [8, 9].

For more complex turbocharging architectures, such as two-stage turbocharging, one investigated strategy is to ascribe the system to the well-known one-stage turbocharging problem which is approached as a single-input single-output (SISO) system. This is achieved by introducing a logic that switches between feedback controlling either the LP or the HP stage depending on the current operating point [15, 16]. For low exhaust gas enthalpy (e.g. at low engine speeds), the small turbocharger is feedback controlled while for high exhaust gas enthalpy (e.g. at high engine speeds), the LP stage is feedback controlled. The NMPC offers the advantage that the entire multiple-input multiple-output (MIMO) control problem including all arising requirements can be tackled in a systematic manner. For this reason, the NMPC framework is a suitable choice for the control algorithm. The following case study shows that the nonlinear system dynamics can be handled with a high control quality and that the NMPC approach is able to respect the constraints on the system states.

12.2 System Setup

Figure 12.1 shows the schematic setup of the air path architecture investigated. Each turbocharging stage consists of a compressor and a turbine that are connected by a common shaft. For an experimental analysis of the system, the architecture depicted was built up and implemented in a demonstrator vehicle with a 1.8 L four-cylinder SI engine with gasoline fuel. The control algorithm was implemented in the vehicle on rapid control prototyping hardware. The experiments conducted in this case study were all recorded with this demonstrator vehicle. A more detailed overview on the system is provided in [5, 10].

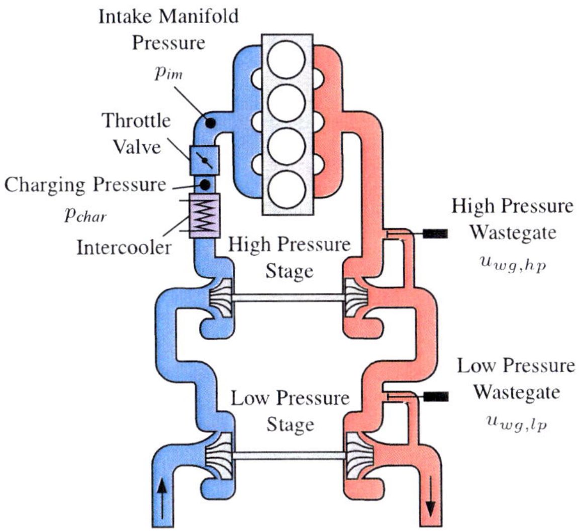

Fig. 12.1 Two-stage turbocharging setup

12.2.1 Sensors

The controlled variable of the system is the charging pressure p_{char}, which is measured with a pressure sensor positioned between the intercooler and the throttle valve. The application of sensors and advanced actuators in the exhaust gas path is unfavorable for a SI engine series configuration. Significant drawbacks in terms of durability and cost result due to the high exhaust gas temperatures in SI engines. Thus, the control approach presented does not include any sensor signal from the exhaust gas path. Besides the controlled variables, several measured variables are employed, i.e. the engine speed ω_{eng}, the ambient pressure p_{amb}, and the ambient temperature ϑ_{amb}. All three values are available in a series-production vehicle. For the purpose of modeling and validation of the control algorithm, additional sensors are installed in the car. However, they are not used in the controller. Especially the turbocharger speeds, the pressure between the compressors and in the exhaust manifold which yield the pressure ratio over the two compressors, are measured in the demonstrator vehicle for modeling and validation purposes, but they are not used as inputs for the control algorithm.

12.2.2 Actuators

As actuators for the control system, a wastegate on the high-pressure $u_{wg,hp}$ and on the low-pressure stage $u_{wg,lp}$ is used. Commonly, electronic wastegates are applied for turbocharging. They have the advantage of containing a position feedback sensor, which allows for an accurate setting of the valve opening area. In this case study, the use of simpler and cheaper pneumatic actuators is investigated. For instance, they do not have any additional sensors for position feedback, which makes the development of the controller more demanding. The wastegate actuation signals are pulse-width-modulated (PWM) signals with an operating range of $u_{wg,hp} = [0 \ldots 100]\%$ and $u_{wg,lp} = [0 \ldots 100]\%$. They allow manipulation of the pilot pressure, influencing the effective opening area of the wastegate. Thus, the amount of exhaust gas that passes the turbine can be adjusted. A value of $u_{wg} = 0\%$ opens the wastegate, while a value of $u_{wg} = 100\%$ closes the wastegate as much as possible. When the wastegate is fully open, the majority of the exhaust gas bypasses the turbine, whereas a fully closed wastegate allows all the exhaust gas to flow through the turbine. The architecture considered also contains a throttle valve which is controlled via a separate controller that is not part of the turbocharger controller. At very high mass flows, the high-pressure stage no longer delivers an increase in the charging pressure. Still, no additional high-pressure bypass was used, thus allowing efficiency losses at high mass flows.

12.2.3 Engine Control Algorithm

The control concept investigated focuses on the closed-loop control of the charging pressure p_{char} by actuating the high-pressure and low-pressure wastegate PWM signals $u_{wg,hp}$ and $u_{wg,lp}$. The engine speed ω_{eng}, the ambient pressure p_{amb}, and the ambient temperature ϑ_{amb} are used as measured disturbances. In today's series applications, the throttle valve is only used in the non-turbocharged region. In the turbocharged region, the throttle valve is set completely open, for reasons of fuel efficiency. As a consequence, the charging pressure p_{char} is equal to the intake manifold pressure p_{im} in the turbocharged region and is thus directly correlated to the torque of the engine. The throttle valve is not investigated further for the turbocharger control concept. The setpoint for the charging pressure is determined based on the requested engine torque, as commonly done in torque-oriented engine control structures; see Fig. 7.7. All other parameters of the engine control structure, such as ignition, injection, and camshaft position, are equal to the standard calibration. The air path NMPC control algorithm and all other engine control functions such as the setpoint calculation and ignition timing are implemented on one single dSPACE MicroAutoBox II system.

12.3 Nonlinear Process Model for Two-Stage Turbocharging

In the following, the derivation of the nonlinear state-space model is introduced which is used as a basis for the controller-internal model of the NMPC algorithm. The resulting model is compared to measurement data obtained from experiments conducted on the vehicle.

12.3.1 Fundamental Equations of Two-Stage Turbocharging

For determining the turbocharger angular speed on the high-pressure $\omega_{tc,hp}$ and on the low-pressure stage $\omega_{tc,lp}$, Newton's second law is used. The equation of motion is described in (12.1) and (12.2), where the powers generated by the turbines $P_{t,hp}$ and $P_{t,lp}$ are considered as well as the powers consumed by the compressors $P_{c,hp}$ and $P_{c,lp}$. The mass moment of inertia of the turbochargers are given by J_{tc}. The values for $J_{tc,hp}$ and $J_{tc,lp}$ are determined via system identification.

$$\dot{\omega}_{tc,hp}(t) = \frac{1}{\omega_{tc,hp}(t)\,J_{tc,hp}}\Big(P_{t,hp}(t) - P_{c,hp}(t)\Big) \tag{12.1}$$

$$\dot{\omega}_{tc,lp}(t) = \frac{1}{\omega_{tc,lp}(t)\,J_{tc,lp}}\Big(P_{t,lp}(t) - P_{c,lp}(t)\Big) \tag{12.2}$$

The expressions (12.3) and (12.4) calculate the consumed power on the two stages of the compressor. Similarly, the relationship for the generated power of both turbine stages is expressed in (12.5) and (12.6).

$$P_{c,hp} = \dot{m}_{c,hp}(t) c_{p,air} \vartheta_{uc,hp} \frac{1}{\eta_{c,hp}} \left(\Pi_{c,hp}^{\frac{\kappa_{air}-1}{\kappa_{air}}} - 1 \right) \tag{12.3}$$

$$P_{c,lp} = \dot{m}_{c,lp}(t) c_{p,air} \vartheta_{uc,lp} \frac{1}{\eta_{c,lp}} \left(\Pi_{c,lp}^{\frac{\kappa_{air}-1}{\kappa_{air}}} - 1 \right) \tag{12.4}$$

$$P_{t,hp} = \dot{m}_{t,hp}(t) c_{p,exh} \vartheta_{ut,hp} \eta_{t,hp} \left(1 - \Pi_{t,hp}^{\frac{1-\kappa_{exh}}{\kappa_{exh}}} \right) \tag{12.5}$$

$$P_{t,lp} = \dot{m}_{t,lp}(t) c_{p,exh} \vartheta_{ut,lp} \eta_{t,lp} \left(1 - \Pi_{t,lp}^{\frac{1-\kappa_{exh}}{\kappa_{exh}}} \right) \tag{12.6}$$

The specific heat capacity is given by c_p, i.e. $c_{p,air} = 1005\,\frac{\text{J}}{\text{kgK}}$ and $c_{p,exh} = 1117\,\frac{\text{J}}{\text{kgK}}$ are used. The heat capacity ratio is denoted by κ, i.e. $\kappa_{air} = 1.4$ and $\kappa_{exh} = 1.33$ are applied. The mass flows through the compressors and the turbines are denoted by $\dot{m}$. The efficiencies are denoted by η and are assumed to be constant. They are determined via system identification. Within the equations, the temperature upstream of the compressors and the turbines are denoted by ϑ. They represent a constant, medium value in their operating range. The pressure ratios relate the upstream and downstream conditions, and they are given by

$$\Pi_t = \frac{p_{ut}}{p_{dt}} \tag{12.7}$$

$$\Pi_c = \frac{p_{dc}}{p_{uc}} \tag{12.8}$$

The compressor is modeled by correlating the rotational kinetic energy of the turbocharger to the pressure ratio over the compressor via an affine mapping. For the low-pressure stage, an affine map with $\Pi_{c,lp}$ as input is used. Due to the large spread of exhaust enthalpy and operating points of the compressor, the high-pressure stage has to be modeled in dependence of $\Pi_{c,hp}$ and the angular speed of the engine ω_{eng}. The constants are identified by a linear regression with measurement data.

$$\omega_{tc,lp}^2 = a_{lp} \Pi_{c,lp} + b_{lp} \tag{12.9}$$

$$\omega_{tc,hp}^2 = \left(a_{hp}\omega_{eng} + c_{hp} \right) \Pi_{c,hp} + \left(b_{hp}\omega_{eng} + d_{hp} \right) \tag{12.10}$$

As described in [1, 14], the model can be simplified by a model order reduction technique called singular perturbation. The application of a singular perturbation confirms that the dynamics of the volumes can be neglected. As a consequence, the mass flows in all cross-section areas of the intake and the exhaust path are assumed to be equal, yielding the relations (12.11)–(12.12). In the exhaust path, the total mass flow is equal to the mass flow through the turbine plus the mass flow through the wastegate.

$$\dot{m}_{asp} = \dot{m}_{c,hp} = \dot{m}_{c,lp} \tag{12.11}$$

$$\dot{m}_{asp} + \dot{m}_{fuel} = \dot{m}_{t,hp} + \dot{m}_{wg,hp} = \dot{m}_{t,lp} + \dot{m}_{wg,lp} \tag{12.12}$$

A simplified model is used to calculate the aspirated mass flow of the engine. It uses a constant volumetric efficiency η_v, the charging pressure p_{char}, and the intake manifold temperature ϑ_{im} which is assumed to be constant and represents a medium value of its operating range, the displacement volume of all the cylinders combined V_{dis}, the specific gas constant R, the engine speed n_{eng} given in revolutions per second, and the number of revolutions per cycle, i.e. $N = 2$. The pressure upstream of the low-pressure compressor is set equal to the measured ambient pressure $p_{uc,lp} = p_{amb}$.

$$\begin{aligned} \dot{m}_{asp} &= \frac{\eta_v p_{char}(t) V_{dis} n_{eng}(t)}{R \vartheta_{im} N} \\ &= \frac{\eta_v p_{amb}(t) \Pi_{c,lp}(t) \Pi_{c,hp}(t) V_{dis} n_{eng}(t)}{R \vartheta_{im} N} \end{aligned} \tag{12.13}$$

The fuel mass flow can be calculated by taking into account the stoichiometric air-to-fuel ratio AFR_s, i.e. $AFR_s = 14.7$ can be used for gasoline combustion. For the direct-injection engine it results:

$$\dot{m}_{fuel} = \frac{\dot{m}_{asp}}{AFR_s} \tag{12.14}$$

The mass flow through the wastegates can be calculated by a compressible flow restriction model, where only the non-choked region is taken into account. The expressions designated by a star $\star$ hold for the high- and the low-pressure stage, i.e. $\star = \{\text{lp}, \text{hp}\}$.

$$\begin{aligned} \dot{m}_{wg,\star} &= c_d A_{wg,\star}(u_{wg,\star}) \frac{p_{dt,\star}}{\sqrt{R \vartheta_{ut,\star}}} \Pi_{t,\star} \psi_{t,\star}\left(\Pi_{t,\star}\right) \\ \psi_{t,\star}\left(\Pi_{t,\star}\right) &= \Pi_{t,\star}^{\frac{-1}{\kappa}} \sqrt{\frac{2\kappa}{\kappa - 1} \cdot f_{\text{lim}}\left(1 - \Pi_{t,\star}^{\frac{-\kappa+1}{\kappa}}\right)} \\ f_{\text{lim}}(x) &= \frac{1}{2}\sqrt{x^2 + \varepsilon} + \frac{1}{2}x \end{aligned} \tag{12.15}$$

The mass flow through the turbine is calculated by a slightly modified version of the compressible flow equation model. The modified version shows a better fit to the measurement data than the standard equations.

$$\dot{m}_{t,\star} = c_d A_{t,\star} \frac{p_{dt,\star}}{\sqrt{R\vartheta_{ut,\star}}} \Pi_{t,\star} \psi_{t,\star}\left(\Pi_{t,\star}\right)$$
$$\psi_{t,\star}\left(\Pi_{t,\star}\right) = \Pi_{t,\star}^{\frac{\kappa-2}{2\kappa}} \sqrt{\frac{2\kappa}{\kappa-1} \cdot f_{\lim}\left(1 - \Pi_{t,\star}^{\frac{-\kappa+1}{\kappa}}\right)} \tag{12.16}$$
$$f_{\lim}(x) = \frac{1}{2}\sqrt{x^2+\varepsilon} + \frac{1}{2}x$$

For the high-pressure and the low-pressure wastegate, different actuation principles are used. At the high-pressure stage, a system is used that actuates the wastegate by means of underpressure delivered by a vacuum pump. The opening area $A_{wg,hp}$ is approximated well by a linear correlation with the actuation signal.

$$A_{wg,hp} = \left(1 - \frac{u_{wg,hp}}{100}\right) A_{wg,hp,max} \tag{12.17}$$

For opening the low-pressure stage wastegate, a system is used that is driven by excess pressure, i.e. the charging pressure. The opening area of the wastegate is modeled in a black-box model fashion by the application of sigmoid functions.

$$A_{wg,lp} = f(p_{amb}\Pi_{c,lp}\Pi_{c,hp}, u_{wg,lp}) \tag{12.18}$$

For several reasons, the implementation of the turbocharger architecture in a vehicle causes a quite considerable dead-time. A major impact results from the working principle of the pneumatic actuation. A change in the actuation signal must first result in a pressure change inside the wastegate actuator before the wastegate position changes. Instead of modeling all these effects individually, they are summarized as one overall dead-time T_D which acts directly on both actuated values. The same dead-time can be used for both actuated values.

$$u_{wg,\star} = u_{wg,\star}(t - T_D) \tag{12.19}$$

12.3.2 Overall State-Space Model

The various equations can be combined to one overall state-space model. The resulting model consists of a set of differential–algebraic equations (DAE). For this purpose, let $\boldsymbol{x}(t) \in \mathbb{R}^{n_x}$ denote the differential states, $\dot{\boldsymbol{x}}(t)$ the differential state derivatives, $z(t) \in \mathbb{R}^{n_z}$ the algebraic variables, while $\boldsymbol{u}(t) \in \mathbb{R}^{n_u}$ represents the actuated values and $\dot{\boldsymbol{u}}(t)$ their derivatives. The system is governed by introducing the differential states $x_1 = \Pi_{c,lp}$, $x_2 = \Pi_{c,hp}$, and the algebraic states $z_1 = \Pi_{t,lp}$, $z_2 = \Pi_{t,hp}$. The output functions are defined by (12.20)–(12.22).

$$p_{char} = p_{amb}\Pi_{c,lp}\Pi_{c,hp} \tag{12.20}$$

$$\omega_{tc,lp} = \sqrt{\left(a_{lp}\Pi_{c,lp} + b_{lp}\right)} \tag{12.21}$$

$$\omega_{tc,hp} = \sqrt{\left(a_{hp}n_{eng} + c_{hp}\right)\Pi_{c,hp} + \left(b_{hp}n_{eng} + d_{hp}\right)} \tag{12.22}$$

The DAE model is of index one and consists of two differential and two algebraic states. The output y_1 corresponds to the charging pressure p_{char}, which forms the tracking objective. The outputs y_2 and y_3 correspond to the low-pressure and the high-pressure turbocharger speeds $\omega_{tc,lp}$ and $\omega_{tc,hp}$, which are to be constrained as part of the NMPC formulation.

12.4 Analysis of the System Dynamics

The implementation of an NMPC algorithm requires a model of the process as well as a suitable formulation of the optimization problem. In the particular case of a two-stage turbocharging unit, the choices need to reflect that the transient behavior should be as fast as possible and that the wastegates should be actuated taking into account the fuel economy, which is influenced by the pumping losses. The pumping losses describe the work required to aspirate gas into and out of the cylinders. These performance requirements can be incorporated in various ways. To derive a suitable optimization problem, the system dynamics are investigated with respect to the control requirements.

12.4.1 Stationary System Behavior

One distinct property of the two-stage turbocharging system is its overactuation. Two actuated values are present to control one output value. A desired charging pressure can be realized by numerous combinations of wastegate positions for the low- and the high-pressure stage. The split over these two stages influences the exhaust gas back pressure. The exhaust gas back pressure itself correlates to the pumping losses and thus to the efficiency and the fuel consumption of the engine. During the exhaust stroke, the cylinder pressure can be assumed to be equal to the exhaust gas back pressure $p_{cyl} = p_{exh}$, whereas during the intake stroke, the cylinder pressure can be assumed to be equal to the charging pressure $p_{cyl} = p_{char}$. Using V_{dis} for the displacement volume of all cylinders, the pumping losses W_p can be calculated by

$$W_p = (p_{exh} - p_{char})V_{dis} \tag{12.23}$$

The pumping loss mean effective pressure ($PMEP$) is calculated as follows:

$$PMEP = p_{exh} - p_{char} \tag{12.24}$$

Figure 12.2 shows the resulting charging pressure and the exhaust gas back pressure in dependence of the wastegate positions. For a constant charging pressure, the pumping losses directly correlate to the exhaust gas back pressure. The dependence of the exhaust gas back pressure on the wastegate position is shown by the isocurves for $p_{char} = 1.5$ bar and $p_{char} = 1.9$ bar. The measurement data depicted in Fig. 12.2 show that for $p_{char} = 1.5$ bar at an engine speed of $n_{eng} = 3500\,\frac{1}{\text{min}}$, the PMEP ranges from a minimum of $PMEP = 0.3$ bar when $u_{wg,hp} = 0$ to a maximum of $PMEP = 0.7$ bar when $u_{wg,lp} = 0$. At $p_{char} = 1.9$ bar, the difference ranges from $PMEP = 0.5$ bar to a maximum of $PMEP = 1.0$ bar. This result implies that choosing the correct control strategy can improve the $PMEP$ by up to 0.5 bar, which corresponds to approximately 3 kW.

The experimental results show that for reducing pumping losses, it is always best to use the combination with a maximum application of the low-pressure turbocharging stage. This can be realized by closing the low-pressure wastegate and opening the high-pressure wastegate as much as possible to realize a certain charging pressure. This characteristic is valid for high as well as for low engine speeds, although at low engine speeds, the difference in the exhaust gas back pressure declines.

12.4.2 *Transient System Behavior*

The two-stage turbocharging system is used to improve the transient response behavior while at the same time allowing for high power outputs. A high-pressure stage thus allows for a faster increase of the charging pressure due to the smaller mass moment of inertia $J_{tc,hp}$ and the significantly lower flow cross-sections of the compressor and the turbine. On the other hand, the low-pressure stage is larger which, due to the larger mass moment of inertia $J_{tc,lp}$, results in a slower increase of the charging pressure.

For the experimental analysis, a vehicle dynamometer was used to measure at several constant engine speeds. Various step inputs were applied to the two wastegates. The value that describes the time required until 95% of the steady-state value is reached for the first time is the characteristic value t_{95}. The value t_{95} is used to compare the transient response. Figure 12.3 shows an exemplary experimental result of the transient behavior of the high- and low-pressure stages. The plot shows a normalized expression of the charging pressure with respect to the steady-state value $\Delta p_{char}/\Delta p_{char,ss}$, the rotational speed of the two turbochargers, and the wastegate signals applied. The colors each correspond to one independent measurement. In both cases, the step is applied at $t = 0$ s. When the step is applied on the low-pressure wastegate, the step response shows a characteristic time value of $t_{95,lp} = 2.8$ s. By contrast, when the step is applied on the high-pressure wastegate, the value can be reduced to $t_{95,hp} = 1$ s. As long as the high-pressure stage is able to deliver the torque required, for the entire operating range investigated, from a transient perspective, it is recommendable to use the high-pressure stage.

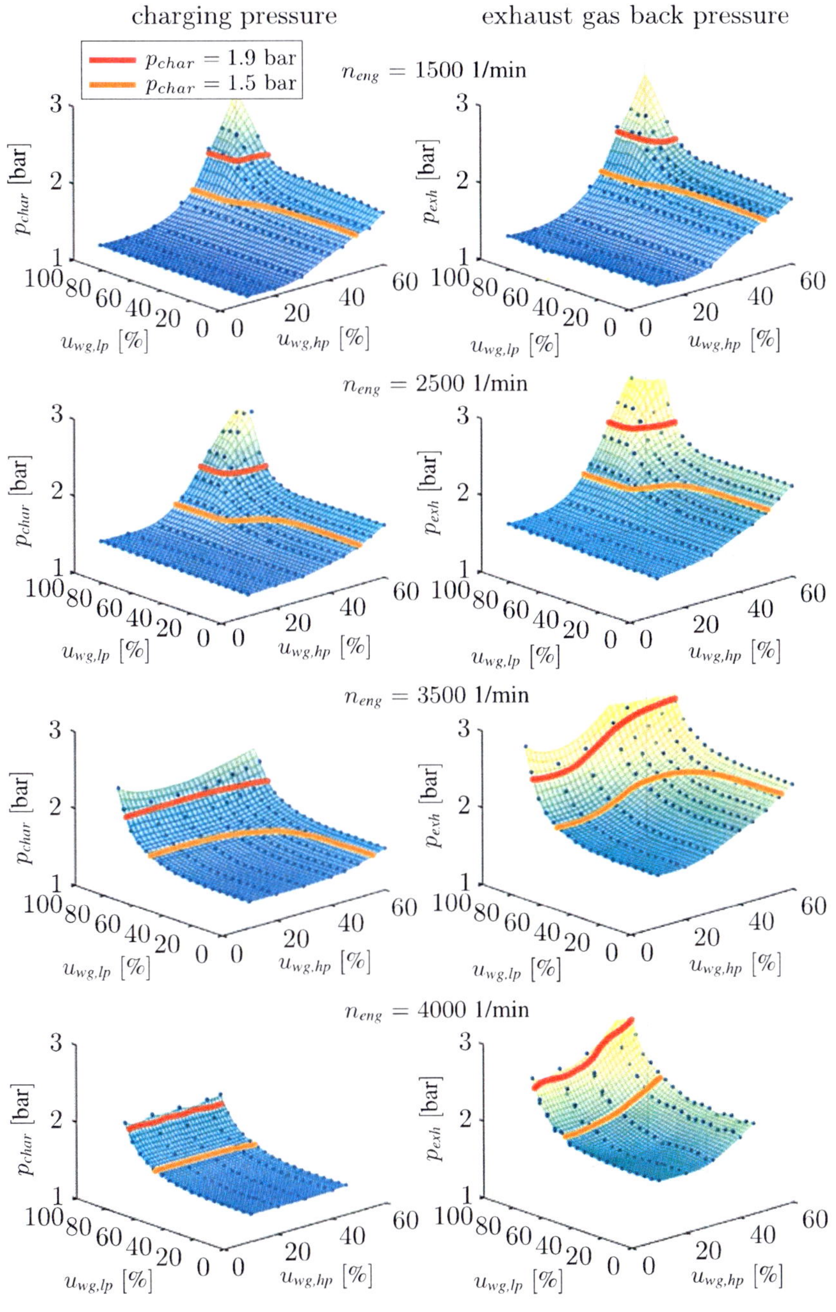

Fig. 12.2 Maps with charging and exhaust gas back pressure for various engine speeds with charging pressure isocurves – Reproduced from [2], originally published open access under a Creative Commons CC BY 4.0 license, https://doi.org/10.3390/en9070530

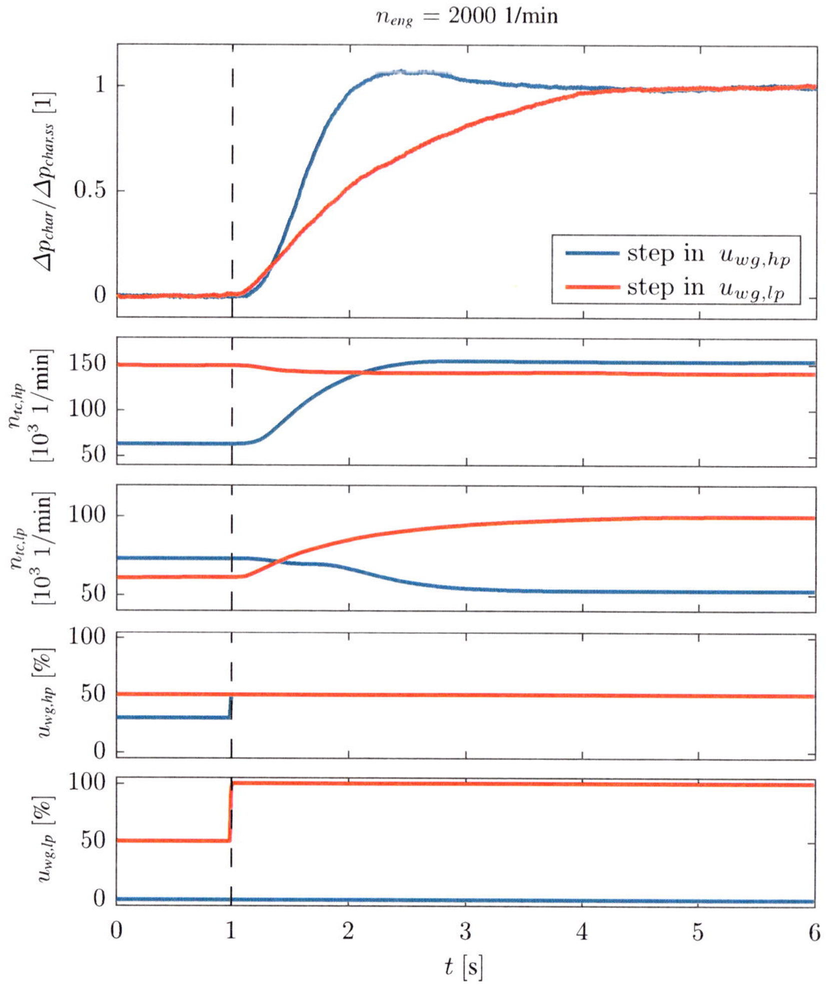

Fig. 12.3 Measured step responses for steps in high- and low-pressure wastegates at constant engine speed – Reproduced from [2], originally published open access under a Creative Commons CC BY 4.0 license, https://doi.org/10.3390/en9070530

12.5 Validation of Reduced-Order Model

In order to validate the proposed model, measurement data was obtained using the demonstrator vehicle at constant engine speed as well as at varying engine speeds. For measurements at constant engine speed, a dynamometer test bench was used, where the engine speed was set to various values between $n_{eng} = 1500\,\frac{1}{\text{min}}$ and $n_{eng} = 3500\,\frac{1}{\text{min}}$.

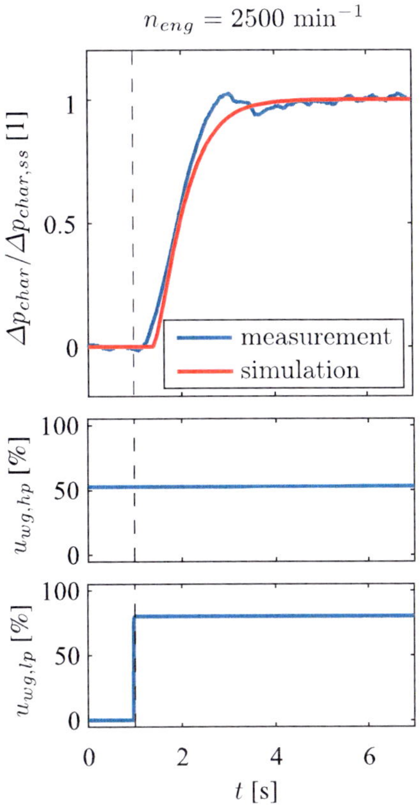

Fig. 12.4 Validation of the dynamic behavior of the model [3], adapted with permission from IFAC-PapersOnLine, © 2015 International Federation of Automatic Control (IFAC)

Figure 12.4 shows a validation of the dynamic behavior. The measurement of a step response on the low-pressure wastegate is compared to a simulated step response. Both are obtained at a constant engine speed of $n_{eng} = 2500\ \frac{1}{\mathrm{min}}$. For comparing the transient behavior, the output, i.e. the charging pressure, is normalized, thus neglecting the stationary offset between these two values. The model is able to reproduce the transient behavior of the system very well, including effects such as the dead-time. Based on the measurement data, the resulting dead-time is estimated to be $T_D = 0.45$ s. This value is quite large and makes the control task more challenging for the NMPC implementation.

Fig. 12.5 Validation of the steady-state behavior of the model – © [2015] IEEE. Reprinted, with permission, from [1]

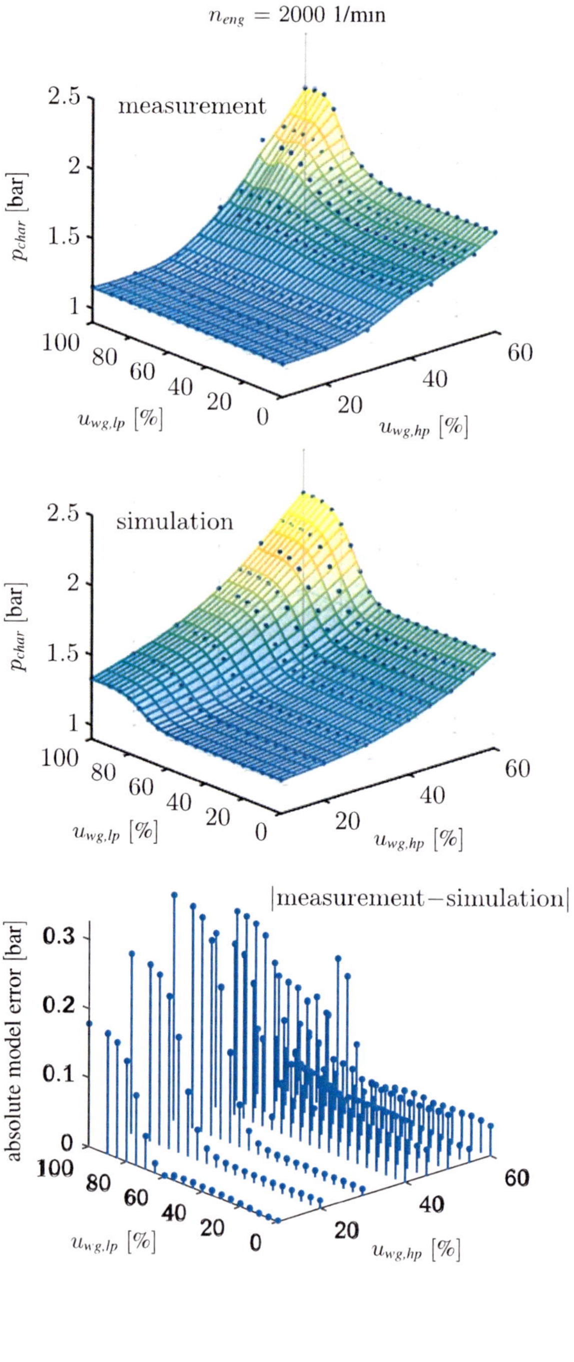

Figure 12.5 shows the result of a validation experiment for steady-state behavior. The upper plot shows the steady-state measurement data for the charging pressure p_{char} recorded on a grid of various values for $u_{wg,lp}$ and $u_{wg,hp}$ at a constant engine speed of $n_{eng} = 2500 \frac{1}{\text{min}}$. The plot in the middle shows the steady-state data for the charging pressure p_{char} obtained by simulating the model. The lower plot shows the resulting model error with absolute values. The model is able to qualitatively and quantitatively reproduce the measurement data quite well.

References

1. T. Albin, D. Ritter, D. Abel, N. Liberda, R. Quirynen, M. Diehl, Nonlinear MPC for a two-stage turbocharged gasoline engine airpath, in *IEEE Conference on Decision and Control* (2015), pp. 849–856
2. T. Albin, D. Ritter, N. Liberda, D. Abel, Boost pressure control strategy to account for transient behavior and pumping losses in a two-stage turbocharged air path concept. Energies **9**(7), 530–545 (2016)
3. T. Albin, D. Ritter, N. Liberda, S. Pischinger, D. Abel, Two-stage turbocharged gasoline engines: experimental validation of model-based control. IFAC PapersOnLine **48**(15), 124–131 (2015)
4. T. Albin, D. Ritter, N. Liberda, R. Quirynen, M. Diehl, In-vehicle realization of nonlinear MPC for gasoline two-stage turbocharging airpath control. IEEE Trans. Control Syst. Technol. **26**(5), 1606–1618 (2018)
5. F. Buchner, S. Wedowski, A. Sehr, S. Glueck, C. Schernus, In-vehicle optimization of 2-stage turbocharging for gasoline engines. Int. J. Autom. Eng. **2**(4), 143–148 (2011)
6. P. Dickinson, K. Glover, N. Collings, Y. Yamashita, Y. Yashiro, T. Hoshi, Real-Time Control of a Two-Stage Serial VGT Diesel Engine Using MPC. IFAC-PapersOnLine **48**(15), 117–123 (2015)
7. P. Drews, K. Hoffmann, R. Beck, R. Gasper, A. Vanegas, C. Felsch, N. Peters, D. Abel, Fast model predictive control for the air path of a turbocharged diesel engine, in *European Control Conference* (2009), pp. 3377–3382
8. J. El Hadef, S. Olaru, P. Rodriguez-Ayerbe, G. Colin, Y. Chamaillard, V. Talon, Explicit non-linear model predictive control of the air path of a turbocharged spark-ignited engine, in *IEEE Conference on Control Applications* (2013), pp. 71–77
9. J. El Hadef, S. Olaru, P. Rodriguez-Ayerbe, G. Colin, Y. Chamaillard, V. Talon, Nonlinear model predictive control of the air path of a turbocharged gasoline engine using Laguerre functions, in *International Conference on System Theory, Control and Computing* (2013), pp. 193–200
10. S. Glück, Charging concepts for a two-stage turbocharging gasoline engine. Ph.D. Thesis, RWTH Aachen University (2013)
11. M. Huang, H. Nakada, K. Butts, I. Kolmanovsky, Nonlinear model predictive control of a diesel engine air path: a comparison of constraint handling and computational strategies. IFAC-PapersOnLine **48**(23), 372–379 (2015)
12. P. Kotman, M. Bitzer, A. Kugi, Flatness-based feedforward control of a two-stage turbocharged diesel air system with EGR, in *IEEE Conference on Control Applications* (2010), pp. 979–984
13. P. Kotman, A. Kugi et al., Flatness-based feedforward control of a diesel engine air system with egr. IFAC Proc. **43**(7), 598–603 (2010)
14. P. Moulin, Air systems modeling and control for turbocharged engines. Ph.D. Thesis, École Nationale Supérieure des Mines de Paris (2010)
15. P. Moulin, O. Grondin, Control design for a second order dynamic system: two-stage turbocharger, in *Proceedings of Advances in Automotive Control* (2013), pp. 460–466

16. P. Moulin, O. Grondin, L. Fontvieille, Control of a two-stage turbocharger on a diesel engine, in *IEEE Conference on Decision and Control* (2009), pp. 5200–5206
17. A. Murilo, M. Alamir, D. Alberer, A general NMPC framework for a diesel engine air path. Int. J. Control **87**(10), 2194–2207 (2014)
18. R. Nitsche, D. Schwarzmann, J. Hanschke, Nonlinear internal model control of diesel air systems. Oil Gas Sci. Technol.-Revue de l'IFP **62**(4), 501–512 (2007)
19. Z. Qiu, M. Santillo, M. Jankovic, J. Sun, Composite adaptive internal model control and its application to boost pressure control of a turbocharged gasoline engine. IEEE Trans. Control Syst. Technol. **23**(6), 2306–2315 (2015)
20. M. Santillo, A. Karnik, Model predictive controller design for throttle and wastegate control of a turbocharged engine, in *American Control Conference* (2013), pp. 2183–2188
21. D. Schwarzmann, R. Nitsche, J. Lunze, Modelling of the air-system of a two-stage turbocharged passenger car diesel engine, in *Proceedings of the MATHMOD* (2006)

Chapter 13
Two-Stage Turbocharging: Control

Abstract This chapter investigates a model predictive control (MPC) algorithm for the two-stage turbocharging concept of an SI engine. Due to the working principle of an SI engine, quick reference tracking is demanded for the air path control while no oscillatory behavior should be present for a step in the reference signal. At the same time, the speed limits for the turbocharger have to be respected. Within the two-stage turbocharging setup, an additional requirement is present due to the overactuation. In stationary operation, the reference is to be tracked in a fuel-efficient manner while quick transients must be achieved. A nonlinear MPC (NMPC) algorithm is investigated that allows all the demanding requirements on the closed-loop system to be satisfied. The present chapter outlines a suitable controller-internal model, a problem formulation, and numerical solution techniques. For the validation of the control algorithm developed, exemplary simulations as well as in-vehicle experiments are presented.

13.1 Nonlinear MPC Algorithm

In the following, an NMPC concept for the two-stage turbocharging system is presented. All necessary components, the controller-internal model, the problem formulation, and the numerical solution techniques are outlined. The NMPC algorithm is based on the process model and the analysis derived in Chap. 12. A special emphasis is placed on the exploitation of the overactuated hardware setup. For this purpose, a tailor-made formulation of the optimization problem is derived.

13.1.1 Controller-Internal Model

The controller-internal model of the NMPC algorithm is based on the process model derived in Sect. 12.3. This base model is defined as an differential–algebraic equation (DAE) system and consists of two inputs $\boldsymbol{u}(t)$, two differential states $\boldsymbol{x}(t)$, two

T. Albin Rajasingham, *Nonlinear Model Predictive Control of Combustion Engines*, Advances in Industrial Control,
https://doi.org/10.1007/978-3-030-68010-7_13

algebraic states $z(t)$, one output that is reference tracked $y_{p_{char}}(t)$, and two outputs that are considered as constraints $\boldsymbol{y}_{\omega_{tc}}(t)$.

For the NMPC algorithm, a discretized model is required. To discretize the continuous-time model, a numerical integration algorithm is applied. Since the model comprises a DAE system, an implicit integration method is required, as discussed in [5]. As the DAE is of index one, the Jacobian matrix $\frac{\partial f(\cdot)}{\partial(z,\dot{x})}$ is invertible, which is a requirement of most simulation methods. In this case, an implicit Runge–Kutta method of fourth order is applied. For the discrete dead-time, $k_d = \lfloor T_d / T_s \rfloor$ is used where T_s is the sampling time. The function $\boldsymbol{f}_d(\boldsymbol{x}(k), \boldsymbol{u}(k-k_d))$ denotes the numerical simulation of the nonlinear states throughout one sampling step, starting from the given state $\boldsymbol{x}(k)$ and using the actuated values $\boldsymbol{u}(k-k_d)$ that are constant over each integration step. The overall discretized model looks as follows, where k is the discrete sampling instant:

$$\begin{aligned} \boldsymbol{x}(k+1) &= \boldsymbol{f}_d(\boldsymbol{x}(k), \boldsymbol{u}(k-k_d)) \\ \boldsymbol{y}(k) &= g(\boldsymbol{x}(k)) \end{aligned} \tag{13.1}$$

The closed-loop controller is required to achieve offset-free tracking even in the presence of disturbances or a model–plant mismatch. For the two-stage turbocharging application, the goal is that the reference of the charging pressure $y_{p_{char}}$ is tracked without any offset for a reference and disturbances that are asymptotically constant. In order to compensate for model–plant mismatch and disturbances, an additional disturbance model is introduced. The calculation of the charging pressure is thus augmented by an additional disturbance state $d(k) \in \mathbb{R}$, which acts on the output, as introduced in Sect. 6.2.2.

$$y_{p_{char},aug}(k) = g_{p_{char}}(\boldsymbol{x}(k)) + d(k) \tag{13.2}$$

For the prediction within the NMPC algorithm, the disturbance state is assumed to be constant over the prediction horizon.

$$d(\cdot|k) = d(k) \tag{13.3}$$

All system, disturbance, and output states are assumed to exhibit zero mean Gaussian white noise. An extended Kalman filter (EKF) is used for the estimation of the system states $\boldsymbol{x}_{es}$ and the disturbance d.

13.1.2 Formulation of the Optimization Problem

The formulation of the optimization problem plays a fundamental role in the closed-loop control performance. The requirements of the control system have to be taken care of appropriately. For the two-stage turbocharging setup, this concerns especially

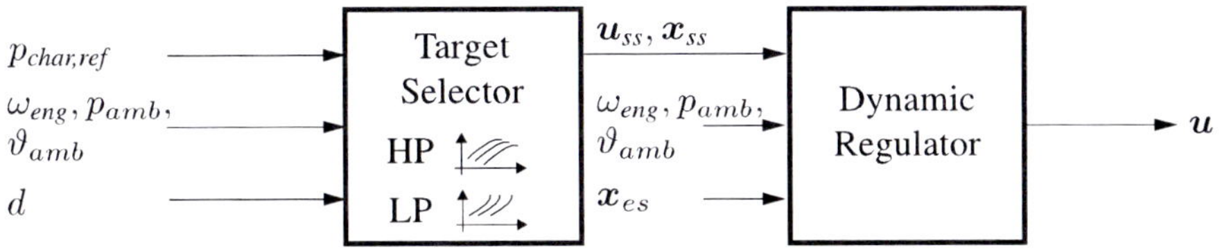

Fig. 13.1 Two-layered approach for handling the multi-objective targets – Reproduced from [1], originally published open access under a Creative Commons CC BY 4.0 license, https://doi.org/10.3390/en9070530

the appropriate exploitation of the overactuation. A fast transient response has to be realized while achieving a high efficiency at steady-state operation. For fast transient response, the high-pressure stage has to be used as much as possible to quickly increase the charging pressure. At steady-state conditions, the low-pressure stage must be used as much as possible. A suitable strategy is using the high-pressure stage in the transients and then shifting to the low-pressure stage. Thus, multi-objective targets need to be satisfied. In the following, a two-layered approach is presented to realize this behavior. It consists of two optimization problems, i.e. the target selector and the dynamic regulator as shown in Fig. 13.1.

Target Selector

The target selector calculates the optimal steady-state values for the actuated values $\boldsymbol{u}_{ss}$ and for the system states $\boldsymbol{x}_{ss}$. Within this optimization problem, the reference value of the charging pressure $p_{char,ref}$, the measured values ω_{eng}, p_{amb}, and ϑ_{amb} as well as the estimated disturbance d serve as inputs. The following optimization problem is solved within the target selector:

$$\min_{\boldsymbol{x}_{ss}, \boldsymbol{u}_{ss}, \boldsymbol{\epsilon}_{ss}} \quad \boldsymbol{u}_{ss}^{\mathsf{T}} \boldsymbol{R}_l \boldsymbol{u}_{ss} + \boldsymbol{S} \boldsymbol{\epsilon}_{ss} \tag{13.4a}$$

$$\text{s.t.} \quad \boldsymbol{0} = \boldsymbol{x}_{ss} - \boldsymbol{f}_d(\boldsymbol{x}_{ss}, \boldsymbol{u}_{ss}), \tag{13.4b}$$

$$p_{char,ref} = g_{p_{char}}(\boldsymbol{x}_{ss}) + d, \tag{13.4c}$$

$$\boldsymbol{u}_{min} \leq \boldsymbol{u}_{ss} \leq \boldsymbol{u}_{max}, \tag{13.4d}$$

$$\boldsymbol{y}_{\omega_{tc},ss} \leq \boldsymbol{\omega}_{tc,max} + \boldsymbol{\epsilon}_{ss}, \tag{13.4e}$$

$$\boldsymbol{0} \leq \boldsymbol{\epsilon}_{ss} \tag{13.4f}$$

The target selector calculates values that strongly rely on the low-pressure stage by appropriate penalization $\boldsymbol{R}_l$ within the cost function such that the goal of high efficiency in steady-state operation is realized. Within the optimization problem, steady-state dynamics are considered as a constraint (13.4b). The reference value is incorporated as equality constraint (13.4c). Within the closed-loop control validations, the reference value has proven to be always feasible; thus, this equality constraint is not softened. Additionally, the physical constraints on the actuated values are considered as (13.4d) shows. The constraints on the turbocharger speeds are implemented, including slack variables (13.4e)–(13.4f). The vector of slack variables are denoted by $\boldsymbol{\epsilon}_{ss} \in \mathbb{R}^2$. The slack variables enter the cost function in a linear fashion. The solver applied uses automatic regularization, thus no additional quadratic terms are applied for penalization of the slack variables.

Dynamic Regulator

The second optimization problem is formed by the dynamic regulator. In this layer, the actuated values are calculated that are applied to the plant. Unlike the target selector, the dynamic regulator also incorporates the transient system behavior. The optimal steady-state values $\boldsymbol{u}_{ss}$, $\boldsymbol{x}_{ss}$, the estimated states $\boldsymbol{x}_{es}$ as well as the measured values ω_{eng}, p_{amb}, and ϑ_{amb} serve as inputs. The optimization problem for the dynamic regulator is set up as given by multiple shooting. An equidistant grid of $N+1$ shooting intervals is considered over the predicted horizon. Within each shooting interval, a piecewise constant control parametrization $\boldsymbol{u}(\tau) = \boldsymbol{u}_i$ for $\tau \in [t_i, t_{i+1})$ is applied. The cost function of the optimization problem for the dynamic regulator reads as follows:

$$J = \sum_{i=0}^{N} \Big[||\boldsymbol{x}(k+k_D+i+1|k) - \boldsymbol{x}_{ss}(k)||^2_{\boldsymbol{Q}_i} + ||\boldsymbol{u}(k+i \mid k) - \boldsymbol{u}_{ss}(k)||^2_{\boldsymbol{R}_{ss,i}} + ||\boldsymbol{u}(k+i \mid k) - \boldsymbol{u}(k+i-1 \mid k)||^2_{\boldsymbol{R}_{d,i}} + \boldsymbol{S}\boldsymbol{\epsilon}(k+i \mid k) \Big] \tag{13.5}$$

The optimization variables are given by the trajectory of actuated values $\boldsymbol{u}(\cdot|k)$, the system states $\boldsymbol{x}(\cdot|k)$, and the slack variables $\boldsymbol{\epsilon}(\cdot|k)$. In the cost function, various terms are considered. The first term takes into account the deviation from the reference value, which is given by $\boldsymbol{x}_{ss}$, weighted by matrix $\boldsymbol{Q}_i$. The reference signal $p_{char,ref}$ and thus also $\boldsymbol{x}_{ss}$ are used as a constant value over the prediction horizon as no future values are known. Second, the deviation from the optimal steady-state actuated values $\boldsymbol{u}_{ss}$ is penalized. The change of the actuated values $\boldsymbol{u}(k+i \mid k) - \boldsymbol{u}(k+i-1 \mid k)$ is represented by the third term, while the final term considers the slack variables. This cost function allows the high-pressure stage to be used in the transient phase plus the subsequent shift to the low-pressure stage approaching the optimal steady-state values.

Within the constraints, the initial value embedding is considered, where the parameter $\boldsymbol{x}_{es} \in \mathbb{R}^{n_x}$ denotes the state estimate (13.6b). Due to the dead-time, the last k_D actuated values applied are also considered as initial values (13.6c). The discretized dynamics are considered by (13.6d). The box constraints for the actuated values are defined in (13.6e). In (13.6f)–(13.6g), the slack variables $\boldsymbol{\epsilon}$ are used to penalize violations of the constraints on the turbocharger speeds. The solver applied uses automatic regularization, thus no additional quadratic terms are applied for penalization of the slack variables. The NLP is given by

$$\min_{u(\cdot),\, x(\cdot),\, \epsilon(\cdot)} \quad J \tag{13.6a}$$

$$\text{s.t.} \quad x(k|k) = x_{es}(k), \tag{13.6b}$$

$$u(k-i|k) = u(k-i),\ i = 1, \ldots, k_D, \tag{13.6c}$$

$$x(k+i+1|k) = f_d(x(k+i|k), u(k-k_D+i|k)), i = 0, \ldots, N+k_D, \tag{13.6d}$$

$$u_{min} \le u(k+i \mid k) \le u_{max}, \qquad i = 0, \ldots, N, \tag{13.6e}$$

$$y_{\omega_{tc}}(k+k_D+i+1 \mid k) \le \omega_{tc,max} + \epsilon(k+i \mid k), \quad i = 0, \ldots, N, \tag{13.6f}$$

$$0 \le \epsilon(k+i \mid k), \qquad i = 0, \ldots, N \tag{13.6g}$$

13.1.3 Numerical Solution of the Optimization Problem

Both NLPs need to be solved at each sampling instant. This case study uses sequential quadratic programming (SQP) to solve the NLPs. For the approximation of the Hessian, the generalized Gauss–Newton method is used. To further reduce the computation time, the real-time iteration (RTI) scheme is applied, i.e. only one SQP iterate is solved in each sampling instant. The sensitivities of the nonlinear functions in the NLP are calculated with algorithmic differentiation. In the algorithm presented, condensing has been used in combination with the QP solver qpOASES [3]. To obtain a real-time feasible NMPC implementation on the embedded control hardware, the ACADO code generation tool has been used as presented in [4, 5].

13.2 Validation of the NMPC Algorithm by Simulations

To validate the NMPC algorithm, simulations with MATLAB®/Simulink® are conducted. The NMPC algorithm presented is referred to as a multi-objective controller, as it tries to find a trade-off between fast transient raise and fuel-efficient steady-state operation. To evaluate the effectiveness of this trade-off, it is compared to two alternative strategies. The first alternative is optimized for the transient behavior and does not consider fuel efficiency. Based on the analysis of the system dynamics, this behavior can be realized by using the high-pressure stage as much as possible during the entire time, while considering its speed limit. As a second alternative, a control strategy is applied which is optimized for reducing the pumping losses and which does not take the transient behavior into account. In contrast to the first alternative, the high-pressure stage here is used as little as possible. In both alternative strategies, the optimization problem and its weightings differ from the multi-objective controller.

The results presented are extracted from [1] where one multi-objective controller is compared to a fuel-efficient controller and a fast transient raise controller. The controllers are comparable to the strategies presented above. The various control strategies are tested in a closed-loop control setting. For these tests, the controller-internal model is identical to the plant model; just measurement noise is added to the outputs. This allows the influence of the control strategy to be validated without dealing with effects such as a model–plant mismatch.

Figure 13.2 shows the closed-loop control performance of the three strategies. Two step-wise changes are applied for the reference, each with a different step height. The control strategy, which is focused on the reduction of pumping losses, uses static actuated values. They are to ensure that the reference is tracked in steady state with a maximum actuation of the low-pressure wastegate. As the focus lies on the reduction of pumping losses, the transient behavior is entirely neglected. The second controller, which is to optimize the transient behavior, uses the high-pressure stage as much as possible. This correlates to the operation at the limit of the high-pressure turbocharger speed. The multi-objective controller uses a strategy in between these two. At the beginning of the transient phase, the actuated values correspond to those of the controller with optimal transients. The actuated values then shift to the solution of the controller with optimally reduced pumping losses. The multi-objective controller is thus able to be fast in the transients. It is only slightly slower in reference tracking than the controller that is purely optimized for the transients. Simultaneously, it reduces pumping losses in the steady-state behavior. At steady state, the actuated values of the multi-objective controller are the same as those of the controller that is optimized for pumping losses.

The multi-objective control strategy is also suited for nonuniform reference trajectories. Figure 13.3 shows a simulation obtained with such a reference trajectory. The NMPC algorithm is able to accurately track the reference. At the same time, the turbocharger speed limits are respected.

13.3 Experimental In-Vehicle Validation of the NMPC Algorithm

The simulation tests of the control algorithm provide important insights. However, various factors require additional experimental validations. The control algorithm has to deal with random reference trajectories that occur in real driving situations, with disturbance signals as well as model–plant mismatches in a system that is highly nonlinear. For the experimental validation of engine control algorithms, various types of experiments usually are conducted. First, the controller is validated on an engine test bench. This has the advantage that adjustments of certain parameters, such as the intake air temperature, are straightforward. The control algorithm then is tested on a vehicle dynamometer. This allows the air path control to be tested with the entire vehicle. Still, certain parameters can be easily adjusted; for instance, the algorithm can be tested at various constant engine speeds, which is especially for air path control very advantageous. As the last step, the control algorithm is tested in the vehicle on a road, where it has to handle all the arising complexities to the full extent.

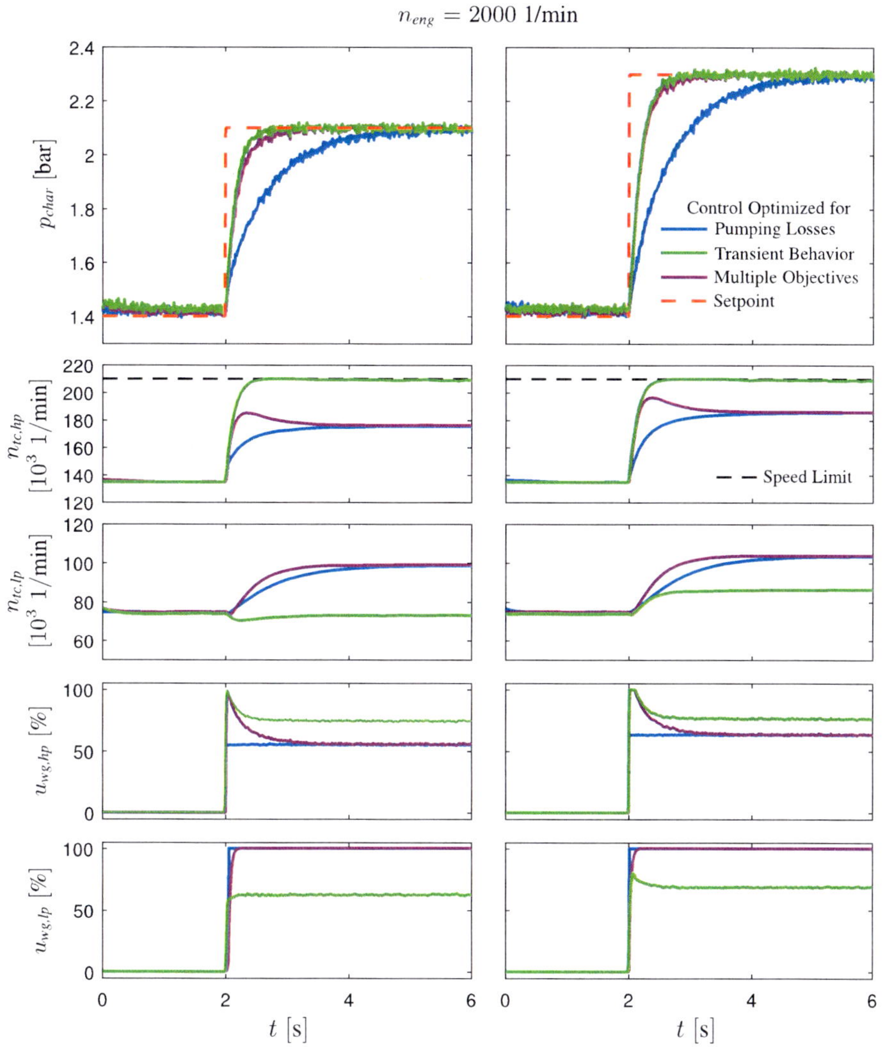

Fig. 13.2 Validation of the control algorithm by simulations (1) Reduction of pumping losses; (2) Fast transients; (3) Multi-objective – Reproduced from [1], originally published open access under a Creative Commons CC BY 4.0 license, https://doi.org/10.3390/en9070530

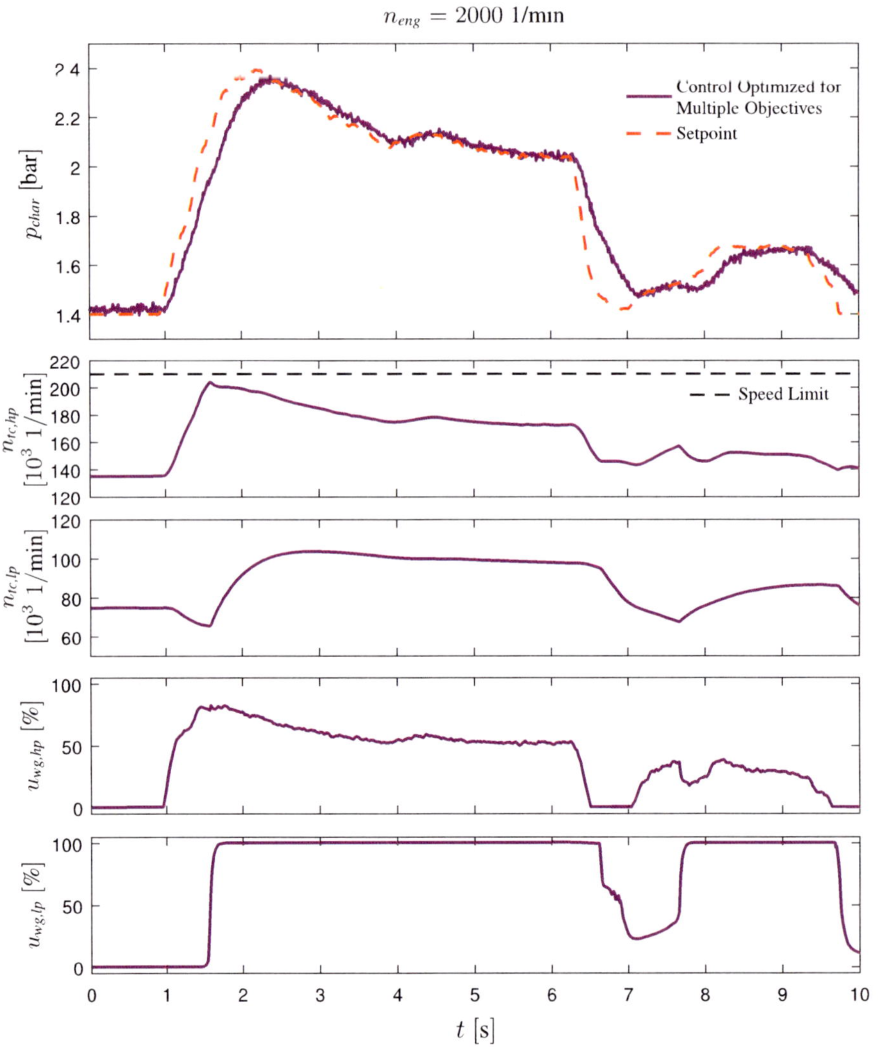

Fig. 13.3 Validation of the control algorithm by simulations for a nonuniform reference profile – Reproduced from [1], originally published open access under a Creative Commons CC BY 4.0 license, https://doi.org/10.3390/en9070530

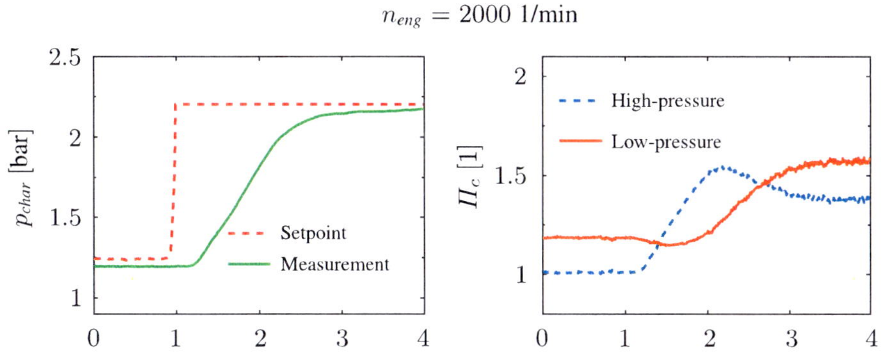

Fig. 13.4 Validation of the control algorithm on a vehicle dynamometer – © [2018] IEEE. Reprinted, with permission, from [2]

In [2], such an NMPC algorithm for two-stage turbocharging is validated with a vehicle dynamometer and with vehicle tests on the road. The road tests with random driving have been performed on an automotive testing track and on public roads. The NMPC deployed is comparable to the one described in this chapter. Two examples of the tests conducted in [2] are reproduced to showcase the capabilities of the NMPC algorithm. The NMPC algorithm is used with a sampling time of $T_s = 0.05$ s.

Figure 13.4 shows a closed-loop control result from a vehicle dynamometer test. The vehicle dynamometer allows testing at a constant engine speed and using uniform as well as nonuniform reference signals. The presented closed-loop control result is obtained at a constant engine speed of $n_{eng} = 2000\frac{1}{min}$. As a reference trajectory, a step input is applied. The reference value for the charging pressure is tracked by the NMPC algorithm without any overshoots. The control algorithm uses the high-pressure stage in the transients to quickly increase the charging pressure. In the progress of the trajectory, there is a shift toward the use of the low-pressure stage. This enables a fuel-efficient stationary operation. The pressure ratios of the two stages show that the control algorithm is capable of engaging and disengaging the two stages without any negative impact on the charging pressure. Thus, the control algorithm is able to exploit the capabilities of the specific hardware setup.

Figure 13.5 shows a control result from tests with the vehicle on an automotive testing center. Due to the real-world driving, the reference profile is rather nonuniform and the engine speed varies continuously. Still, the controller is able to track the reference trajectory very accurately. A certain lag is present between the reference trajectory and the system output. This lag cannot be reduced due to the physics of the system, such as the dead-time present and the inertia of the turbochargers. The consideration of the limit values on the turbocharger speeds allows reduction of overstepping.

In conclusion, the experiments show that the NMPC algorithm developed is able to deliver high performance for all driving profiles in the entire operating range. It can cope with all complexities arising for an in-vehicle implementation and it meets

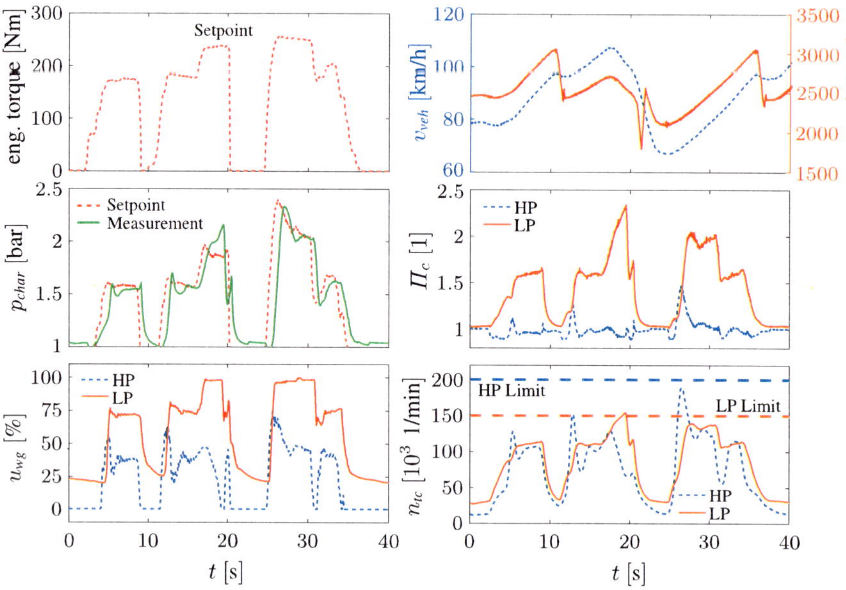

Fig. 13.5 In-vehicle validation of the control algorithm on the road – © [2018] IEEE. Reprinted, with permission, from [2]

all requirements specified. The optimization-based method allows to account for the specifics of the two-stage turbocharged system such that the goals associated with the hardware design can be realized. Additionally, the demanding system properties can be handled by using a very systematic development approach. A critical point for using NMPC is its real-time feasibility, which is achieved by using a reduced-order model in combination with tailor-made numerical solution methods.

References

1. T. Albin, D. Ritter, N. Liberda, D. Abel, Boost pressure control strategy to account for transient behavior and pumping losses in a two-stage turbocharged air path concept. Energies **9**(7), 530–545 (2016)
2. T. Albin, D. Ritter, N. Liberda, R. Quirynen, M. Diehl, In-vehicle realization of nonlinear MPC for gasoline two-stage turbocharging airpath control. IEEE Trans. Control Syst. Technol. **26**(5), 1606–1618 (2018)
3. H.J. Ferreau, C. Kirches, A. Potschka, H.G. Bock, M. Diehl, qpOASES: a parametric active-set algorithm for quadratic programming. Math. Program. Comput. **6**(4), 327–363 (2014)
4. B. Houska, H.J. Ferreau, M. Diehl, An auto-generated real-time iteration algorithm for nonlinear MPC in the microsecond range. Automatica **47**(10), 2279–2285 (2011)
5. R. Quirynen, M. Vukov, M. Zanon, M. Diehl, Autogenerating microsecond solvers for nonlinear MPC: a tutorial using ACADO integrators. Optim. Control Appl. Methods **36**, 685–704 (2014)

Part IV
In-Depth Case Studies: Combustion Control

Chapter 14
Fundamentals of CI Engine Combustion Control and Modeling

Abstract This chapter presents the basics of the combustion control of compression ignition (CI) engines. In this context, a control-oriented model is presented that can be used for control purposes, e.g. as a controller-internal model for optimization-based control. A single-zone model is detailed with a focus on the high-pressure working process. The simplifications are chosen such that a computationally efficient model results that is still capable of reproducing the cylinder pressure trace with reasonable accuracy. The basic conservation equations for mass and energy are derived. Additionally needed submodels such as the one for wall heat losses are introduced as well. The chapter also details the two fundamental concepts for combustion control. First, the cycle-to-cycle control of cycle-integral combustion parameters is presented. The cycle-integral parameters are derived from the cylinder pressure trace and are used as surrogate parameters for the combustion performance. Second, the combustion rate shaping is presented. For combustion rate shaping, the heat release rate is controlled in a highly time-resolved manner, thus offering a high number of degrees of freedom for combustion control.

14.1 Introduction to Combustion Control

The combustion control plays an important role in CI engines as the combustion process influences the overall performance, e.g. the level of pollutant and noise emissions as well as the efficiency of operation. In CI engines, a multitude of parameters is available that can actively control the combustion process. There exist several air path parameters, such as the pressure and the burnt gas ratio, in the intake manifold. They influence the combustion process only on a slow time scale. On the other side, there are several parameters available from the fuel path. They allow for influencing the combustion process on a combustion cycle time basis. Modern injection systems that are applied in series production are characterized by increased rail pressures and

T. Albin Rajasingham, *Nonlinear Model Predictive Control of Combustion Engines*, Advances in Industrial Control,
https://doi.org/10.1007/978-3-030-68010-7_14

decreased dwell times, i.e. the minimum time distance between two injection events. They offer a high flexibility in terms of many injection events per combustion cycle. Each injection event is parametrized by the time of injection and the duration of the injection, which yields twice as many actuated values as the number of injections. Recent injection systems are capable of delivering up to nine injection events per cycle [10, 26]. Thus, a high number of degrees of freedom are present for achieving the highest performance concerning fuel efficiency, noise, and pollutant emissions.

In general, there exist two fundamental approaches for the control of the combustion process. In the first approach, a limited number of cycle-integral combustion parameters are used as controlled variables. The actuation is updated on a cycle-to-cycle basis. The control concept along with the most important cycle-integral combustion parameters is introduced in Sect. 14.3. An alternative procedure is the so-called combustion rate shaping. To its fullest extent, the entire quasi-continuous trace of the heat release is controlled based on a cycle-to-cycle frequency. While the cycle-to-cycle-based control of cycle-integral combustion characteristics can be considered as state of the art, the combustion rate shaping is a topic of current research. The combustion rate shaping is detailed in Sect. 14.4. For the understanding of the two combustion control concepts, first, a brief summary of the control-relevant fundamentals of the thermodynamics of the engine process is given. For this purpose, a simplified process model is introduced. Readers interested in a detailed description of the fundamentals are referred to books such as [12, 18].

14.2 Control-Oriented Process Model

In the following, the engine process of a four-stroke CI engine with direct injection of fuel is investigated. In this case, a full working cycle is given by four strokes, which corresponds to a rotation of the crankshaft by 720 °CA. Each stroke refers to the the full upward respectively the full downward movement of the piston inside the cylinder. They are given by

1. Intake stroke
2. Compression stroke
3. Power stroke
4. Exhaust stroke

The intake and the exhaust stroke determine the charge exchange process, i.e. fresh air is aspirated into the cylinder and the exhaust gas is pushed out. Due to the low pressure inside the cylinder, these two strokes are called low-pressure cycle. Within the compression and the power stroke, the air–fuel mixture is compressed and subsequently, combustion takes place which results in expansion and work done on the piston. The two strokes are called high-pressure cycle due to the high pressure levels inside the cylinder.

Figure 14.1 shows relevant parameters of the CI combustion process around the top-dead center during the high-pressure cycle. The phases of compression, fuel injection, combustion, and expansion are shown. The digital injector actuation signal act_{inj} is shown along with the fuel mass flow into the cylinder $\dot{m}_{fuel,inj}$. The pressure inside the combustion chamber p is shown for the fired and for the motored operation, i.e. without any fuel injection and combustion. Additionally, the accumulated heat release Q_{comb} from combustion and its time derivative, the heat release rate $\dot{Q}_{comb}$ from combustion are depicted.

The progress in the variables is shown in dependence of the crank angle ϕ. The crank angle ϕ can be related to the time t using the engine speed n_{eng} in $\frac{1}{s}$ or the angular velocity of the engine ω_{eng}. Assuming the engine speed to be constant during one working cycle, i.e. $d\omega_{eng} = dn_{eng} = 0$, the relationship between time and crank angle can be described by:

$$\phi = \omega_{eng}t = 2\pi n_{eng}t \tag{14.1}$$

$$d\phi = \omega_{eng}dt \tag{14.2}$$

Setting the actuation signal of the injector to $act_{inj} = 1$ leads, with some time delay, to a mass flow of fuel into the cylinder. The time point where $act_{inj} = 1$ is set is called start of energizing SOE. Setting the actuation signal of the injector again to $act_{inj} = 0$ leads, with some time delay, to a decrease of the fuel mass flow until it entirely stops. The time period where $act_{inj} = 1$ holds is called the duration of energizing DOE. A single injection event is parameterized with the two values SOE and DOE. While the fuel is injected, already a high pressure is present inside the cylinder. Thus, the fuel autoignites during the injection process. The combustion of the fuel occurs after the so-called ignition delay. A significant measure for the progress in combustion is the accumulated heat release Q and its derivative the heat release rate $\dot{Q}$. In the heat release rate trace depicted, two distinctive peaks are visible. The initial phase with the characteristic first peak results from the combustion of a premixed air–fuel mixture. The subsequent phase of the combustion, with the second peak in it, is dominated by diffusive combustion. Finally, the heat release rate decays exponentially, which is the so-called burn-out phase. The heat release increases the temperature of the charge and the cylinder pressure.

In the following, the fundamentals of modeling the in-cycle, i.e. crank-angle-resolved CI engine process, are detailed. A lumped-parameter process is presented, where the volume is assumed to be ideally mixed. Thus, the gas states such as temperature, pressure, and mixture composition are not spatially distributed. Additionally, it is assumed that the gas is well described by ideal gas laws. This modeling approach allows for low computation times. These kinds of models are typically called zero-dimensional or single-zone models [12]. They offer reasonable model quality for reproducing the cylinder pressure trace. However, they are not suited for reproducing pollutant emissions, as their formation results mainly from local effects in the combustion chamber.

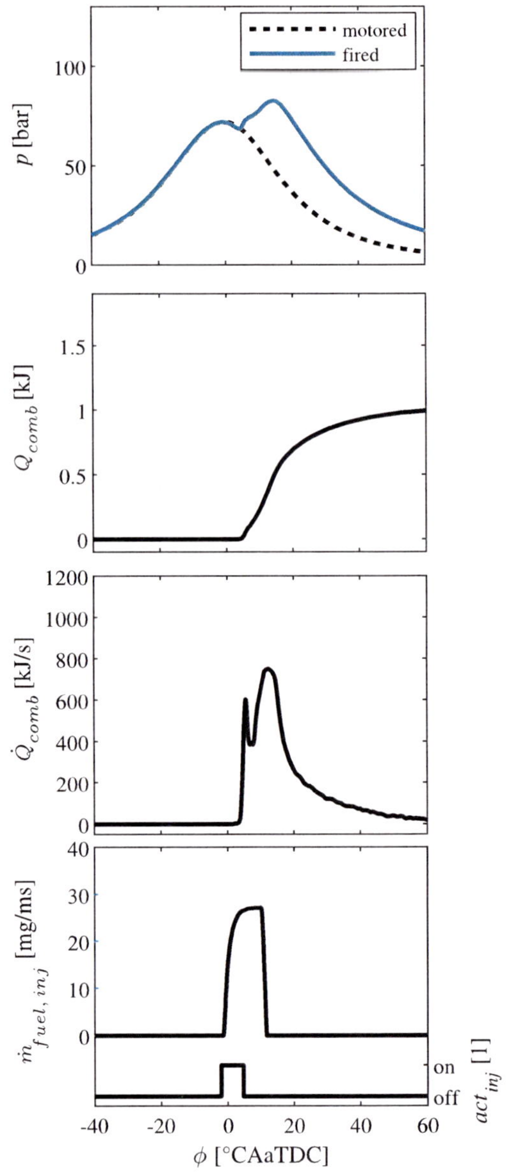

Fig. 14.1 CI combustion process around the top-dead center during the high-pressure cycle

14.2.1 Mass Balance

An important quantity of the process is the mass m of the total charge inside the combustion chamber. The change of the total mass with respect to the time $\frac{\mathrm{d}}{\mathrm{d}t}m$ is given by the sum of all mass flows flowing inside and outside of the combustion chamber:

$$\frac{\mathrm{d}}{\mathrm{d}t}m = \dot{m}_{int} + \dot{m}_{exh} + \dot{m}_{fuel,inj} \tag{14.3}$$

The mass flows entering and exiting through the intake and exhaust valves are denoted by $\dot{m}_{int}$ and $\dot{m}_{exh}$. The mass flow of the fuel injected is given by $\dot{m}_{fuel,inj}$. To simplify the model, it is assumed that the injected fuel evaporates instantaneously when entering the combustion chamber. There exist additional mass flows due to leakage effects. For modern CI engines, they can be neglected without significantly impairing the model quality [18]:

$$\dot{m}_{leak} = 0 \tag{14.4}$$

Low-Pressure Cycle
During the low-pressure cycle, the gas exchange process takes place. As no fuel is injected in the low-pressure cycle, $\dot{m}_{fuel,inj} = 0$ holds. Within the intake stroke, the exhaust valves are closed, resulting in $\dot{m}_{exh} = 0$. On the other side, the intake valves are closed during the exhaust stroke, resulting in $\dot{m}_{int} = 0$. The mass flow through the intake valves during the intake stroke and the mass flow through the exhaust valves during the exhaust stroke can be described by the compressible flow restriction model. The mass flow through the valves can be calculated by

$$\dot{m}_{valve}(t) = c_d A(t) \frac{p_{us}(t)}{\sqrt{R \vartheta_{us}(t)}} \psi\left(\Pi(t)\right) \tag{14.5}$$

The flow function $\psi(\Pi)$ for the compressible flow restriction model is determined by

$$\psi(\Pi) = \begin{cases} \sqrt{\kappa \cdot \left(\frac{2}{\kappa+1}\right)^{\frac{\kappa+1}{\kappa-1}}}\,, & \Pi \leq \left(\frac{2}{\kappa+1}\right)^{\frac{\kappa}{\kappa-1}} \\ \Pi^{\frac{1}{\kappa}} \sqrt{\frac{2\kappa}{\kappa-1} \cdot \left(1 - \Pi^{\frac{\kappa-1}{\kappa}}\right)}\,, & \Pi > \left(\frac{2}{\kappa+1}\right)^{\frac{\kappa}{\kappa-1}} \end{cases} \tag{14.6}$$

A continuously differentiable alternative and a more detailed description of the parameters involved can be found in Sect. 9.4.2. For calculation of the mass flow through the valves, the factor $c_d A(t)$, i.e. the multiplication of the discharge coefficient c_d and the reference flow area A, has to be modeled carefully. The factor needs to take the motion of the valves into account. A model for the reference flow area $A(t)$ that considers the valve motion is presented in [18].

For more detailed models, besides the total mass, also the mixture composition plays an important role. In this case, the gas exchange model calculates the total in-cylinder charge mass and the composition at intake valve closing (IVC). This is used subsequently as the initial condition for the calculations of the high-pressure cycle. The mixture composition is especially of relevance when a detailed caloric model is used.

High-Pressure Cycle
During the high-pressure cycle, the exhaust and the intake valves are closed. It results

$$\dot{m}_{int} = 0 \tag{14.7}$$

$$\dot{m}_{exh} = 0 \tag{14.8}$$

Thus, the total mass balance (14.3) can be simplified. The change of total mass is directly determined by the mass flow of fuel inside the combustion chamber:

$$\frac{\mathrm{d}}{\mathrm{d}t} m = \dot{m}_{fuel,inj} \tag{14.9}$$

Typically, a model is needed that relates the actuation signal such as start of energizing SOE and duration of energizing DOE to the mass flow through the injector $\dot{m}_{fuel,inj}$. One data-based approach that is suitable for the purpose of combustion rate shaping is presented in Sect. 15.3.

14.2.2 Energy Balance

The governing equations for the cylinder pressure p can be derived by the use of the energy balance via the first law of thermodynamics. Typically, the kinetic and potential energies are neglected, such that the change of the total energy $\mathrm{d}E$ is assumed to be equal to the change of the internal energy $\mathrm{d}U$:

$$\mathrm{d}E = \mathrm{d}U \tag{14.10}$$

For the open system, the change of the internal energy with respect to differential changes in the crank angle $\mathrm{d}\phi$ can be calculated by [21]

$$\begin{aligned}\frac{\mathrm{d}}{\mathrm{d}\phi} U = & \frac{\mathrm{d}}{\mathrm{d}\phi} W_{mech} + \frac{\mathrm{d}}{\mathrm{d}\phi} Q_{wall} + \frac{\mathrm{d}}{\mathrm{d}\phi} Q_{comb} \\ & + \frac{\mathrm{d}}{\mathrm{d}\phi} H_{int} + \frac{\mathrm{d}}{\mathrm{d}\phi} H_{exh} + \frac{\mathrm{d}}{\mathrm{d}\phi} H_{leak} + \frac{\mathrm{d}}{\mathrm{d}\phi} H_{fuel,inj}\end{aligned} \tag{14.11}$$

Various processes affect the change in the internal energy. These are the change of heat release by combustion of the fuel $\frac{\mathrm{d}}{\mathrm{d}\phi} Q_{comb}$, the change in performed mechanical work $\frac{\mathrm{d}}{\mathrm{d}\phi}W_{mech}$ that results from cylinder volume change work, the change in heat transfer to the combustion chamber walls $\frac{\mathrm{d}}{\mathrm{d}\phi}Q_{wall}$ as well as the change of enthalpy flows into and out of the combustion chamber, i.e. through the intake and exhaust valves $\frac{\mathrm{d}}{\mathrm{d}\phi}H_{int}$ and $\frac{\mathrm{d}}{\mathrm{d}\phi}H_{exh}$, by fuel injection $\frac{\mathrm{d}}{\mathrm{d}\phi}H_{fuel,inj}$, and due to leakage $\frac{\mathrm{d}}{\mathrm{d}\phi}H_{leak}$. The heat necessary for the evaporation of the liquid fuel is neglected.

For the process control, only the cylinder pressure trace of the high-pressure cycle is of relevance. For the high-pressure cycle, further going simplifications can be made in order to obtain a computationally simple model. First, a closed system can be assumed, where all mass flows into and out of the combustion chamber are neglected. As the intake and exhaust valves are closed, the mass flows are given by $\dot{m}_{int} = \dot{m}_{ext} = 0$ and thus $\frac{\mathrm{d}}{\mathrm{d}\phi}H_{int} = \frac{\mathrm{d}}{\mathrm{d}\phi}H_{exh} = 0$. The leakage mass flow and the fuel mass flow are neglected as well for the internal energy, resulting in $\frac{\mathrm{d}}{\mathrm{d}\phi}H_{fuel,inj} = 0$ and $\frac{\mathrm{d}}{\mathrm{d}\phi}H_{leak} = 0$. Compared to the other terms in (14.11), these enthalpy flows have a minor effect. A simplified relation for the internal energy results:

$$\frac{\mathrm{d}}{\mathrm{d}\phi}U = \frac{\mathrm{d}}{\mathrm{d}\phi}W_{mech} + \frac{\mathrm{d}}{\mathrm{d}\phi}Q_{wall} + \frac{\mathrm{d}}{\mathrm{d}\phi}Q_{comb} \tag{14.12}$$

The term $\frac{\mathrm{d}}{\mathrm{d}\phi}U$ is reformulated to express it in terms of measurable quantities. For this purpose, first the following relation can be applied for the internal energy:

$$\frac{\mathrm{d}}{\mathrm{d}\phi}U = \frac{\mathrm{d}}{\mathrm{d}\phi}(mu) = m\frac{\mathrm{d}}{\mathrm{d}\phi}u + u\frac{\mathrm{d}}{\mathrm{d}\phi}m \tag{14.13}$$

As the change of mass is neglected, the change of internal energy is due to changes in the specific internal energy $\frac{\mathrm{d}}{\mathrm{d}\phi}u$. The specific internal energy u can be expressed as a function that depends on the temperature ϑ and the specific heat at constant volume c_v, i.e. $u(\vartheta) = c_v\vartheta$. Thus, it follows for the change in internal energy:

$$\frac{\mathrm{d}}{\mathrm{d}\phi}U = mc_v\frac{\mathrm{d}}{\mathrm{d}\phi}\vartheta \tag{14.14}$$

The gas is modeled as an ideal gas. Thus, the following equation holds:

$$pV = mR\vartheta \tag{14.15}$$

The equation can be expressed in differential form, taking into account that the mass is assumed to be constant $\mathrm{d}m = 0$. Additionally, it can be assumed that the specific gas constant stays constant as well, i.e. $\mathrm{d}R = 0$. It follows for the differentials:

$$p\mathrm{d}V + V\mathrm{d}p = mR\mathrm{d}\vartheta \tag{14.16}$$

Inserting (14.16) in (14.14) allows to derive an expression with measurable quantities:

$$\frac{\mathrm{d}}{\mathrm{d}\phi}U = \frac{c_v}{R}\left(p\frac{\mathrm{d}}{\mathrm{d}\phi}V + V\frac{\mathrm{d}}{\mathrm{d}\phi}p\right) = \frac{1}{\kappa - 1}\left(p\frac{\mathrm{d}}{\mathrm{d}\phi}V + V\frac{\mathrm{d}}{\mathrm{d}\phi}p\right) \tag{14.17}$$

The ratio between c_v and R is expressed in terms of the isentropic exponent κ using the specific heat at constant pressure c_p [12]:

$$\kappa = \frac{c_p}{c_v} \tag{14.18}$$

$$R = c_p - c_v \tag{14.19}$$

For a computationally efficient model approach, the caloric properties, i.e. the specific heats, the isentropic exponent, and the specific gas constant, are assumed to be constant. However, for a more accurate modeling, the caloric properties need to be modeled in dependence of the temperature and the gas composition. Often, the temperature dependency of the constant pressure specific heat is modeled using the so-called NASA polynomials [5, 8]. A full approach using the NASA polynomials is depicted in [15].

The inclusion of (14.17) in the energy balance (14.12) results in

$$\frac{1}{\kappa - 1}\left(p\frac{\mathrm{d}}{\mathrm{d}\phi}V + V\frac{\mathrm{d}}{\mathrm{d}\phi}p\right) = \frac{\mathrm{d}}{\mathrm{d}\phi}W_{mech} + \frac{\mathrm{d}}{\mathrm{d}\phi}Q_{wall} + \frac{\mathrm{d}}{\mathrm{d}\phi}Q_{comb} \tag{14.20}$$

The change in performed mechanical work $\frac{\mathrm{d}}{\mathrm{d}\phi}W_{mech}$ is given by the change in volumetric work:

$$\frac{\mathrm{d}}{\mathrm{d}\phi}W_{mech} = -p\frac{\mathrm{d}}{\mathrm{d}\phi}V \tag{14.21}$$

Using this equation, the final relation for the energy balance can be derived:

$$\frac{1}{\kappa - 1}V\frac{\mathrm{d}}{\mathrm{d}\phi}p + \frac{\kappa}{\kappa - 1}p\frac{\mathrm{d}}{\mathrm{d}\phi}V = \frac{\mathrm{d}}{\mathrm{d}\phi}Q_{comb} + \frac{\mathrm{d}}{\mathrm{d}\phi}Q_{wall} \tag{14.22}$$

Typically, the gross heat release Q_{gross} and the net heat release Q_{net} are distinguished [12]. The gross heat release Q_{gross} describes the total heat release by combustion. Its change is calculated by

$$\frac{\mathrm{d}}{\mathrm{d}\phi}Q_{gross} = \frac{\mathrm{d}}{\mathrm{d}\phi}Q_{comb} \tag{14.23}$$

The net heat release Q_{net} describes the effective work done on the piston. It takes all loss terms into account. For instance, for the simplified approach described, it follows

$$\frac{\mathrm{d}}{\mathrm{d}\phi}Q_{net} = \frac{\mathrm{d}}{\mathrm{d}\phi}Q_{comb} + \frac{\mathrm{d}}{\mathrm{d}\phi}Q_{wall} \tag{14.24}$$

14.2.3 Applications of the Single-Zone Model

The single-zone model can be used for the purpose of engine process simulation and for heat release analysis. Within the *engine process simulation*, the pressure of the combustion chamber p is determined. For this application, the fuel injection profile is used as input. For optimization-based control, the model is used in the same way. Based on (14.22), the equation for the pressure change can be determined.

$$\frac{\mathrm{d}}{\mathrm{d}\phi}p = -\frac{\kappa}{V}p\frac{\mathrm{d}}{\mathrm{d}\phi}V + \frac{\kappa - 1}{V}\left(\frac{\mathrm{d}}{\mathrm{d}\phi}Q_{comb} + \frac{\mathrm{d}}{\mathrm{d}\phi}Q_{wall}\right) \tag{14.25}$$

For the prediction of the pressure trace, various submodels are needed. A submodel is needed that calculates the volume in dependence of the crank angle as detailed in Sect. 14.2.4. Additionally, a submodel is needed that calculates the loss terms. A model for the wall heat losses is detailed in Sect. 14.2.5. Another important submodel is the combustion model that calculates the heat release by combustion Q_{comb}. A data-based approach that is suitable for combustion rate shaping is explained below in Sect. 15.2.

The second application of the model is the *heat release analysis*. In this case, the heat release is calculated and the pressure trace serves as input. For instance, the heat release analysis can be used to calculate the net heat release based on a measured cylinder pressure profile. For calculation of the net heat release, the following equation results:

$$\frac{\mathrm{d}}{\mathrm{d}\phi}Q_{net} = \frac{1}{\kappa - 1}V\frac{\mathrm{d}}{\mathrm{d}\phi}p + \frac{\kappa}{\kappa - 1}p\frac{\mathrm{d}}{\mathrm{d}\phi}V \tag{14.26}$$

In fact, the net heat release can be calculated, when the pressure trace, the volume, and the isentropic coefficient are known. For the validity of the heat release analysis, the quality of the measured cylinder pressure trace data needs to be high. Thus, the data needs to be recorded appropriately and the data needs to be preprocessed; see [1] for details. For calculation of the gross heat release, additional submodels for the heat losses are needed.

14.2.4 Combustion Chamber Volume

For the process simulation (14.25) as well as for the heat release analysis (14.26), the change of the volume $\mathrm{d}V$ needs to be calculated. The relevant volume V is given by the combustion chamber, where the boundaries are given by the piston, the cylinder liner walls, and the cylinder head. The volume V is changing in dependence of the crank angle ϕ due to the movement of the piston, defined by the crank kinematics. The distance from the top-dead center (TDC) to the piston is denoted by $s(\phi)$. At $\phi = 0$, the position TDC is present where the combustion chamber volume equals the minimum volume, the so-called compression volume V_c. At $\phi = 180°$, the bottom-dead center (BDC) is reached where the combustion chamber volume is at its maximum, i.e. it equals the so-called displacement volume V_d. The position of the piston s can be calculated by [12]

$$s(\phi) = r_{cs} \cdot \left(1 + \frac{1}{\lambda_{cr}} - \cos(\phi) - \frac{1}{\lambda_{cr}}\sqrt{1 - \lambda_{cr}^2 \sin^2(\phi)}\right) \tag{14.27}$$

In this equation, r_{cs} defines the radius of the crank shaft and $\lambda_{cr} = \frac{r_{cs}}{l_{cr}}$ is the ratio of the radius of the crank shaft r_{cs} and the length of the connecting rod l_{cr}.

Taking into account the cross-section area of the cylinder A_{cyl} and the bore diameter d_b, the combustion chamber volume V is well approximated by

$$V(\phi) = V_c + A_{cyl} s(\phi) \quad \text{with} \quad A_{cyl} = \frac{\pi d_b^2}{4} \tag{14.28}$$

With ω_{eng} being the engine speed, the derivative of the combustion chamber volume with respect to the crank angle $\frac{\mathrm{d}V}{\mathrm{d}\phi}$ is given by

$$\frac{\mathrm{d}V}{\mathrm{d}\phi}(\phi) = A_{cyl} r_{cs} \omega_{eng} \cdot \left(\sin(\phi) \frac{\lambda_{cr}}{2} \frac{\sin(2\phi)}{\sqrt{1 - \lambda_{cr}^2 \sin^2(\phi)}}\right) \tag{14.29}$$

Alternatively, the position can be approximated by a simplified relation [9]:

$$s(\phi) = r_{cs} \cdot \left(1 - \cos(\phi) + \frac{\lambda_{cr}}{2} \sin^2(\phi)\right) \tag{14.30}$$

The combustion chamber volume V can be expressed in terms of the compression ratio ϵ. The compression ratio is defined by

$$\epsilon = \frac{V_d + V_c}{V_c} \tag{14.31}$$

Considering that the full stroke has a length of $2r_{cs}$, the combustion chamber volume V results to be [9]

$$V(\phi) = V_d \frac{1}{\epsilon - 1} \frac{s(\phi)}{2r_{cs}} \tag{14.32}$$

The mean piston velocity c_m is a characteristic value that is used for various purposes such as the calculation of the heat transfer through the combustion chamber walls; see Sect. 14.2.5. It is calculated by

$$c_m = \frac{1}{\pi} \int_0^{\pi} \frac{\mathrm{d}s}{\mathrm{d}\phi}(\phi) \frac{\mathrm{d}\phi}{\mathrm{d}t} \mathrm{d}\phi = \frac{2 r_{cs} \omega_{eng}}{\pi} \tag{14.33}$$

14.2.5 *Heat Transfer Through Combustion Chamber Walls*

For the process simulation of the cylinder pressure as well as for the determination of the gross heat release, a submodel is needed that captures the heat transfer between the in-cylinder charge and the combustion chamber walls. The physical processes involved are quite complex, and in-detail models have high computational demands. Simplified approaches only model the convective effect of the heat transfer by a Newtonian approach of the following form:

$$\dot{Q}_{wall} = \alpha A \left(\vartheta_{wall} - \vartheta\right) \tag{14.34}$$

In this equation, α denotes the heat transfer coefficient, ϑ_{wall} the wall temperature, and A the surface area of heat transfer. The value ϑ denotes the in-cylinder charge temperature which can be expressed in terms of the pressure by the ideal gas state equation (14.15). All terms in this equation depend on the time or crank angle, respectively. In the following, simplified relations for the various terms are described.

Heat Transfer Surface Area

The heat transfer surface area A depends on the position of the piston $s(\phi)$. For simplification, only three areas can be considered, namely the cylinder head area A_{head}, the area of the piston A_{pis}, and the area of the cylinder liner $A_{liner}(\phi)$ that is a function of the crank angle due to the piston motion. It results

$$A(\phi) = A_{head} + A_{pis} + A_{liner}(\phi) \tag{14.35}$$

The cylinder liner area is given by

$$A_{liner}(\phi) = \pi d_b s(\phi) \tag{14.36}$$

For determination of A_{head} and A_{pis}, the geometry of the combustion chamber is needed. However, typically rough estimates are sufficient that rely on simplified geometrical relations.

Heat Transfer Coefficient Based on Woschni Modeling Approach

A multitude of approaches are available to model the heat transfer coefficient α; see for instance [3, 6] or [13]. A commonly used approach is the one developed by

Woschni [27, 28] and reproduced in [2]. The heat transfer coefficient is determined by the following empirical relation:

$$\alpha_{woschni} = 130 d_b^{-0.2} p^{0.8} \vartheta^{-0.53} \left(C_1 c_m + C_2 \frac{V_d \vartheta_{ivc}}{p_{ivc} V_{ivc}} (p - p_m) \right)^{0.8} \tag{14.37}$$

In this equation, the term d_b denotes the bore diameter, p the combustion chamber pressure, ϑ the temperature of the charge in the cylinder, c_m is the mean piston velocity, and p_m the pressure for motored operation. The values ϑ_{ivc}, p_{ivc}, V_{ivc} define the conditions at the time point intake valve closing, which serves as a reference point. The constants C_1, C_2 have to be determined based on simulations or experiments. According to [9], if the effect of the swirl can be neglected, the parameters C_1, C_2 can be chosen to be

$$C_1 = 6.18 \tag{14.38}$$

$$C_2 = 0.00324 \text{ [m/(sK)]} \tag{14.39}$$

Wall Temperature

The wall temperature ϑ_{wall} is typically assumed to be quasi-stationary, i.e. no dynamic effects are modeled. One approach presented in [14] follows an empirical approach:

$$\vartheta_{wall} = 360 + 9\lambda^{0.4}\sqrt{n_{eng} d_b} \text{ [K]} \tag{14.40}$$

The air-to-fuel ratio of the in-cylinder charge is denoted by λ, n_{eng} is used for the engine speed in $\frac{1}{\text{min}}$, while d_b denotes the bore diameter.

14.3 Control of Cycle-Integral Combustion Parameters

The state-of-the-art combustion control algorithms rely on the use of a limited number of cycle-integral surrogate parameters. Within the control algorithm, these parameters, typically, between two and four, are used as controlled variables. They reflect the quality of the combustion process within the individual combustion cycle, e.g. they are used as surrogates for performance parameters such as the torque generated, the process efficiency, or the combustion noise emissions. Sensor data, especially those from an in-cylinder pressure sensor, from the preceding cycle are used to determine these parameters. The values obtained are utilized by the control algorithm to calculate the actuated values for the subsequent cycle. In order to account for the MIMO dynamics, approx. the same number of parameters is needed from the actuation side. These consist of parameters from the air and fuel path. For the control purpose, only a limited number of injection events is applied. The publication in [24] describes the use of two pilot, one main, and one post injection. The actuation is updated on a cycle-to-cycle basis. In the following, the most relevant cycle-integral parameters are introduced.

Indicated Mean Effective Pressure

An important surrogate parameter for combustion control is the indicated mean effective pressure (IMEP). The IMEP correlates with the torque generated and thus can be used as a controlled variable in order to satisfy the desired engine torque which is calculated by the torque manager as described in Sect. 7.3.4. The IMEP is defined as the ratio between the total performed mechanical work during one engine cycle W_t, i.e. the volume change work which is conducted on the piston and the displacement volume V_d [12]:

$$IMEP = \frac{W_t}{V_d} \tag{14.41}$$

Using the indicated pressure p and the volume V, it can be calculated with the following integral over one working cycle:

$$IMEP = \frac{1}{V_d} \oint p\mathrm{d}V \tag{14.42}$$

Using the relation given in (14.28), this can be reformulated to

$$IMEP = \frac{1}{V_d} \oint p(\phi) A_{cyl} \mathrm{d}s(\phi) \tag{14.43}$$

For the mean indicated torque M_{ind}, it follows

$$M_{ind} = \frac{V_d}{4\pi} IMEP \tag{14.44}$$

Accordingly, the indicated power P_{ind} can be calculated using the angular speed of the engine ω_{eng}:

$$P_{ind} = M_{ind} \omega_{eng} \tag{14.45}$$

Combustion Average

An important performance parameter of the combustion process is the indicated efficiency η_i. The efficiency η_i can be calculated by the ratio between the mechanical work and the total energy of the fuel injected, where $m_{fuel,tot}$ is the total fuel mass injected and H_{lhv} is the lower heating value [12]:

$$\eta_i = \frac{IMEP\ V_d}{m_{fuel,tot}\ H_{lhv}} \tag{14.46}$$

A widely used surrogate parameter for the indicated efficiency is the combustion average [23]. The combustion average is defined as the crank angle position where 50% of the total fuel mass injected is burnt. It is denoted as CA50:

$$Q_{comb}(\phi = CA50) = 50\ \%\ m_{fuel,tot}\ H_{lhv} \tag{14.47}$$

The characteristic value CA50 works well as a surrogate parameter for efficiency in the case of conventional CI and SI engine operation. In other applications, the correlation to efficiency is not that unanimous. This concerns especially operation modes of CI engines with high shares of post injections [15] or combustion modes such as premixed charge compression ignition (PCCI) with very early injections [11]. One example is shown in Fig. 8.6.

Peak in-Cylinder Pressure

The peak in-cylinder pressure PMAX serves as an important parameter for engine control. It is given by the maximum pressure arising within the combustion cycle:

$$PMAX = (p(\phi))_{max} \tag{14.48}$$

The engine control has to ensure that a critical value for PMAX is not surpassed as the peak pressure correlates with the mechanical load of the engine. The value of the limit depends on the hardware design. Modern CI engines typically operate at peak pressures of 180–220 bar [23]. However, it is desirable to achieve a PMAX that is as high as possible within the specified limit. High peak pressures approximate the ideal isochoric working process, thus increasing the indicated efficiency η_i.

Combustion Noise Indicators

Another performance parameter that has to be taken care of in the combustion control algorithm is the CI engine noise. The most significant reason for noise can be found in the excitation of the combustion chamber walls due to the acting gas forces. A good surrogate parameter for the combustion noise is the maximum in-cylinder pressure gradient DPMAX [15]:

$$DPMAX = \left(\frac{\mathrm{d}p}{\mathrm{d}\phi}\right)_{max} \tag{14.49}$$

In order to maintain a good noise, vibration, and harshness (NVH) behavior, an upper limit for DPMAX has to be sustained by the combustion control algorithm.

14.4 Combustion Rate Shaping

The combustion rate shaping enables highly accurate control of the combustion process. Instead of controlling a few cycle-integral parameters that indirectly reflect the quality of the combustion process, the combustion process itself is controlled in a highly time-resolved manner. The actuated values are updated with a cycle-to-cycle frequency.

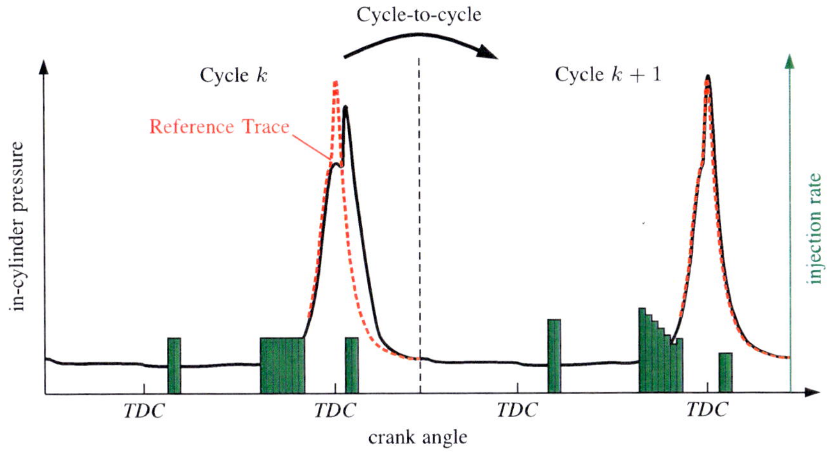

Fig. 14.2 Combustion rate shaping for PCCI

The scheme of combustion rate shaping is depicted in Fig. 14.2. In order to be able to shape the combustion process, a high number of degrees of freedom on the actuation side have to be available that have an effect on the combustion process on a cycle-to-cycle basis. This requirement makes the injector the main actuator for combustion rate shaping. The necessary high number of degrees of freedom is given in modern injection systems. They offer the possibility of multiple injection events in one combustion cycle with a high flexibility concerning the timings and the durations of the injection events. As a result, the manipulation of injection events is used as the main actuated values to conduct combustion rate shaping.

The combustion rate shaping is a promising concept for advanced combustion concepts and in applications with demanding requirements. A combustion concept where the combustion rate shaping is very promising is the PCCI; see Sect. 8.3. Combustion rate shaping can reduce the emissions of noise, particulate matter, and nitrogen oxide while maintaining high levels of efficiency and similar levels with respect to the emissions of carbon monoxide and unburned hydrocarbon [7].

In general, the combustion rate shaping can be used for two scenarios. It can be used as a feedback algorithm where the actuated values, i.e. the injection rate profile, are adjusted on a cycle-to-cycle basis during real operation. This allows to reject disturbances and ensures a high-performance quality of the combustion process during the lifetime of the combustion engine. Disturbance parameters that can be compensated for are the fuel quality and the wear and tear of the injector [22].

The second possibility is the use of combustion rate shaping in order to derive feedforward maps. In this case, combustion rate shaping is applied in the development phase at the engine test bench. The feedforward maps derived are used for real operation without any feedback portion. Due to the high amount of calibration parameters in modern injection systems, their tuning becomes very complex. The combustion rate shaping allows to systematically calibrate these parameters for various influencing variables, e.g. engine speed and rail pressure [17]. This allows the full exploitation of the many degrees of freedom that are present.

To realize combustion rate shaping, various control strategies have been developed, which differ in terms of the control algorithms used and the level of detail concerning the resolution of the controlled combustion rate. The fact that the MIMO system dynamics are considered is common to all these concepts. One possibility is given by the introduction of new combustion metrics in addition to CA50, such as CA30, where 30% of the heat is released [19]. These extended metrics can be controlled using multivariable PI controllers [20]. Alternatively, the highly time-resolved combustion process can be controlled. In this case, the quasi-continuous in-cylinder pressure trace or the quasi-continuous heat release rate trace can be used as controlled variables. Thus, an accurate control of the time evolution of the combustion process can be achieved. In [25], pressure profile tracking is presented by the use of MPC. In [17], a combustion rate shaping approach based on an invertible model is applied. The model consists of simplified physics-based relations. Another possibility is the use of iterative learning control methods in order to control the profile in an iterative fashion [16, 17, 29]. The iterative learning control allows to track a reference profile of the controlled values by cycle-to-cycle adjustment of the profile of the actuated values [4].

References

1. C. Barba, Erarbeitung von verbrennungskennwerten aus indizierdaten zur verbesserten prognose und rechnerischen simulation des verbrennungsablaufes bei pkw-de-dieselmotoren mit common-rail-einspritzung, *Ph.D. Thesis, ETH Zurich*, 2001
2. M. Bargende, Wärmeübergang und wärmebelastung im dieselmotor, *Handbuch Dieselmotoren* (Springer, 2016), pp. 471–496
3. G. Borman, K. Nishiwaki, Internal-combustion engine heat transfer. Prog. Energy Combust. Sci. **13**(1), 1–46 (1987)
4. D. Bristow, M. Tharayil, A. Alleyne, A survey of iterative learning control. IEEE Control Syst. Mag. **26**(3), 96–114 (2006)
5. M.W. Chase, *NIST-JANAF Thermochemical Tables* (American Institute of Physics, 1998)
6. M. Chiodi, M. Bargende, Improvement of engine heat-transfer calculation in the three-dimensional simulation using a phenomenological heat-transfer model, SAE Technical Paper, 2001
7. E.G. Giakoumis, *Driving and Engine Cycles* (Springer, Berlin, 2017)
8. S. Gordon, B.J. McBride, Computer program for calculation of complex chemical equilibrium compositions, rocket performance, incident and reflected shocks, and Chapman-Jouguet detonations, in *National Aeronautics and Space Administration* (1971)
9. L. Guzzella, C.H. Onder, *Introduction to Modeling and Control of Internal Combustion Engine Systems* (Springer, Berlin, 2010)
10. J. Hammer, M. Raff, D. Naber, Advanced diesel fuel injection equipment – a never ending BOSCH story, in *Internationales Stuttgarter Symposium* (2014), pp. 31–45
11. S. Hänggi, G. Moretto, T. Albin, C. Onder, The potential of heat release rate and cylinder pressure feedback control for conventional and premixed charge compression ignition combustion. Int. J. Engine Res., p. 1468087420948314 (2020)
12. J.B. Heywood, *Internal Combustion Engine Fundamentals* (McGraw-Hill Education, 2018)
13. G.F. Hohenberg, Advanced approaches for heat transfer calculations, SAE Transactions, pp. 2788–2806 (1979)

14. R. Isermann, *Engine Modeling and Control* (Springer, Berlin, 2014)
15. C. Jörg, Development of a combustion rate shaping controller for transient engine operation on a direct injection compression ignition engine, *Ph.D. Thesis, RWTH Aachen University*, 2018
16. C. Jörg, T. Schnorbus, S. Jarvis, B. Neaves, K. Bandila, D. Neumann, Feedforward control approach for digital combustion rate shaping realizing predefined combustion processes. SAE Int. J. Engines **8**(3), 1041–1054 (2015)
17. C. Jörg, J. Schaub, D. Neumann, S. Pischinger, Diesel combustion control via rate shaping. MTZ Worldwide **79**(4), 16–21 (2018)
18. U. Kiencke, L. Nielsen, *Automotive Control Systems* (Springer, Berlin, 2005)
19. X. Luo, M. Donkers, B. de Jager, F. Willems, Multi-pulse fuel injection controller design using a quadratic model, in *IEEE Conference on Control Applications* (2016), pp. 305–310
20. X. Luo, M. Donkers, B. de Jager, F. Willems, Systematic design of multivariable fuel injection controllers for advanced diesel combustion. IEEE Trans. Control Syst. Technol. **99**, 1–12 (2018)
21. G. Merker, C. Schwarz, G. Stiesch, F. Otto, *Simulating Combustion: Simulation of Combustion and Pollutant Formation for Engine-Development* (Springer, Berlin, 2014)
22. D. Neumann, P. Muthyala, C. Frenken, J. Schaub, C. Jörg, M. Kötter, Flex-fuel capability via advanced digital combustion rate shaping and airpath control. Internationaler Motorenkongress **2019**, 333–350 (2019)
23. T. Schnorbus, M. Lamping, T. Körfer, S. Pischinger, New challenges for combustion control in advanced diesel engines. MTZ Worldwide **69**(4), 18–26 (2008)
24. F. Tschanz, S. Zentner, C.H. Onder, L. Guzzella, Cascaded control of combustion and pollutant emissions in diesel engines. Control Eng. Pract. **29**, 176–186 (2014)
25. G. Turesson, L. Yin, R. Johansson, P. Tunestål, Predictive pressure control with multiple injections. IFAC-PapersOnLine **51**(31), 706–713 (2018)
26. T. Wintrich, S. Rothe, K. Bucher, H.-J. Hitz, Diesel injection system with closed-loop control. MTZ Worldwide **79**(9), 54–59 (2018)
27. G. Woschni, Die Berechnung der Wandverluste und der thermischen Belastung der Bauteile von Dieselmotoren. Motortechnische Zeitschrift **31**(12), 338–353 (1970)
28. G. Woschni, A universally applicable equation for the instantaneous heat transfer coefficient in the internal combustion engine, in *SAE transactions* (1968), pp. 3065–3083
29. R. Zweigel, F. Thelen, D. Abel, T. Albin, Iterative learning approach for diesel combustion control using injection rate shaping, in *European Control Conference* (2015), pp. 3168–3173

Chapter 15
Combustion Rate Shaping Control

Abstract The current state-of-the-art combustion control concepts are based on cycle-to-cycle actuation and the control of cycle-integral values. A more advanced combustion control concept, the so-called combustion rate shaping, aims to control the combustion process in a highly time resolved manner. In case of the fullest extent, the crank-angle-resolved in-cylinder pressure signals are controlled. The concept relies on the actuation side on injection systems that are highly flexible as given for modern injection systems. In this chapter, two aspects of the combustion rate shaping are detailed. First, the modeling of the combustion process is investigated. A low-order model is required that is able to reproduce the combustion process in case of highly flexible injection events. A suitable approach based on data-based models is presented. Second, the digitalization of a continuous injection profile is examined. Digitalization is needed when a series-production injector is applied as they are only capable of realizing digital injection events. In both parts, optimization-based methods are depicted.

15.1 Introduction

The goal of combustion rate shaping is the control of a highly time-resolved combustion profile. One prerequisite is the usage of an injector that is highly flexible. In general, there exist two injection systems that can be applied for this purpose. Fuel injectors can be applied that are capable of injecting a continuously shaped injection rate profile [14]. Thus, as actuated values the trace of the injection shape profile can be manipulated. These kinds of fuel injectors are only available as prototypes for research [11]. Alternatively, conventional fuel injection systems can be applied that are used in series production. The electric actuation signal turns the injector "on" and "off" for each injection event, making it "digital". The "on" and "off" switching can be conducted multiple times within one combustion cycle. Thus, a multi-pulse

T. Albin Rajasingham, *Nonlinear Model Predictive Control of Combustion Engines*, Advances in Industrial Control,
https://doi.org/10.1007/978-3-030-68010-7_15

fuel injection strategy results with a flexible number of injection events [13, 16]. As actuated values the timings and the duration of the various injection events can be chosen. Often, these two approaches are referred to as continuous vs. digital combustion rate shaping. The maximum achievable control quality is higher for continuous combustion rate shaping, but it poses higher demands on the injection system. The achievable quality of digital rate shaping is strongly dependent on the injection system applied. This concerns hardware-specific parameters such as the minimum amount of fuel that can be injected per injection event and the minimum time distance between two injection events, the so-called dwell time. For the calculation of a suitable injection profile, the consideration of these limitations is essential to fully exploit the capabilities of the injection system applied [23].

Figure 15.1 shows a possible control structure for the realization of digital combustion rate shaping which is reproduced from previous collaborative work of the author in [23]. In this concept, a continuously shaped trace of heat release rate is used as a controlled variable. Thus, the desired trace of heat release rate is needed which is used as a reference profile. The desired trace of heat release rate needs to fulfill criteria such as providing the requested torque and achieving low pollutant emissions and also low noise emissions. Based on the reference trace and the current measurements, a controller calculates a continuous injection rate profile. Due to the nonlinearities of the system to be controlled, model-based approaches are advantageous for this task. In turn, a model of the combustion process is necessary. For digital combustion rate shaping, an injector is applied that cannot inject a continuous injection rate profile. Thus, subsequently, an injection rate digitalization is necessary. The goal of the digitalization process is to obtain the digital actuated signals of the injector that resemble as close as possible the continuous injection rate profile.

Two aspects of the combustion rate shaping are depicted. First, a data-based approach is presented for the modeling of the combustion process. This model can be used within the combustion controller that transforms the desired heat release trace along with the measurements into a continuous injection profile. Second, an

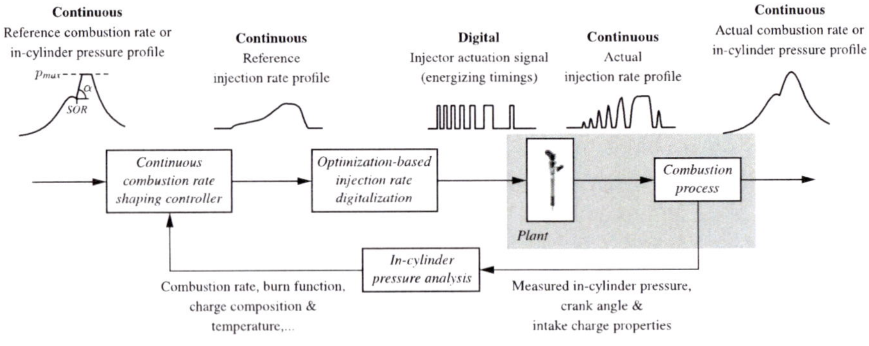

Fig. 15.1 Control structure for digital combustion rate shaping – reproduced from [23] with permission of the American Automatic Control Council

optimization-based approach for injection rate digitalization is presented. The data-based injector model is presented along with the optimization problem.

15.2 Combustion Modeling

For model-based combustion rate shaping, a low-order model of the combustion process is needed. The combustion model uses the injection rate profile as input variables, and the relevant output is the accumulated heat release from combustion $Q_{comb}(\phi)$. Typically, the progress is represented in dependence of the crank angle ϕ as an independent variable. The main challenge arises as the inputs and the outputs are high-dimensional, as they represent quasi-continuous traces, and the input–output correlation is strongly nonlinear. Multiple approaches exist to model the combustion process. They vary in terms of the detail level and the computational complexity.

15.2.1 Overview on Combustion Modeling Approaches

The most detailed models are given by *Reactive Computational Fluid Dynamics* models. They describe the physical and chemical processes involved by first-principle models, e.g. for the fuel spray vaporization, the turbulent flow dynamics, and the chemical reactions [17, 21]. Typically, the mathematical model is given by partial differential equations (PDEs). As the effect of the engine geometry can be predicted, they are, for instance, well suited for applications such as the design of the combustion chamber [10]. However, they are many orders of magnitude too complex for real-time optimization-based control.

There exist two alternatives that can be used for optimization-based control. One possibility is given by *data-based models* which use global empirical relations in order to approximate the measured system behavior. They can approximate the behavior very well with low computational demands. However, they are only well suited if enough measurement data is available, i.e. they have little capability of extrapolation.

Another possibility is given by simplified physics-based models, often referred as *phenomenological models*. They are better suited in terms of extrapolation capability but usually are more complex than data-based models. Phenomenological models use various submodels in order to describe the heat release by combustion. The submodels are represented by simplified physical relations or by data-based models. Thus, phenomenological models cover a broad spectrum regarding computational complexity and detail level [18]. Depending on the level of detail described, subprocesses like spray development and mixture formation can be included [22]. A computationally inexpensive model is described in [6, 7]. The model is based on the idea that the heat release rate is proportional to the recent fuel mass available for combustion. To account for this, the difference between the energy of the total fuel mass injected into the cylinder up until that moment $Q_{fuel}(\phi)$ and the energy of the already burnt fuel mass which is equivalent to the accumulated heat release $Q_{comb}(\phi)$ is considered in the differential equation:

$$\frac{\mathrm{d}Q_{comb}}{\mathrm{d}\phi}(\phi) = C(\phi)\left(Q_{fuel}(\phi) - Q_{comb}(\phi)\right) \tag{15.1}$$

The progress-dependent proportionality factor $C(\phi)$ can be interpreted in a physical sense, i.e. it represents the local density of the turbulent kinetic energy. Other researchers as [5, 16, 19] have extended this approach in order to include the ignition delay. In this case, a time-variant ignition delay time $\tau_{ign}(\phi)$ is added. Using H_{lhv} as lower heating value and $m_{fuel,inj}$ as the injected fuel mass, it results:

$$Q_{fuel}(\phi) = H_{lhv} m_{fuel,inj}(\phi - \tau_{ign}(\phi)) \tag{15.2}$$

As the input for these kind of models is given by the profile of injected fuel mass, they are applicable for combustion rate shaping. Several approaches for combustion rate shaping have been presented that are based on phenomenological models. The models are, for instance, well suited for feedforward control. Combustion rate shaping approaches are detailed in publications such as [16, 19].

15.2.2 Data-Based Combustion Models

Data-based combustion models can be used within single-zone models as presented in Sect. 14.2. They offer the possibility to reproduce the combustion process with low computational demands. They use basis functions for the approximation of the system behavior. These basis functions are also called replacement burn functions. The parameters of the model are given as a function that depends on global variables describing the current operating point such as the engine speed. The suitability of a certain replacement burn function depends heavily on the combustion type investigated. For instance, different functions are needed for SI and CI combustion and for single-pulse injection and multi-pulse injection. Thus, the replacement burn function has to be chosen carefully according to the given system setup.

In the following, the most common replacement burn function, the Vibe function, is introduced. More specifically, the classical Vibe function is introduced along with an extension to a Vibe mixture model. The Vibe mixture model is suited for modeling the CI combustion process arising for CI combustion rate shaping.

Vibe Combustion Model

The Vibe function is the most common replacement burn function that is used for data-based modeling of the combustion process [12]. The original publication dates back to 1956 [26]. The Vibe function describes the accumulated heat release from combustion $Q_{comb}(\phi)$ with the progress in crank angle ϕ as an independent variable. The analytic function which is defined for $\phi \geq \phi_{SOC}$ is given by

$$Q_{comb}(\phi) = Q_{tot} \cdot \left(1 - \mathrm{E}^{-a\left(\frac{\phi-\phi_{SOC}}{\Delta\phi_{bd}}\right)^{m+1}}\right) \tag{15.3}$$

The value $Q_{tot} = m_{fuel,tot} H_{lhv}$ denotes the total energy of the fuel injected over the combustion cycle. The instant of start of combustion is denoted by ϕ_{SOC}. The duration of combustion is given by $\Delta\phi_{bd}$. The two parameters a and m determine the shape of the resulting heat release profile.

The differentiation of (15.3) with respect to ϕ yields the analytic function for the heat release rate $\frac{\mathrm{d}Q_{comb}}{\mathrm{d}\phi}(\phi)$:

$$\frac{\mathrm{d}Q_{comb}}{\mathrm{d}\phi}(\phi) = a\,Q_{tot} \cdot (m+1) \cdot \left(\frac{\phi - \phi_{SOC}}{\Delta\phi_{bd}}\right)^{m} \mathrm{E}^{-a\cdot\left(\frac{\phi-\phi_{SOC}}{\Delta\phi_{bd}}\right)^{m+1}} \tag{15.4}$$

Figure 15.2 shows an exemplary variation of a and the resulting profiles for the heat release rate and the accumulated heat release. Similarly, Fig. 15.3 shows an exemplary variation of the parameter m.

The classical Vibe function is well suited for the approximation of the combustion process of an SI engine. In fact, it was developed for reproducing the SI engine behavior. However, it is not well suited for CI combustion. The single Vibe function cannot reproduce the characteristic phases of CI combustion, i.e. the premixed com-

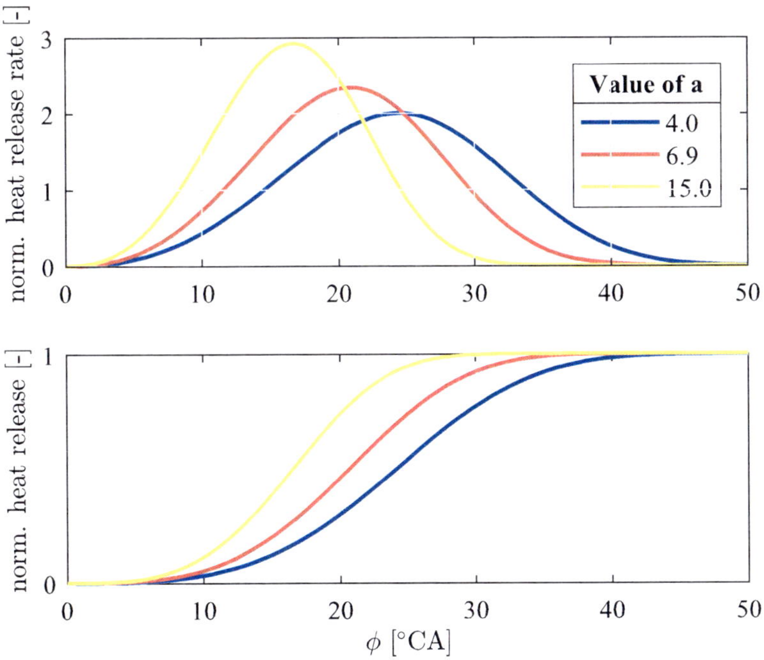

Fig. 15.2 Vibe function model for various values of the shape parameter *a*

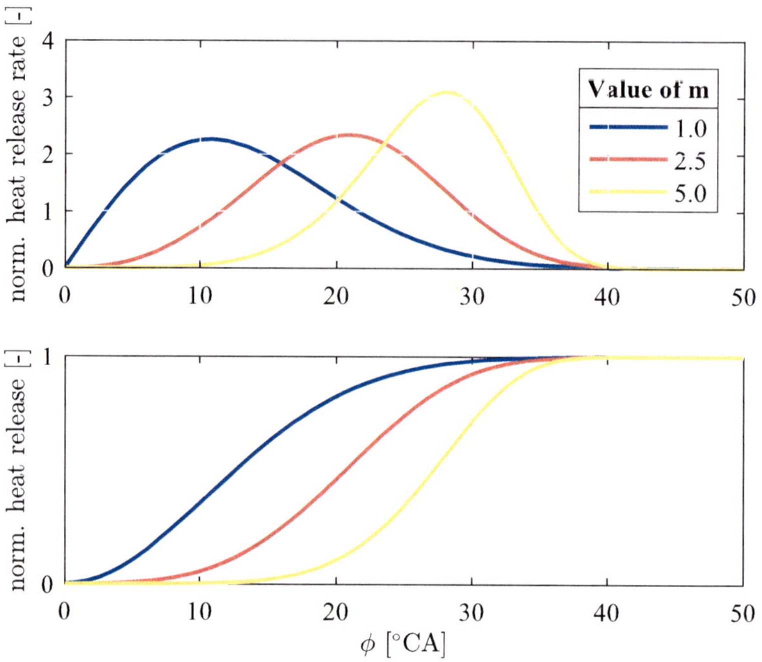

Fig. 15.3 Vibe function model for various values of the shape parameter m

bustion phase, the subsequent diffusive combustion phase, and the burn-out phase. For data-based modeling of the CI combustion, several adaptations have been proposed. This includes the use of multiple Vibe functions [18] as well as the use of other replacement burn functions, such as the use of hyperbolic profiles or polygon-hyperbola functions [3, 25]. In the following, the use of multiple Vibe functions, so-called Vibe mixture models, is introduced.

Vibe Mixture Combustion Model

Vibe mixture combustion models have been proposed for accurately reproducing the behavior of CI combustion for combustion rate shaping applications [24]. The main idea is the representation of the heat release rate by the linear superposition of multiple Vibe functions. The basic principle is depicted in Fig. 15.4 where the sum of three Vibe functions is calculated to derive the overall heat release rate. The linear superposition of various Vibe functions allows to reproduce complex combustion profiles.

One Vibe mixture combustion model approach was presented in collaborative work of the author [24], which is detailed in the following. In this case, the continuous fuel injection profile is discretized into N_u samples, i.e. $\boldsymbol{u} = [u_1 \ \dots \ u_{N_u}]^\mathsf{T}$. The unit of each discretized input shall be given by the injected energy, i.e. $u_i = m_{fuel,i} H_{lhv}$. The sum of all discretized injected energy packages is equal to the total energy injected in one cycle:

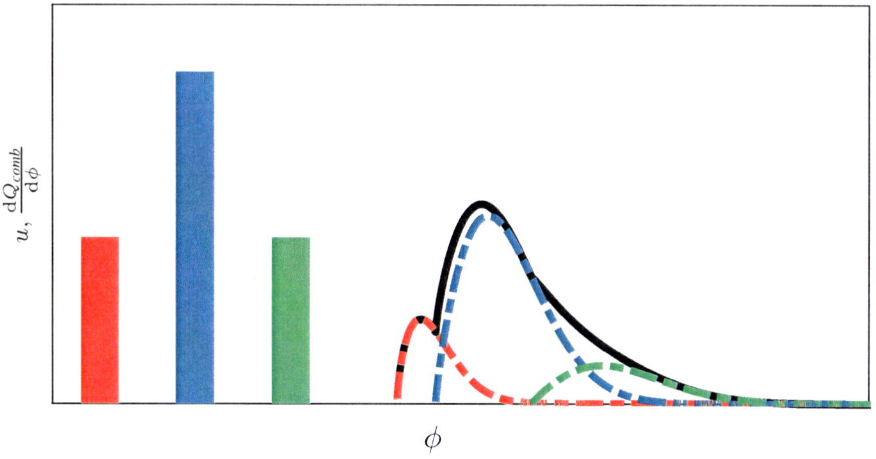

Fig. 15.4 Linear superposition of various Vibe functions, adapted by permission from Springer Nature: [24], copyright 2020

$$Q_{tot} = \sum_{i=1}^{N_u} u_i \tag{15.5}$$

The Vibe mixture model relates the i fuel packages u_i to the accumulated heat release denoted by $\boldsymbol{Q}_{comb,i}(\phi)$. For the Vibe mixture model, it is assumed that the injected fuel is burnt entirely and that each injected fuel package results in a distinctive heat release $\boldsymbol{Q}_{comb,i}(\phi)$. The overall heat release is given by the sum of all the individual heat release functions.

$$\boldsymbol{Q}_{comb}(\phi) = \sum_{i=1}^{N_u} \boldsymbol{Q}_{comb,i}(\phi) = \sum_{i=1}^{N_u} u_i \cdot \left(1 - \mathrm{E}^{-a_i\left(\frac{\phi-\phi_{SOC,i}}{\Delta\phi_{bd,i}}\right)^{m_i+1}}\right) \tag{15.6}$$

For each of the N_u Vibe functions, the parameters m_i, $\Delta\phi_{bd,i}$, and $\phi_{soc,i}$ need to be determined. The parameters a_i are typically fixed to a constant value. In [24], a value of $a = 6.908$ is chosen, which results in 99.9% burnt fuel after the duration of combustion $\Delta\phi_{bd,i}$, which is a reasonable assumption.

The identification of the various parameters within the Vibe mixture model can be conducted by numerical solution of an optimization problem where measured data is used as input. Within the optimization problem, constraints can be considered such that a reasonable solution results. The overall optimization problem can be formulated as follows, where $Q_{comb,meas,k}$ denotes the measured accumulated heat release at the discrete-time point k and $Q_{comb,k}$ denotes the one predicted by the Vibe mixture model at the discrete-time point k.

$$\min_{\boldsymbol{m},\, \boldsymbol{\phi}_{soc} \boldsymbol{\Delta\phi}_{bd}} \quad \sum_{k=1}^{N} \| Q_{comb,meas,k} - Q_{comb,k}(\boldsymbol{m},\, \boldsymbol{\Delta\phi}_{soc},\, \boldsymbol{\phi}_{bd}) \|^2 \tag{15.7a}$$

$$\text{s.t.} \quad \boldsymbol{\phi}_{soc,min} \leq \boldsymbol{\phi}_{soc} \leq \boldsymbol{\phi}_{soc,max}, \tag{15.7b}$$

$$\boldsymbol{\Delta\phi}_{bd,min} \leq \boldsymbol{\Delta\phi}_{bd} \leq \boldsymbol{\Delta\phi}_{bd,max}, \tag{15.7c}$$

$$\boldsymbol{m}_{min} \leq \boldsymbol{m} \leq \boldsymbol{m}_{max}, \tag{15.7d}$$

$$\phi_{soc,i} \leq \phi_{soc,i+1} i = 1, .., N_u - 1, \tag{15.7e}$$

$$\phi_{soi,i} \leq \phi_{soc,i} \quad i = 1, .., N_u - 1 \tag{15.7f}$$

In the optimization problem, a least-squares cost function is used which penalizes the deviation from the measured data. Box constraints are defined for the parameters $\boldsymbol{m}$, $\boldsymbol{\phi}_{soc}$, $\boldsymbol{\Delta\phi}_{bd}$; see (15.7b)–(15.7d). These constraints limit the parameter range to realistic values. Additionally, the constraint in (15.7e) ensures that the order of combustion is maintained, i.e. $\phi_{soc,i} \leq \phi_{soc,i+1}$. The constraint in (15.7f) forces the start of combustion $\phi_{soc,i}$ of a package i to be after the start of injection of the corresponding fuel injection package, which is denoted by $\phi_{soi,i}$.

One possibility to solve this optimization problem is given by sequential quadratic programming (SQP). As a least-squares objective is used, the Gauss–Newton approximation can be applied. Figure 15.5 shows a validation plot of a Vibe mixture model. In this case, the fuel injection was realized with a series-production injector. In the combustion cycle shown, four injections have been realized. The continuous injection profile has been discretized using $N_u = 60$ packages. Each of the $N_u = 60$ packages results in a specific Vibe function. Within the figure, the sum of all individual Vibe

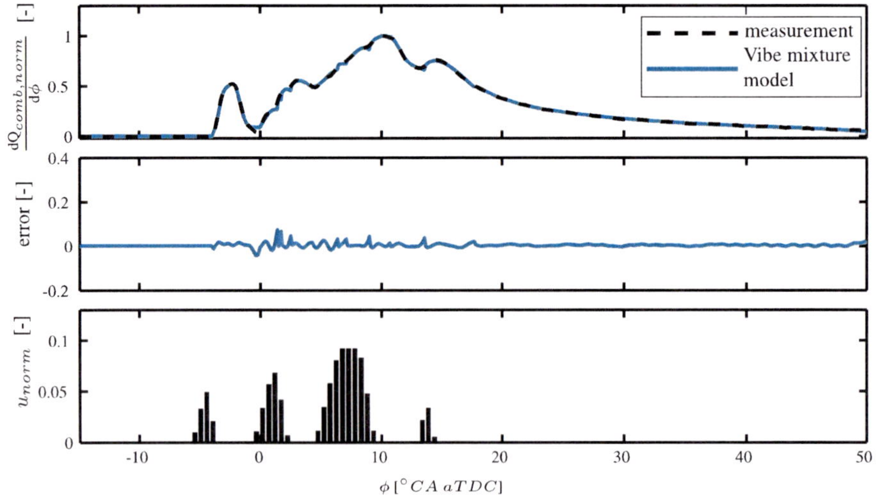

Fig. 15.5 Validation of the Vibe mixure model against measured data, adapted by permission from Springer Nature: [24], copyright 2020

functions is shown. The figure shows that the complex combustion profile can be reproduced very accurately using the Vibe mixture model.

The speed of the SQP algorithm even allows to solve the optimization problem during the runtime of the process. This can be exploited for the purpose of deriving feedforward maps at the combustion engine test bench. For instance, a learning-based control algorithm can be applied. In this case, in each iteration, the parameters of the model are adapted, i.e. by model parameter identification, to improve the model accuracy. The updated model itself is used subsequently to generate the actuation profile for the next iteration. This algorithm relies on fast updates of the model parameters, which can be accomplished with the SQP algorithm.

15.3 Optimization-Based Fuel Injection Rate Digitalization

Several approaches have been proposed to "digitalize" the continuous fuel injection profile. For instance, in [15], an iterative approach is presented. Boundary conditions such as the dwell time which specifies the minimum time between two subsequent injection events and the minimum fuel quantity injected are considered using heuristic rules. An alternative possibility was presented in the collaborative work of the author [23]. It uses real-time optimization for the digitalization task. The formulation as an optimization problem has the advantage that the nonlinear model of the injection system can be taken into account. Additionally, limitations of the hardware such as the dwell time can be considered directly as constraints of the optimization problem. Thus, the optimal digitalized injection strategy can be found that is effectively realizable with the given injection system. The real-time optimization-based approach from [23] is detailed below. First, the data-based fuel injection model is introduced along with its validation. Based on this, the formulation of the optimization problem is detailed. Finally, results from the digitalization process are highlighted.

15.3.1 Data-Based Fuel Injection Model

For the optimization-based fuel injection rate digitalization, a model is needed that correlates the actuated values start of energizing SOE_i and duration of energizing DOE_i of the n_{inj} injection events to the output which is the overall volume flow of injected fuel $\dot{V}_{fuel,inj}(t)$. The model needs to be of low order for usage within optimization-based control. At the same time, a continuously differentiable model makes the application of gradient-based optimization methods possible.

There exist several approaches to model the injector fuel volume flow. For instance, physical models can be used, such as proposed in [8, 9]. All the components of the injection system are modeled, such as the fuel flow dynamics, mechanical displacement of moving parts, and pressure wave propagation. A very detailed model results that captures the most important physical effects occurring in the injection system. However, the resulting model is of high order which makes it too complex to be used for real-time optimization. Alternatively, data-based models can be used which are typically computationally cheap such that they are suitable for real-time optimization. Usually, a sum of multiple instances of a base-function is used with the individual energizing timing SOE_i and DOE_i as inputs. The overall fuel volume flow $\dot{V}_{fuel,inj}$ is given by the sum of each individual injection event i:

$$\dot{V}_{fuel,inj}(t) = \sum_{i=1}^{n_{inj}} \dot{V}_{fuel,inj,i}(t, SOE_i, DOE_i) \tag{15.8}$$

The superposition reproduces the real behavior appropriately well as long as the injection events are not influenced by a preceding injection event. This assumption holds true when the time between two injection events is sufficiently long, i.e. longer than the dwell time τ_{dwell}.

Various base-functions have been applied to reproduce the behavior of the injection system. For instance, in [16, 20], the use of trapezoidal shapes has been proposed. The trapezoidal base-function can reproduce the system behavior quite well; however, it has the disadvantage that it is not continuously differentiable. In [23], the use of second-order step response functions with dead-time as base-function is suggested which are smoothly connected such that the overall model is continuously differentiable. Additionally, an empirical function is used to describe the injection dead-time τ_{inj} between the start of energizing SOE and the start of injection SOI. Physically, this dead-time is caused by the delay of the piezoelectric stack and the needle [4]. The model can reproduce the fuel volume flow very accurately. An exemplary validation from [23] is shown in Fig. 15.6.

15.3.2 *Formulation of the Optimization Problem*

The optimization variables are given by the start timings and the durations of energizing for the various injection events $SOE_i, DOE_i \in \mathbb{R}$ for $i = \{1, \ldots, n_{inj}\}$. The number of injection events $n_{inj} \in \mathbb{N}$ itself is an optimization variable, as the optimal number can vary based on the reference trace and the rail pressure applied. The inclusion of integer optimization variables makes the optimization problem a mixed-integer problem. For appropriate injection rate digitalization, the optimization problem has to consider various aspects. The main goal of the digitalization method is to calculate the optimization variables that reproduce as good as possible the continuous reference signal $\dot{V}_{fuel,ref}(t)$. The rail pressure is considered as an external parameter. At the same time, the limitations of the hardware are considered as constraints.

Formulation of the Continuous-Time Optimal Control Problem

A continuous-time finite-horizon mixed-integer nonlinear optimal control problem is set up in order to achieve the goals mentioned above. Within the cost function (15.9a),

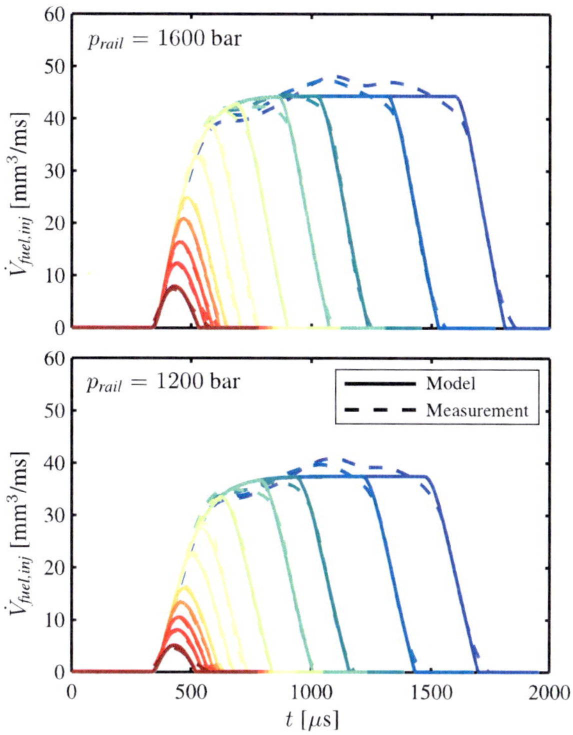

Fig. 15.6 Validation of injection rate model for various rail pressures p_{rail} and durations of energizing *DOE* – reproduced from [23] with permission of the American Automatic Control Council

the least-squares deviation between the integrated fuel volume flow and the integrated reference trace is used. The consideration of the integrated fuel volume flow has advantages compared to the non-integrated volume flow. The integration has a low-pass characteristic, thus a higher sampling time can be used which decreases the computational complexity. Additionally, the integration preserves the information of the preceding fuel injection events, which avoids undersampling where small injection events are crossed over.

$$\min_{\substack{SOE_i, DOE_i, n_{inj} \\ \text{for } i \in \{1, \ldots n_{inj}\}}} \int_{T_{N_1}}^{T_{N_2}} \left\| V_{fuel,inj}(t) - V_{fuel,ref}(t) \right\|^2 \mathrm{d}t \tag{15.9a}$$

s.t.

$$\dot{V}_{fuel,inj}(t) = \sum_{i=1}^{n_{inj}} \dot{V}_{fuel,inj,i}(t, SOE_i, DOE_i), \tag{15.9b}$$

$$V_{fuel,inj}(T_{N_1}) = 0, \tag{15.9c}$$

$$V_{fuel,inj}(T_{N_2}) = V_{fuel,ref}(T_{N_2}), \tag{15.9d}$$

$$SOI_i(SOE_i) \geq EOI_{i-1}(SOE_{i-1}, DOE_{i-1}) + \tau_{dwell},\ i \in \{2, \ldots, n_{inj}\}, \tag{15.9e}$$

$$SOI_1(SOE_1) \geq T_{N_1}, \tag{15.9f}$$

$$EOI_{n_{inj}}(SOE_{n_{inj}}, DOE_{n_{inj}}) \leq T_{N_2}, \tag{15.9g}$$

$$DOE_{min} \leq DOE_i \leq DOE_{max},\ i \in \{1, \ldots, n_{inj}\}, \tag{15.9h}$$

$$1 \leq n_{inj} \leq n_{inj,max} \tag{15.9i}$$

The data-based model of the injector is added as equality constraint (15.9b). The boundary conditions for the beginning and the end of the integrated fuel volume flow are incorporated in (15.9c)–(15.9d). The consideration of the terminal constraint is important in order to maintain the amount of injected fuel volume respectively fuel mass and thus the generated torque. Additionally, hardware-related limitations of the injection system are incorporated. The inequality constraint (15.9e) ensures that the dwell time is considered. The start of injection *SOI* and the end of injection *EOI* are given by data-based models that depend on the optimization variables. For the injections timings, a minimum and a maximum value are defined, which are included by (15.9f)–(15.9g). For the durations of injections, minimum and maximum values are defined as well by (15.9h). The upper limit for the number of injection events n_{inj} is considered by (15.9i).

Discretization and Numerical Solution of the Resulting Mixed-Integer NLP

For discretization, a single shooting approach is used, due to the structure of the model, i.e. the injection events are decoupled in the model. All in all, N discretization intervals are used with a sampling time of $T_s = \frac{T_{N_2} - T_{N_1}}{N}$. The numerical integration of the fuel volume flow is conducted with n_{int} integration steps per interval using a third-order explicit Runge–Kutta integration scheme, resulting in

$$V_{fuel,inj}(k+1) = \sum_{i=1}^{n_{inj}} f_{dis}(V_{fuel,inj}(k), SOE_i(k), DOE_i(k)) \tag{15.10}$$

The following optimization problem results:

$$\min_{\substack{SOE_i, DOE_i, n_{inj} \\ \text{for } i \in \{1, \ldots n_{inj}\}}} \sum_{k=1}^{N} \left\| V_{fuel,inj}(k) - V_{fuel,ref}(k) \right\|^2 T_s \tag{15.11a}$$

s.t.

$$V_{fuel,inj}(k+1) = \sum_{i=1}^{n_{inj}} f_{dis}(V_{fuel,inj}(k), SOE_i(k), DOE_i(k)),\ k \in \{0, \ldots, N-1\}, \tag{15.11b}$$

$$V_{fuel,inj}(k=0) = 0, \tag{15.11c}$$

$$V_{fuel,inj}(k=N) = V_{fuel,ref}(k=N), \tag{15.11d}$$

$$SOI_i(SOE_i) \geq EOI_{i-1}\left(SOE_{i-1}, DOE_{i-1}\right) + \tau_{dwell},\ i \in \{2, \ldots, n_{inj}\}, \tag{15.11e}$$

$$SOI_1(SOE_1) \geq T_{N_1}, \tag{15.11f}$$

$$EOI_{n_{inj}}(SOE_{n_{inj}}, DOE_{n_{inj}}) \leq T_{N_2}, \tag{15.11g}$$

$$DOE_{min} \leq DOE_i \leq DOE_{max},\ i \in \{1, \ldots, n_{inj}\}, \tag{15.11h}$$

$$1 \leq n_{inj} \leq n_{inj,max} \tag{15.11i}$$

For numerical solution of the optimization problem, the problem is split up into several NLPs. More specifically, for each integer possibility of number of injection events $n_{inj} = 1, \ldots, n_{inj,max}$, a single NLP is solved. Each of this NLPs is solved by an SQP algorithm. The overall minimizer can be obtained by comparing the various solutions, i.e. the one with the lowest cost is the overall minimizer. As a least-squares tracking term is given in the cost function, the Gauss–Newton Hessian approximation can be employed. CasADi is used for automated calculation of the derivatives [1].

15.3.3 Validation of the Fuel Injection Rate Digitalization Concept

In the following, the validation of the fuel injection rate digitalization method is detailed. For this purpose, a reference trace for the continuous injection rate is digitalized with the optimization-based method. Figure 15.7 shows the results of an exemplary digitalization. The reference trace is a so-called "boot"-shaped injection profile which is commonly used for advanced CI combustion. From a combustion point of view, it offers the advantage that the peak of the premixed portion is reduced, and a combustion with a quite constant pressure rise gradient results [2]. At the same time, it is a profile that is challenging to be reproduced with digital injection events, due to the flat slope at the beginning of the injection. The digitalization method is applied for three rail pressures. In all three cases, the approach detailed is able to find suitable injection profiles that reproduce the continuous injection very well. For quantification of the performance, the root mean squared error (RMSE) is used. The optimal number of injections depends on the rail pressure applied as the maximum fuel volume flow depends on the rail pressure setting. As well, all the profiles comply with the constraints formulated, such as the dwell time and the total fuel volume injected over all injections.

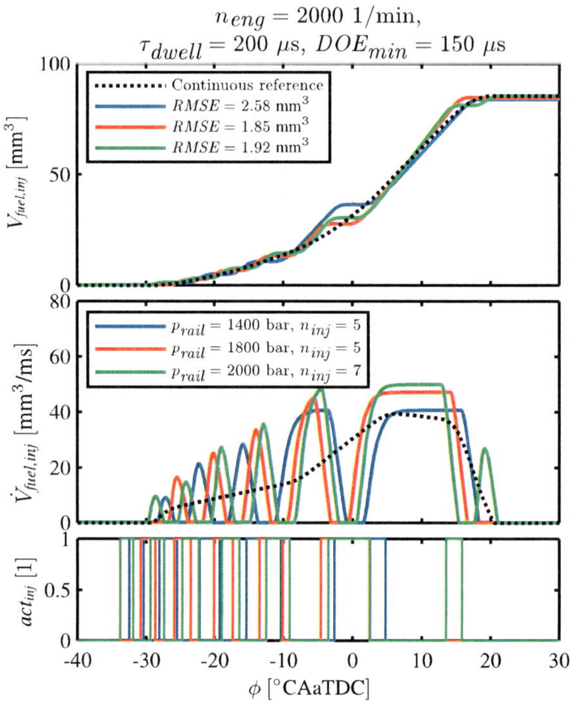

Fig. 15.7 Validation of injection rate digitalization for various rail pressures p_{rail} – reproduced from[23] with permission of the American Automatic Control Council

References

1. J. Andersson, J. Akesson, M. Diehl, CasADi: a symbolic package for automatic differentiation and optimal control, in *Recent Advances in Algorithmic Differentiation* (Springer, Berlin, 2012), pp. 297–307
2. F. Atzler, O. Kastner, A. Rotondi, A. Weigand, Multiple injection and rate shaping – Part 1: Emissions reduction in passenger car diesel engines. SAE Technical Paper, no. 2009-24-0004 (2009)
3. C. Barba, Erarbeitung von verbrennungskennwerten aus indizierdaten zur verbesserten prognose und rechnerischen simulation des verbrennungsablaufes bei pkw-de-dieselmotoren mit common-rail-einspritzung. Ph.D. Thesis, ETH Zurich (2001)
4. G. Bianchi, S. Falfari, F. Brusiani, P. Pelloni, G. Osbat, M. Parotto, Numerical investigation of critical issues in multiple-injection strategy operated by a new CR fast-actuation solenoid injector. SAE Technical Paper, no. 2005-01-1236 (2005)
5. A. Catania, R. Finesso, E. Spessa, Predictive zero-dimensional combustion model for DI diesel engine feed-forward control. Energy Convers. Manag. **52**(10), 3159–3175 (2011)
6. F. Chmela, G. Orthaber, Rate of heat release prediction for direct injection diesel engines based on purely mixing controlled combustion. SAE Trans. 152–160 (1999)
7. F. Chmela, G. Pirker, A. Wimmer, Zero-dimensional ROHR simulation for DI diesel engines - a generic approach. Energy Convers. Manag. **48**(11), 2942–2950 (2007)
8. C. Dongiovanni, M. Coppo, D. Siano, Accurate modelling of an injector for common rail systems. Fuel Inject. 95–119 (2010)
9. R. Garrappa, P. Lino, G. Maione, F. Saponaro, Model optimization and flow rate prediction in electro-injectors of diesel injection systems. IFAC-PapersOnLine **49**(11), 484–489 (2016)

10. H.-W. Ge, Y. Shi, R.D. Reitz, D. Wickman, W. Willems, Engine development using multi-dimensional cfd and computer optimization. SAE Technical Paper (2010)
11. P. Grzeschik, H.-J. Laumen, U. Schlemmer Kellig, FEV HiFORS injector with continuous rate shaping: influence on mixture formation and combustion process. Energy Environ. **1**(9), 123–135 (2016)
12. L. Guzzella, C.H. Onder, *Introduction to Modeling and Control of Internal Combustion Engine Systems* (Springer, Berlin, 2010)
13. J. Hammer, M. Raff, D. Naber, Advanced diesel fuel injection equipment – a never ending BOSCH story, in *Internationales Stuttgarter Symposium* (2014), pp. 31–45
14. J. Hinkelbein, F. Kremer, M. Lamping, T. Körfer, J. Schaub, S. Pischinger, Experimental realisation of predefined diesel combustion process using advanced closed-loop combustion control and injection rate shaping. Int. J. Engine Res. **13**(6), 607–615 (2012)
15. C. Jörg, Development of a combustion rate shaping controller for transient engine operation on a direct injection compression ignition engine. Ph.D. Thesis, RWTH Aachen University (2018)
16. C. Jörg, T. Schnorbus, S. Jarvis, B. Neaves, K. Bandila, D. Neumann, Feedforward control approach for digital combustion rate shaping realizing predefined combustion processes. SAE Int. J. Eng. **8**(3), 1041–1054 (2015)
17. S. Kumar, M.K. Chauhan et al., Numerical modeling of compression ignition engine: a review. Renew. Sustain. Energy Rev. **19**, 517–530 (2013)
18. G. Merker, C. Schwarz, G. Stiesch, F. Otto, *Simulating Combustion: Simulation of Combustion and Pollutant Formation for Engine-Development* (Springer, Berlin, 2014)
19. D. Neumann, C. Jörg, N. Peschke, J. Schaub, T. Schnorbus, Real-time capable simulation of diesel combustion process for HiL applications. Int. J. Eng. Res. **19**(2), 214–229 (2018)
20. R. Payri, J. Gimeno, R. Novella, G. Bracho, On the rate of injection modeling applied to direct injection compression ignition engines. Int. J. Eng. Res. **17**(10), 1015–1030 (2016)
21. R. Reitz, C. Rutland, Development and testing of diesel engine cfd models. Progress Energy Combust. Sci. **21**(2), 173–196 (1995)
22. D. Rether, M. Grill, A. Schmid, M. Bargende, Quasi-dimensional modeling of CI-combustion with multiple pilot- and post injections. SAE Int. J. Eng. **3**(1), 12–27 (2010)
23. D. Ritter, M. Korkmaz, H. Pitsch, D. Abel, T. Albin, Optimization-based fuel injection rate digitalization for combustion rate shaping, in *American Control Conference* (2019)
24. J. Schilliger, N. Keller, S. Hänggi, T. Albin, C. Onder, Data-based modeling for the crank angle resolved ci combustion process, in *Data Analysis for Direct Numerical Simulations of Turbulent Combustion* (Springer, Berlin, 2020), pp. 197–213
25. G. Stiesch, *Modeling Engine Spray and Combustion Processes* (Springer Science & Business Media, Berlin, 2003)
26. I.I. Vibe, Semi-empirical expression for combustion rate in engines, in *Proceedings of Conference on Piston Engines, USSR Academy of Sciences* (1956), pp. 185–191

Index

T. Albin Rajasingham, *Nonlinear Model Predictive Control of Combustion Engines*, Advances in Industrial Control,
https://doi.org/10.1007/978-3-030-68010-7